The Reston Encyclopedia of
Biomedical Engineering Terms

The Reston Encyclopedia of Biomedical Engineering Terms

Rudolf F. Graf

George J. Whalen

Reston Publishing Company, Inc.
A Prentice-Hall Company
Reston, Virginia

Library of Congress Cataloging in Publication Data

Graf, Rudolf F
  The Reston encyclopedia of biomedical engineering
terms.

  1. Biomedical engineering—Dictionaries.
I. Whalen, George J., joint author.   II.   Title.
[DNLM: 1.   Biomedical engineering—Dictionaries.
QT13 G736e]
R856.G68   610'.28   76-41417
ISBN 0-87909-728-0

© 1977 by
Graf Whalen Corp.

10 9 8 7 6 5 4 3 2 1

Printed in the United States of America

To Jeffrey:
Inheritor of the Legacy

Preface

Growing interest in the field of biomedical engineering has brought technology increasingly closer to its ancient and fundamental mandate: to harness and utilize the forces of nature in the service of mankind. And yet, this hybrid, interdisciplinary field has long been burdened by the weight of two languages: that of the *physician* and that of the *engineer*. For medicine and engineering have evolved as separate arts, only recently converging because necessity has dictated an increased understanding between the two disciplines. Perhaps the greatest barrier to understanding between medical personnel and engineers has been the language which each uses when communicating with his peers. Physicians are trained to communicate with physicians; engineers, to communicate with engineers. Not surprisingly, the interface between these arts has often proven to be a gulf of misunderstanding.

Progress in biomedical engineering requires that health care specialists and technical personnel learn to speak each other's language. For, in the unhindered exchange of ideas, concepts, problems, and solutions between these two important disciplines, exciting new breakthroughs will become possible, leading the way to better health care for us all.

This book, then, is dedicated to the task of improving communications between the provider of health care and the technologist. It expresses the language of biomedical engineering in the tongues of both the physician and the engineer, with painstaking emphasis on simplifying the definitions of specialized words from both disciplines. Our objective has been to write definitions of terms which are satisfyingly faithful to present understanding within each discipline, while being simple enough to allow the meaning of each term to cross over to the reader in the other discipline. In effect, we

vii

have attempted to provide a *common* reference work for medical and technical personnel, to aid in the transfer of information between these two great disciplines.

In preparing this work, we have been very fortunate to have had the aid, advice, and encouragement of countless sources in the fields of medicine and technology. The contributions of these sources are gratefully acknowledged, as are the meritorious efforts of Mrs. John J. Dillon, whose skills at the typewriter first made this work a reality.

Rudolf F. Graf
George J. Whalen

The Reston Encyclopedia of Biomedical Engineering Terms

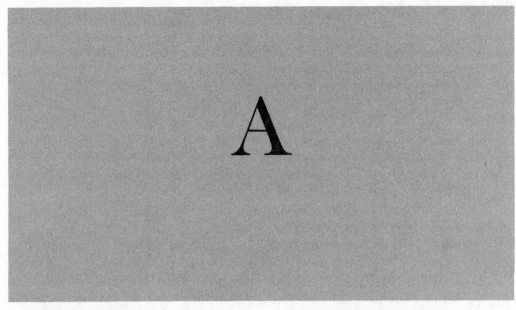

abducens nerve The sixth cranial nerve, arising in the pons and innervating the musculature of the eye.

abduct To move a part of the body away from the midline.

ablation Removal or detachment of a part.

-able Suffix meaning capable of.

abnormal morphology A significant difference in shape or appearance of an oscilloscopic or graphic trace from the norm or expected waveshape. Typically, an electrocardiographic waveform in which an irritable focus causes myocardial contraction, rather than the sinoatrial node, thus giving rise to an ECG complex of abnormal appearance in the clinical electrocardiogram.

ABO grouping Term descriptive of the types of human blood: A, B, AB, and O. An individual's blood type is determined by the presence or absence of identifiable constituents. For example, the agglutinogens, A and B can be absent, or present, or one can be missing. Many distinct sub-groups have been identified in the Standard or Universal grouping, after the system devised by Landsteiner.

abortion The termination of a pregnancy before fetal viability has been reached.

abrasion An area of the body rubbed bare of skin or mucous membrane.

abscess A collection of pus in a cavity formed by the disintegration of tissue.

absolute address 1. An address used to identify work in a program with respect to its location in storage, but without regard to its position in that program; it may be either symbolic or nonsymbolic. 2. A binary number that is permanently assigned as the address of a storage location in a computer.

absolute coding Coding written in machine language. It does not require processing before it can be understood by the computer.

absolute pressure A measurement of pressure with respect to true zero pressure. It is the sum of the atmospheric (barometric) and gauge pressures.

1

absolute-value computer A computer that processes all data expressed in full values of all variables at all times. Contrasted with incremental computer which handles changes in variables as well as absolute values.

absorbed dose 1. Energy imparted by any ionizing radiation absorbed by a unit mass of material at a particular point of interest. 2. The portion of an administered medication which is not excreted by the body.

absorbed dose rate The dose per unit of time, measured in rads per second, minute, or hour.

absorptiometer An instrument for quantitative analysis of liquids or gases by measurement of radiation absorbed by a sample.

absorption 1. The process by which a medication or radiated energy is taken into the body. 2. The process by which the number of particles or photons entering a body is reduced by interaction of the particle or radiation with the matter. Similarly, the reduction of the energy of a particle while traversing a body of matter. 3. Penetration of a substance or pathogen into the body of another.

ac Abbreviation for alternating current.

acapnia The absence of carbon dioxide from the blood. In cases of hyperventilation, too much CO_2 may be lost from the bloodstream, inducing syncope (fainting).

ac bias The alternating current, usually of a frequency several times higher than the highest signal frequency, that is fed to a magnetic recording head, in addition to the signal current. This ac bias serves to linearize the magnetic recording process.

accelerated life test Any set of test conditions designed to reproduce in a short time the deteriorating effect obtained under normal service conditions.

accelerators 1. Machines that accelerate either protons or electrons to high energies. Types of accelerators are: betatrons, cyclotrons, Van de Graaff devices, linear accelerators, and other newer machines, the energy limits of which range between 10 million and several billion electron volts. Accelerators are used medically in the treatment of various types of malignant tumors. 2. An agent used to hasten a reaction, to reduce curing, or hardening time of a thermosetting resin, by entering into the reaction. The term *accelerator* is often used interchangeably with the term *promoter*. An accelerator or promoter is often used along with a catalyst, hardener, or curing agent.

access In data processing, the act of summoning data from, or placing data in, storage.

accessory nerves See *cranial nerves*.

access time 1. The time it takes a computer to locate data or an instruction word in its memory or storage section, and transfer it to its arithmetic unit where the required computations are performed. 2. The time it takes to transfer information which has been operated on, from the arithmetic unit to the location in memory where the information is to be stored. Also called *waiting time*. 3. The time interval between the instant at which data is to be stored and the instant at which storage is completed (i.e., the *write time*).

accident prone Special susceptibility to accidents due to psychological causes.

ac coupling Signal transfer from one circuit to another circuit through a capacitor or other device which passes the desired signal, but not the static characteristics of an electrical signal.

accretion Growth, accumulation, or enlargement.

accumulator 1. Register and associated equipment in the arithmetic unit of the computer in which arithmetical and logical operations are performed. 2. Main computational register of a computer. 3. Unit in a digital computer where numbers are totaled—that is, accumulated. Often the accumulator stores one operand, and upon receipt of any second operand, it forms and stores the result of performing the indicated operation on the first and second operands. Related to adder. 4. Means of storing energy or electric power with such techniques as lifted weights and chemical changes.

accuracy 1. The ability of a measurement system to determine the true level or state of a variable in terms according to standards of reference. 2. That property of a well-described measurement process that sets forth the proximity of a given output of the process to a true value, agreed upon by workers in the field. 3. The limit of an electrical indicating instrument usually expressed as a percentage of rated full-scale value, which the errors inherent in a meter will not exceed when used at reference conditions. These errors include full-scale error, balance error, scale error, and magnetic panel error. While accuracy is the composite effect of these errors, not all of these errors are necessarily present in every meter. 4. The measure of a meter's ability to indicate a value corresponding to the absolute value of electrical energy applied. Accuracy is expressed as a percentage of the meter's rated full-scale value. 5. The maximum angular difference between the shaft angle input to an encoder and the indicated shaft angle as read from the code output. Accuracy includes both transition error and quantizing error. 6. The ratio of the error to the full-scale output, or the ratio of the error to the output expressed as a percent.

acetone A colorless liquid with a pleasant ethereal odor. It is found in small quantities in normal urine and is used as a solvent for fats and resins.

acetylcholine A chemical agent which aids in the transmission of nerve impulses across junctional synapses between nerves and muscles.

achalasia 1. An inability to relax the muscles of a hollow organ. 2. A spasm; typically, of the esophageal musculature at the entrance to the stomach, bringing about an uncomfortable sensation of fullness.

achlorhydria An abnormally low hydrochloric acid level in the stomach, resulting from a disease state or from the process of aging in some cases.

acid A chemical substance that, in solution, releases hydrogen ions. Acids taste sour, turn blue litmus red, unite with bases to form salts, and in general have chemical properties opposed to those of alkalis. Many thousands of acids are known to chemistry; hundreds are useful to medicine and scores are present in the human body.

acid-base balance Balance between acidic and basic component which determines the pH of the body fluids.

acid-fast In bacteriology, a term applied to certain bacteria which retain the red carbo-fuchsin stain after the application of an acid solution. Other organisms are decolorized. Tubercle bacilli are acid-fast.

acidophils Those cells situated in the anterior pituitary gland which secrete hormone directly influencing growth.

acidosis A tendency toward a higher acid reaction in the blood stream and usually a lower alkaline reaction. The body's alkaline reserve is depleted. This condition occurs in advanced diabetes or under other circumstances, such as prolonged diarrhea, in which the body loses sodium bicarbonate. Retention of carbon dioxide in the blood, as occurs in drowning, when the lungs are underventilated, also induces acidosis. Patients with severe acidosis may gasp for breath, go into a coma, even die. This is because the carbon dioxide content of the blood importantly regulates the breathing process.

acne An inflammatory disease of the sebaceous glands.

acoustic impedance The degree of resistance to transmitted sound imparted by the characteristic elasticity of a given substance.

acoustic nerve The eighth cranial nerve, arising in the pons. This paired nerve innervates the sensory components of the ears and is associated with hearing and the state of balance.

acoustic shock The physical pain, dizziness, and sometimes nausea caused by hearing a sudden loud sound. The threshhold of pain is about 120 dB.

acoustic telemetry The utilization of sound energy for the transmission of information. It differs from others in that information derived from the received signal is encoded by the transmitting source.

ac set and reset inputs See *synchronous inputs*.

actin The major component of the thin filaments of living muscle tissue. It is a protein which acts with other components of the muscle structure to bring about contraction.

actino electricity Electricity produced by the action of radiant energy upon crystals.

actinometer An instrument that measures the intensity of radiation by determining the amount of fluorescence produced by that radiation.

actinotherapy A treatment method involving exposure of the body to ultraviolet light rays or infrared radiation.

action current 1. A brief and very small electric current flowing in a nerve following stimulus. 2. The local flow of current into the depolarized region of the cell membrane during generation of the action potential. Since the cell membranes involved are normally those of nerve or muscle, they are polarized until the flow of action current, whereupon a state of depolarization is attained.

action potential The voltage variations that occur when a nerve or muscle cell is excited or fired by an appropriate stimulus. After a brief interval, the cell recovers its normal *resting potential,* which is typically about 80 millivolts. The interior of the cell is negative with respect to the outside.

active 1. Elements which control flow of energy from a separate supply. 2. The general class of devices that requires a power supply separate from the control source.

active circuit Any network or electric circuit which requires an energy source for its operation and is characterized by its production of an output derived by controlling the flow of energy from that source. Amplifiers, oscillators, and switching circuits are *active*. Resistive, capacitive, and inductive circuits are typically classified as *passive*.

active component An electrical or electronic element which can control voltages or currents to produce gain or switching action in a circuit (e.g., transistor, diode, IC, vacuum tube, or saturable reactor). Also called *active device* or *active element*.

active computer   When two or more computers are installed, the active computer is the one that is on-line and processing data.

active device   See *active component*.

active electric network   Electric network containing one or more sources of energy.

active element   A device which modifies or amplifies the input and contains one or more sources of energy. Example: ICs, transistors, vacuum tubes, relays, ferromagnetic cores and saturable reactors. See also *active component*.

active network   An electrical network which includes a source of energy.

active pull-up   Similar to a pull-up resistor, except that a transistor replaces the resistor connected to the positive supply voltage. This allows low output impedance without high power consumption.

active RC filter   A filter which uses both negative and positive feedback, provided by solid-state amplifiers, to enhance its characteristics. Excellent selectivity can be provided at very low frequencies.

active substrate   A substrate for an integrated component in which parts of the substrate display transistance. Examples of active substrates are single crystals of semiconductor materials, within which transistors and diodes are formed, and ferrite substrates, within which flux is steered to perform logical, gating, or memory functions. For other functions, it can also serve as a passive substrate.

active transducer   A transducer whose output waves receive their power from a separate power source, controlled by the input waves to the transducer (e.g., strain-gauge pressure transducer, thermistor, or photo-sensitive resistor).

acuity   Sharpness or clearness, especially of the visual sense.

acute exposure   Term used to denote radiation exposure of short duration.

acute myocardial infarction   A condition resulting from diminished or blocked flow of oxygenated blood to a region of the myocardium, owing to occlusion of a vessel supplying that region. Atherosclerotic deposits are the usual cause and necrosis of the tissue, the usual result. The condition is life-threatening, because spontaneous ectopic beats (either stimulated by the ischemic tissue, or caused by conduction delay) can disrupt normal cardiac contraction and bring on fibrillation. Prompt counteractive drug therapy can reduce the possibility of this occurrence, although other steps must also be taken. The condition requires immediate, intensive, and highly skilled treatment, and modern practice has scored resounding successes in defusing the lethal nature of this once major killer.

ad   Prefix meaning to or toward.

adapter   A device for connecting two parts of an apparatus, which would not be directly connectable because of incompatible dimensions, terminations, voltages, currents, frequencies, etc.

adaptive communications   Communications system capable of automatic change to meet changing inputs or changing characteristics of the device or process being controlled. Also called *self-adjusting communications* or *self-optimizing communications*.

adaptive system   A system which provides reliability and stability by sensing the environment and controlling the mechanism through proportional feedback.

adaptometer Instrument for measuring time required for the retina to adapt; indicates night blindness, deficiency of vitamin A.

adc See *A/D converter*.

A/D (analog-to-digital) converter A circuit that translates an analog signal to a digital code which is representative of the value of the signal, usually at a given instant. Many digital codes may be generated at different times.

adder Switching circuits which combine binary bits to generate the *sum* and *carry* of these bits. Takes the bits from the two binary numbers to be added (*addend* and *augend*) plus the carry from the preceding less significant bit and generates the sum and the carry.

addiction The state of being given to some habit; the habitual use of drugs.

Addison's disease A cortisone deficiency which disrupts the working of the circulatory and digestive systems. Skin discoloration is a manifest symptom, as is the abnormal activities of the skeletal musculature.

address 1. An identification, as represented by a name, label, or number, for a register, location in storage, or other data source or destination. See also *tag*. 2. Loosely, any part of an instruction that specifies the location of an operand for the instruction.

address field The portion of an instruction which specifies the location in a computer memory where a particular piece of information is located.

adduct To move a part of the body toward the midline.

A/D encoder Analog-to-digital encoder for changing an analog quantity to equivalent digital representation.

adenine An amino purine base important for the synthesis of nucleic acids.

adenine flavine dinucleotide A cofactor in the electron transport system associated with the biochemical pathways of respiration.

adenoids Spongy lymphoid tissue masses located on the upper surface of the palate in the floor of the nasopharynx.

adenoma A tumor, usually benign, of epithelial cells.

adenopathy Effects of any of several glandular diseases, most particularly involving morbid enlargement or swelling of the lymph nodes.

adenosis Any disease of the glands, particularly one involving the lymph nodes.

adenoviruses A group of viruses capable of causing adenoidal, conjunctival, and pharyngeal infections, often with pneumonia. They average 80 millimicrons in diameter, are icosahedral in shape, and comprise a DNA core surrounded by capsomeres.

adhesion The abnormal joining of two surfaces, especially after surgery.

adipose tissue Fatty tissue found beneath the skin and within the abdomen. The cells contain fat globules.

adjustable speed/torque drive system An adjustable drive system consisting of a drive unit and a control unit. The drive unit may be a motor or gearmotor while the control may be encased or the chassis type. The units work together through a closed loop feedback circuit to control speed and torque electronically.

adolescence Between the age of puberty and adulthood.

ADR Abbreviation for air dose rate.

adrenal glands A pair of endocrine

glands, also called suprarenal glands, situated adjacent to the upper pole of each kidney. They consist of two regions, the cortex and the medulla.

adrenalin (adrenin, epinephrine) Hormone secreted, together with closely related noradrenalin, by the medulla of the adrenal gland. Both substances are also secreted at many nerve endings of sympathetic nervous system, and this accounts for similarity of the action of adrenal medullary hormone to the effects of massive stimulation of sympathetic system (increased work of heart, blood pressure, and blood-sugar; dilation of blood vessels of muscles, heart and brain, and contraction of those of skin and viscera; widening of pupil; erection of hair, etc.). The two hormones have similar but not identical action. Adrenalin is amino-hydroxyphenyl-propionic acid.

adrenal nerves Cholinergic, preganglionic sympathetic nerve fibers supplying the adrenal medulla. Stimulation causes the release of adrenalin into the circulation.

adrenergic Of a motor nerve fiber, secreting at its end adrenalin and noradrenalin when nerve-impulse arrives there. These substances stimulate the effector innervated by the nerve fiber. Many sympathetic motor nerve fibers are adrenergic.

adrenocorticosteroids More than 20 synthetic adrenal hormones and their chemical derivatives. They are used primarily in the treatment of collagen diseases (e.g., rheumatoid arthritis and various forms of allergy), and in the suppression of certain inflammatory diseases or immunological processes.

adsorbent Substance causing adsorption.

adsorption The property possessed by certain porous substances, e.g., charcoal, of taking up other substances.

adynamic ileus An obstruction of the intestine resulting from inhibition of bowel motility, usually following abdominal surgery. Electrical stimulators have been applied to the intestinal musculature in an attempt to increase intestinal motility in cases of adynamic ileus.

aeroembolism The formation of gaseous bubbles in the blood and other body tissues consisting primarily of nitrogen, carbon dioxide and water vapor. This occurs at altitudes where release of gases from the body tissues form bubbles in accordance with Henry's Law.

aerosol Very fine particles of liquid or solid substances suspended in air or other gas.

afferent In neurology, referring to the direction of an impulse which is traveling toward a specified center, such as the cell body of a neuron.

afibrinogenemia The absence of fibrinogen from the blood. This substance plays a vital role in blood clotting. Thus, its lack can result in excessive bleeding from small wounds. It is possible to infuse fibrinogen into the blood to control this abnormality. Also called *hypofibrinogenemia*. May be found in obstetrics if thromboplastins are absorbed into the bloodstream from the damaged placenta or retained blood clot. The clotting mechanism of the blood is impaired and there may be uncontrollable hemorrhage.

afterglow Persistence of luminosity in a gas-discharge tube after the voltage has been removed or on the screen of a cathode-ray tube after the electron beam has moved. In many types of monitors, this property is exploited to trace a representative waveshape of a physiologic event (e.g., ECG, blood pressure, carotid pulse or peripheral pulse), so that the briefly persisting image of the wave on an oscilloscopic screen can be studied by an observer.

afterpotential A relatively slow variation of membrane potential following the spike.

agammaglobulinemia An inherited or acquired deficiency of the important substance, gammaglobulin. Presence of this protein in the blood is essential to the functioning of the body's immune response. Thus, where it is lacking, the power of the individual to resist or overcome infection is seriously impaired.

AGC Automatic gain control. A circuit which stabilizes a large variation in the strength of an incoming signal. A circuit function in a tuner or receiver which maintains the mean amplitude of the modulated signal at almost a constant level. AGC is also used to describe the audio compression system employed in automatic tape recorders, in which the recording signal level is adjusted to compensate for widely differing sound levels.

agglutination The sticking together of cells, e.g., red blood cells, or bacteria, due to an alteration of surface charge. Usually brought about by the effects of antibodies.

agglutinins Antibodies such as those found in the blood serum of persons suffering from typhoid or paratyphoid fever. They have the property of causing bacteria to clump together or agglutinate. See also *blood grouping*.

agnosia A disease of the perceptive faculties through which people and objects are recognized. It may be optical, auditory, tactile, or involving the senses of smell and taste.

-ago Suffix meaning disease.

-agra Suffix meaning seizure, catching, or rough.

air dose X-ray or gamma ray dose expressed in roentgens delivered at a point in free air. In radiologic practice, it consists of the radiation of the primary beam and that radiation scattered from surrounding air.

air gap 1. Airspace between two objects which are electrically or magnetically related. 2. A small gap left in the magnetic circuit of a closed-core transformer which increases the magnetic reluctance and prevents saturation of the core.

airway resistance A measurement encountered in pulmonary function studies, usually made by pneumotachography. The measurement reveals the degree of opposition offered by the major components of the respiratory system (i.e., trachea, bronchi, bronchioles and alveoli) to the flow of air.

-al Suffix meaning pertaining to.

albuminuria The presence of protein in the urine in the form of leukocytes.

aldolase Enzyme in muscle concerned in the conversion of glycogen to lactic acid.

aldosterone A hormone secreted by the cortex of the adrenal glands; a mineralocorticoid.

alexia Inability or difficulty in understanding the written or printed word, owing to a lesion of the brain.

algae A family grouping of simple, photosynthetic plants with unicellular organs of reproduction. Pond scums and most seaweeds are algaes. More precise taxonomy is now based on structure, pigments, and chemical character of cellular wall.

-algia Suffix meaning pain.

algorithm 1. An ordered sequence of mathematical steps that always produces the correct answer to a problem, though the solution may be more lengthy than necessary. 2. A series of equations, some of which may state inequalities which cause decisions to be made and the computational process to be altered based on these decisions.

algorithmic Pertaining to a constructive calculating process usually assumed to lead to the solution of a problem in a finite number of steps.

algorithmic language An arithmetic language by which numerical procedures may be precisely presented to a computer in a standard form. The language is intended not only as a means of directly presenting any numerical procedure to any suitable computer for which a compiler exists, but also as a means of communicating numerical procedures among individuals. The language itself is the result of international cooperation to obtain a standardized algorithmic language. The International Algebraic Language is the forerunner of ALGOL.

algorithm translation A specific, effective, essentially computational method for obtaining a translation from one language to another.

align 1. To adjust the tuning of a multistage device so that all stages are adjusted to the same frequency, or so that they work together properly. 2. To adjust relative positions of optical components in a system. 3. To adjust two or more components of a system, so as to gain synchronism in their functions, or so that they work properly with one another.

alignment 1. Process of adjusting the tuned circuits of an electronic device so that the device will be properly responsive to a given frequency or range of frequencies. 2. Process of adjusting two or more components of a system so their functions are properly synchronized.

alimentary Pertaining to the absorption of nourishment. The alimentary canal is the whole digestive tract, extending from the mouth to the anus.

-alis Suffix meaning pertaining to.

alive 1. Energized. 2. Connected to a source of electrical voltage. 3. Charged so as to have an electrical potential different than that of the earth. 4. Reverberant, as a room in which sound reflects and echoes.

alkali The opposite of acid; a chemically basic substance (base) that forms hydroxyl ions (OH) when in solution and turns red litmus blue. Among the alkalis are potassium (potash), sodium, and ammonia. Mixed with fatty acids, alkalis form soluble soaps.

alkaline battery Any of several types of primary or storage batteries which use an alkaline electrolyte, as opposed to an acid electrolyte. The alkaline electrolyte is usually a solution of potassium hydroxide.

alkaloids Group of nitrogen-containing, basic organic compounds present in plants of a few families of Dicotyledons (e.g., Solanaceae, Papaveraceae); possibly end-products of nitrogen metabolism. Of great importance because of their poisonous and medicinal properties, e.g., atropine, cocaine, morphine, nicotine, quinine, strychnine.

alkalosis An excess of alkali in fluids and tissues of the body. It is the opposite of acidosis. It may occur after excessive consumption of acid-neutralizing substances, such as sodium bicarbonate.

all-diffused monolithic integrated circuit Microcircuit consisting of a silicon substrate into which all of the circuit parts (both active and passive elements) are fabricated by diffusion and related processes.

See also *compatible monolithic integrated circuit.*

alligator clip   A test clip having long, narrow, serrated jaws.

allocate   To assign storage locations in a computer to main routines and subroutines; hence, fixing the absolute values of symbolic addresses.

all-or-none law   The principle that the response to a stimulus is either a full-sized impulse or no impulse, no intermediate response being possible.

alloy   A substance made up of two or more chemical elements, of which at least one is a metal, possessing metallic properties, and combined in such a way that its ingredients are not readily identifiable with the unaided eye.

alloy junction   A junction produced by alloying one or more impurity metals to a semiconductor. A small button of impurity metal is placed at each desired location on the semiconductor wafer, heated above its melting point, and cooled. The impurity metal alloys with the semiconductor material to form a P or N region, depending on the impurity used.

all-purpose computer   A computer combining the specific talents heretofore assigned solely to a general-purpose or special-purpose computer, scientific or business.

alnico   An alloy of iron with aluminum, nickel, and cobalt used to make permanent magnets.

alpha   Brain wave signals, detectable in the EEG having frequencies of approximately 8 to 13 Hz. The associated mental state is relaxation, heightened awareness, elation, and in some cases, dreamlike.

alphameric   A contraction of the term alphanumeric.

alphanumeric   Pertaining to a character set that contains both letters and digits and usually other characters such as punctuation marks. Synonymous with *alphameric.*

alpha particle   Nucleus of a free helium atom which contains two neutrons and two protons (an electronic charge of 2 and a mass number of 4). The particle gives up a large quantity of energy during interaction scattering electrons which in turn results in a high density of ionization.

alpha rays   A stream of fast-moving alpha particles (each having two protons and neutrons) which produce intense ionization in gases through which they pass, are easily absorbed by matter, and produce a glow on a fluorescent screen. The lowest in frequency of the radioactive emissions. Alpha rays have little penetrating capability (a 1 MeV alpha ray will penetrate less than a millimeter of air).

alpha rhythm   A slowly changing bioelectric component of the electroencephalogram (EEG) associated with a state of physical and mental relaxation, usually with the eyes closed. Alpha frequencies range from 8 to 13 Hz and their origin is thought to be the visual cortex of the brain.

alpha wave detector   A device that detects the 8 to 13 Hz alpha wave content of brain wave output. Used in biofeedback. Also called *alpha wave meter.*

alpha wave meter   See *alpha wave detector.*

alternating current (ac)   1. Electric current, the flow of which reverses (or alternates) at regular intervals. Two current reversals are termed one cycle. The number of complete cycles per second is the frequency. (Most American homes are supplied with alternating current with a frequency of 60 cycles per second.) 2. Current that is continually changing in magnitude and reversing in polarity. 3. A periodic current, the average value of which over a period is zero, and the plot of which, over time, is a sine wave.

alternation One-half of an alternating current cycle. The complete rise and fall of a current traveling in one direction, from zero to maximum and back to zero.

alveolar oxygen pressure The oxygen pressure in the alveoli (typically, 105 mmHg).

alveolus 1. An air cell of the lungs formed by the terminal dilations of the bronchioles. 2. A small cavity or pit, resembling a honeycomb cell. 3. The socket into which a tooth is set.

ambulating Walking.

ambulatory Relating to walking; moving about. The ambulatory treatment of fractures enables the patient to remain up and at work. The limb is immobilized in plaster of Paris.

ameboid cell See *leukocyte*.

amenorrhea Absence or abnormal cessation of menstruation.

amino acids A variety of organic compounds containing one, or more than one, of the basic amino groups, and one or more of the acidic carboxyl groups, polymerized to form proteins and peptides.

ammeter Instrument for measuring the strength of an electric current in amperes. The type usually employed consists of a pivoted coil placed between the poles of a permanent magnet. Current to be measured flows through the coil and creates a field which interacts with the field of the permanent magnet pivoting the coil within the magnet's field. A pointer is attached to the movable coil and registers amperes

on a calibrated scale which is fitted to the face of the instrument.

amnesia Loss of memory.

amnion A thin membrane forming a closed sac surrounding the embryo. It contains a thin, watery fluid, the amniotic fluid, in which the embryo is immersed.

amniotic See *amnion*.

A-mode (amplitude mode) A method of echographic data presentation in which the horizontal axis of the oscilloscope trace represents time (interpreted as depth), while the vertical axis displays amplitude of the echo field.

amp Abbreviated form of ampere.

amperage The rate of current flow in a circuit, expressed in amperes.

ampere 1. A unit of electrical current or rate of flow of electricity. One volt across one ohm resistance causes a current flow of one ampere. A flow of one coulomb per second equals one ampere. Also the unvarying current, through an aqueous solution of silver nitrate of standard concentration at a fixed temperature, that deposits silver at the rate of .001118 gram per second. Equivalent to the passage of 6.25×10^{18} electrons per second through a given point in a circuit. 2. The constant current which, if maintained in two straight parallel conductors of infinite length, of negligible circular sections and placed 1 meter apart in a vacuum, will produce between these conductors a force equal to 2×10^{-7} newton per meter of length.

ampere-hour The quantity of electricity represented by a current of one ampere that flows for one hour.

ampere-hour capacity Number of ampere-hours that can be delivered by a storage battery or other battery under specified conditions.

ampere-turn A term used to express the magnetomotive force surrounding a coil. The total force is derived by multiplying the number of turns of wire around a coil by the current (in amperes) flowing through the coil and expressing the product in ampere turns.

amp-hr Abbreviation for ampere-hour(s).

ampicillin A broad-spectrum semisynthetic penicillin used in the treatment of many types of meningitis and severe intestinal, respiratory, and other disorders.

amplification 1. Increase in signal magnitude from one point to another, or the process causing this increase. 2. Of a transducer, the scalar ratio of the signal output to the signal input.

amplification factor The symbol used to signify this is μ. It is the measurement of amplification in a circuit usually in voltage or current terms calculated by dividing the output by the input. This figure is often converted to *decibels* (which see).

amplifier Device which enables an input signal to control power from a source independent of the signal and thus to be capable of delivering an output which bears some relationship to, and is generally greater than, the input signal. Amplifiers are both linear and nonlinear. Linear types are frequently used for increasing a low-level signal without appreciably distorting its waveform. Nonlinear types are most often used to secure greatest power output where distortion of the signal wave is of secondary importance. Typical amplifying elements are transistors, integrated circuits, electron tubes, and magnetic circuits.

amplitude The magnitude of a simple wave or the simple part of a complex wave. It is the largest value measured from zero.

amplitude distortion The distortion of a signal by the undesirable influence of a circuit or component and takes place when there is a change of amplitude in part of the input signal bringing about harmonic distortion and intermodulation distortion. This part is expressed as a percentage of the whole waveform at any given period of time.

amplitude-frequency distortion The distortion which occurs when the various frequency components of a complex wave are not amplified, attenuated, or transmitted equally well.

amplitude modulation A process whereby the amplitude of a single-frequency carrier wave is varied in step with the instantaneous value of a complex modulating wave. In amplitude modulation, two *sidebands* are created, one consisting of the sum of the carrier and modulating frequencies (upper sideband), and the other consisting of the difference between the carrier frequency and the modulating frequencies.

amplitude response The maximum output amplitude obtainable at various points over the frequency range of an instrument operating under rated conditions.

ampoule A sealed phial containing a drug or solution sterilized ready for use.

amyotonia A lack of muscular tone, often congenital in origin.

an Prefix meaning absence of.

anabolism The phase of metabolism associated with the building up (synthesis) of material by biochemical reaction in living tissue.

anacrotic Refers to the upstroke ascending limb of a tracing of pulse wave.

anaemia Diminished oxygen-carrying capacity of the blood, due to a reduction in the numbers of red cells, or in their content of haemoglobin, or both.

anaerobic Capable of living without oxygen, such as anaerobic bacteria. Opposite of aerobic.

anaerobic conditions Conditions in which oxygen is excluded. Normal life, depending on oxygen, is not possible here. Some bacteria are able, however, to live in these conditions.

anaesthetic An agent which produces insensibility. As an adjective, it means insensible to touch.

anal character In psychoanalysis a pattern of behavior in an adult that originates in the anal eroticism of infancy and is characterized by such traits as excessive orderliness, miserliness, and obstinacy.

analgesia Absence of sensitivity to pain.

analgesic A pain-relieving agent.

analog 1. Continuous, cursive, or having an infinite number of connected points. The instrumentation industry uses the words *analog* and *digital* where the more precise language would be *continuous* and *discrete*. Short for analogous. 2. The use of physical variables, such as voltage, distance, or rotation, to represent either a physiological parameter or a numerical value. We can therefore let a voltage represent the blood pressure in a certain vessel or we can let it approximate a certain exact number, such as π. In the latter sense, the word analog can be contrasted with the term digital.

analog computer 1. Computer in which quantities are represented by physical variables. Problem parameters are translated into equivalent mechanical or electrical circuits as an analog for the physical phenomenon being investigated without the use of a machine language. An analog computer measures continuously; a digital computer counts discretely. 2. A machine

which performs mathematical operations on voltages or other characteristics which are analogous to the quantities in a problem. Analog computers must be designed or modified for the particular type of problem they are to handle. They are most frequently used as models of real systems, or as controllers which constantly calculate the changes necessary to obtain the desired output from a process. 3. A computing machine that works on the principle of measuring, as distinguished from counting. 4. A computer that solves problems by setting up equivalent electric circuits and making measurements as the variables are changed in accordance with the corresponding physical phenomena. An analog computer gives approximate solutions, whereas a digital computer gives exact solutions. 5. A nondigital computer that manipulates linear (continuous) data to measure the effect of a change in one variable on all other variables in a particular problem.

analog data 1. A physical representation of information such that the representation bears an exact relationship to the original information. The electrical signals on a telephone channel are an analog data representation of the original voice. 2. Data represented in a continuous form, as contrasted with digital data represented in a discrete (discontinuous) form. Analog data is usually represented by physical variables, such as voltage, resistance, or rotation.

analog device A mechanism that represents numbers by physical quantities (e.g., by lengths, as in a slide rule, or by voltages as in a differential analyzer, or a computer of an analog type).

analog multiplier A device that accepts two or more inputs in analog form and then produces an output proportional to the product of the input quantities.

analog network A circuit or circuits that represent physical variables in such a manner as to permit the expression and solution of mathematical relationships between the variables, or to permit the solution directly by electric or electronic means.

analog recording A method of recording in which some characteristic of the record current, such as amplitude or frequency, is continuously varied in a manner analogous to the time variations of the original signal.

analog representation A representation that does not have discrete values but is continuously variable.

analog signal A nominally continuous electrical signal that varies in amplitude or frequency in response to changes of sound, light, heat, position, or pressure.

analog switch 1. A device that either transmits an analog signal without distortion, or completely blocks it. 2. Any solid-state device, with or without a driver, capable of bilaterally switching voltages or current. It has an input terminal, output terminal and, ideally, no offset voltage, low ON resistance, and extreme isolation between signal being gated and control signals.

analog-to-digital converter A device which converts analogous physical motion, electrical potentials, or other representative inputs, into a coded electrical signal which is acceptable to a digital computer.

analog-to-digital conversion The process of converting a continuously variable (analog) signal to a digital signal (binary code) that is a close approximation of the original signal.

analog voltage A voltage that varies in a continuous fashion in accordance with the magnitude of a measured variable.

analyzer 1. Instrument or device designed to examine the functions of components, circuits, or systems, and their relations to each other, as contrasted with an instrument designed to measure some specific parameter of such a system or circuit. 2. Of computers, a routine whose purpose is to analyze a program written for the same or a different computer. This analysis may consist of summarizing instruction references to storage and tracing sequences of jumps.

anaphylactic Serving to decrease the immunity or susceptibility to an infection instead of increasing it; may be due to the introduction of a foreign protein into the body following an infection.

anaphylaxis An unusual or exaggerated reaction of the body tissues to a foreign protein or other substance, sometimes following sensitization produced by the injection of agents prepared with horse serum, as in tetanus antitoxin.

anatomy The study and classification of the parts of the body.

ancillary equipment Support equipment not required for operation of a system, but necessary for maintenance, transport, etc.

AND A Boolean logic expression used to identify the logic operation wherein given two or more variables, all must be logical 1 for the result to be logical 1. The AND function is graphically represented by the dot (·) symbol.

AND-circuit See *AND gate.*

AND device A device which has its output in the logical 1 state if and only if all the control signals assume the logical 1 state.

AND gate A signal circuit with two or more input wires which has the property that the output wire gives a signal if and only if all input wires receive coincident signals.

AND/NOR gate A single logic element whose operation can be interpreted by 2 AND gates with outputs feeding into a NOR gate. Since this is a single logic element, no access is provided for the internal logic elements (i.e., no connection is provided at the output of the AND gates).

AND-NOT gate A circuit that has inputs A, B, may have as an output A or else B or B or else A.

AND-OR circuit Gating circuit that produces a prescribed output condition when several possible combined input signals are applied. Exhibits the characteristics of the AND gate and the OR gate.

androsperm The Y chromosome carrying male sperm cell. When united with an ovum, the resulting cell will produce an individual having male characteristics.

anelectrotonus The reduced sensitivity produced in a nerve or muscle in the region of contact with the anode when an electric current is passed through it. Anelectronic is the adjective.

anelectronic See *anelectrotonus*.

anemia 1. Decrease in the number of circulating red blood cells or in their hemoglobin (oxygen-carrying pigment) content. Can result from excessive bleeding or blood destruction (either inherited or disease caused) or from decreased blood formation (either nutritional deficiency or disease). In one form, sickle-cell anemia, it is inherited, occurring almost wholly among blacks. Pernicious anemia is caused by inability to absorb vitamin B_{12}. 2. A condition in which the blood is deficient in hemoglobin or red blood cells.

anesthesia Loss of consciousness and sensation, induced by a drug for the purpose of carrying out an operation painlessly. Anesthesia, or rather analgesia (loss of pain sense), brought on deliberately, may involve loss of consciousness (general anesthesia) or merely the loss of the ability to feel pain in some parts of the body (local anesthesia). Until the discovery of the use of ether and chloroform about 1846, the only anesthetic was alcohol in large amounts. For nearly fifty years, ether and chloroform, with laughing gas (N_2O) for short operations, held sway. Then cocaine came into use for local anesthesia. The latter is now used only for eye surgery. Less poisonous drugs, of which the best known is procaine, are now used by injection into parts to be operated on, or into nerves, or into the space around the spinal cord (spinal anesthesia). For general anesthesia, ether is still the safest drug given by inhalation, but cyclopropane or gas and oxygen are also much used. To avoid the unpleasant stage of inhalation anesthesia, solutions of barbiturates with a rapid effect are frequently given by injection into a vein, so that the patient falls asleep and other anesthetics may then be administered.

aneurysm A permanent dilation of an artery, usually with rupture of the internal and middle coats. It may be (a) fusiform or (b) sacculated. The thoracic aorta and the innominate artery are those usually affected, more rarely the abdominal aorta.

angina Suffocating pain; thus, any severe pain associated with swelling of the walls of the air passages may be termed angina. Ludwig's angina and Vincent's angina are examples. The most common use of the term ·is in *angina pectoris* (pain in the chest), a typical symptom of cardiac ischema.

angina pectoris Syndrome usually characterized by pain in the chest and left arm occurring on exertion. Due to insufficiency of the blood supply to the myocardium.

angiocardiogram A diagnostic radiographic technique which makes visible the blood flow through the chambers and valves of the heart. A radiopaque dye is injected into the blood and a series of X-ray films are exposed over the cardiac area. The resulting sequence reveals cardiac defects as the contractions of the heart act upon the now-visible blood.

angiocardiography An opaque liquid is injected into a vein and X-ray photographs are taken when the liquid circulates through the heart, to provide radiographic images of the cardiovascular circulation and the patency of its vessels.

angiogram X-ray film showing blood vessels outlined by radiopaque materials (such as indocynanine green). The objective is to reveal atherosclerotic deposits which are preventing normal blood perfusion of the cardiac musculature.

angioplacentography Radiography of placental blood vessels by injection of radiopaque dye.

angle of lag Phase angle difference between two waves or sinusoidal curves having the same frequency.

angle of lead 1. Time or angle by which one alternating electrical quantity leads another of the same cyclic period. 2. Angle through which the commutator brushes of a generator or motor must be moved from the normal position to prevent sparking.

angstrom (Å) A unit of length particularly for measuring electromagnetic wavelengths; one angstrom $= 10^{-10}$ meters $= 10^{-4}$ microns $= 3.937 \times 10^{-9}$ inches. The unit is named in honor of the distinguished Swedish scientist, Andres Jonas Angstrom (1814–1874), who worked at the University of Uppsala.

anion 1. An atom with a surplus of electrons, thereby having a negative charge. So called because it is attracted to the positive, anode electrode. 2. An ion bearing a negative electrical charge. 3. One of the negative ions that moves towards the anode in a discharge tube, electrolytic cell or similar apparatus.

anisotropic Exhibiting electrical or optical properties having different values when measured along axes in different directions.

anisotropy Phenomenon noted in true and liquid crystals in which certain physical properties (e.g., refractive index) depend on the direction in which they are measured. Calcite crystals are anisotropic.

ankylosis The fusion of the bones of a joint with resultant immobility.

anodal break excitation Excitation occurring after the end of an anodal stimulus pulse.

anodal current stimulus An applied current passing positively inward through the membrane.

anode 1. Positive pole or element. 2. The outermost positive element in a vacuum tube, also called the plate. 3. The positive element of an electrochemical cell. 4. A galvanic anode. 5. The electrode to which anions are attracted. 6. In a cathode ray tube, the elements which accelerate, concentrate, and focus the electron beam. 7. The less noble and/or higher potential electrode of an electrolytic cell, at which corrosion occurs. This may be an area on the surface of a metal or alloy, the more active metal in a cell composed of two dissimilar metals, or the positive electrode of an impressed-current system.

anode terminal The semiconductor–diode

terminal that is positive with respect to the other terminal when the diode is biased in the forward direction.

anogenital Pertaining to the area around the anus and genitalia.

anomaloscope Instrument which tests color vision by presenting a variable intensity yellow light, in the presence of fixed intensity red and green lights.

anomalous (rhythm) 1. Deviating from normal rhythmicity; irregular. 2. A disturbance of the natural heart rhythm which manifests as an irregular beat.

anomaly Any part of the body which is abnormal in form, position, or structure.

anorexia Loss of appetite or of the desire for food.

anoxemia A diminished supply of oxygen in the blood.

anoxia Absence of oxygen. Often implies insufficient oxygen available for normal respiratory metabolism, e.g., hypoxia.

ante Prefix meaning before.

antecubital Situated in front of the forearm.

antenna A conductive system capable of converting current flow at radio frequencies into electromagnetic waves for transmission, or able to reverse the process, for reception. In its simplest form, the *dipole,* the antenna consists of two linear elements, the lengths of which, when summed, are equal to one-half wave length at the frequency of interest. These are driven at their adjacent ends, so that the two elements are electrically 180° out of phase. Countless variations of this antenna design, both vertically polarized and horizontal, are in use.

anterior Situated before or toward the front.

anterior commissure A bundle of nerve fibers crossing the midline in front of the third ventricle and serving to connect certain parts of the two cerebral hemispheres.

antero Prefix meaning in front of.

anterograde amnesia The inability to remember recent events.

anti Prefix meaning against.

antibiotic 1. Opposed to life; drugs, derived from living cells, especially fungi, which prevent microorganisms from multiplying (e.g., penicillin). 2. A substance produced by microorganisms which destroys or inhibits the growth of other microorganisms.

antibody Protein substance, usually circulating in the blood, which neutralizes corresponding antigens.

antidote The corrective to a poison.

antigen Any substance which, when introduced into the body, brings about the formation of antibodies which react against it.

antimitotic Agent which inhibits cell division.

antipyretic A drug which reduces fever.

antisepsis The prevention of infection by inhibiting the growth of microorganisms.

antiseptic Any substance that will inhibit the growth of microorganisms, preventing decay or putrefaction.

antithrombin A substance in the blood having the power of retarding or preventing coagulation.

antrum A cavity or chamber within a bone or visceral organ, such as the pyloric antrum of the stomach.

anuria Lack of production of urine by the kidneys.

anxiety Apprehension, the source of

which is largely unknown or unrecognized. It is different from fear, which is the emotional response to a consciously recognized and usually external danger.

anxiety neurosis　A neurosis in which fear and apprehension mainly control the patient's behavior and ideas.

aorta　Main artery of the human body. It starts in an upward arch from the left side of the heart and is approximately one inch in diameter. This great artery gives out many sub–branches which again divide and send the blood into the body organs. Through the heartbeat (contraction) the blood is pumped into the entire system of blood circulation. See *heart*.

aortic　Pertaining to the aorta.

aortic incompetence　A condition in which blood from the aorta regurgitates back into the left ventricle, due to inefficiency of the aortic valve.

aortic stenosis　Narrowing of the aortic valve due to malformation and/or disease, resulting in inadequate flow.

aortic valve　Three leaflet semilunar valve guarding the entrance from the left ventricle to the aorta and preventing the backward flow of the blood. See *heart valve*.

aortography　Radiography of the aorta visualized by the injection of radiopaque dye.

AP　Abbreviation for anteroposterior.

aperiodic　Irregular. Without periodic vibrations. Describing circuits which are not resonant.

aperiodic damping　Condition of a system when the amount of damping is so large that, when the system is subjected to a single disturbance, either constant or instantaneous, the system comes to a position of rest without passing through that posi-

tion. While an aperiodically damped system is not strictly an oscillating system, it has such properties that it should become an oscillating system if the damping were sufficiently reduced. Also called *overdamping*.

aperture　An opening, in some instances called an *orifice*.

apex　1. Top, extreme point, summit. 2. The pointed portion of a conical structure, as in the apex of the lung or heart.

apex beat　The heart beat as felt at its most forcible point on the chest wall. This corresponds approximately with the position of the apex of the left ventricle.

apexcardiography　The process of recording and interpreting movement of the chest directly over the apex of the heart. These are low-frequency vibrations in the range of 0.1 to 20 Hz and the recording (apexcardiogram) of these displacements provides diagnostically important information about the functioning of the chambers and valves of the heart.

apex of the heart　Narrow end of heart enclosing left ventricle.

aphasia　Loss of the ability to speak language, or sometimes to understand the spoken word.

aphemia　The inability to articulate words or coherent sentences, usually caused by a central lesion. Also called *motor aphasia*.

apical beat　The pulsation of the heart, as it is detected over the apex.

aplasia　Nondevelopment of an organ or tissue.

aplastic anaemia　Anaemia resulting from destruction of bone marrow cells.

apnea　A usually temporary cessation of breathing, brought about by a decrease in the carbon dioxide level in the blood and resulting loss of stimulation to the respiratory center in the brain. Breathing may resume spontaneously, as blood carbon dioxide rises to the stimulus level.

apoplexy A stroke. The effects of a cerebrovascular accident.

A positive (A + or A plus) 1. Positive terminal of an A-battery or positive polarity of other sources of filament voltage. 2. Denoting the terminal to which the positive side of the filament voltage source should be connected.

apparent power The value of power in an alternating current circuit, obtained by multiplying the voltage by the current. The real power is obtained by multiplying the apparent power by the power factor. See also *power factor*.

application In data processing, the system or problem to which a computer is applied. NOTE: Reference is often made to an application as being either of the computational type, wherein arithmetic computations predominate, or of the data processing type, wherein data handling operations predominate.

application packages A combination of required hardware, including remote inputs and outputs, plus programming of the computer memory to produce the specified results.

application schematic diagram Pictorial representation in which symbols and lines are used to illustrate the interrelation of a number of circuits.

applications program A program used to perform some logical or computational task which is important to the user rather than some internal computer function.

aqueous humour Fluid in the eye between the cornea and the iris and the lens.

-ar Suffix meaning pertaining to.

arachnoid Spider-like. An arachnoid membrane surrounds the brain and spinal cord. It is between the dura and pia mater.

arborizations A form resembling a tree in properties, growth, structure, or appearance, or such a form and arrangement.

arc 1. A discharge of electricity through a gas, characterized by a change in space potential in the immediate vicinity of the cathode which is approximately equal to the ionization potential of the gas. 2. Flow of electric current in a flame-like stream of incandescent gas particles.

arc-back 1. In an electron tube, a failure of rectifying action which results in electron flow from anode to cathode. 2. To fail in rectifying, resulting in a reverse flow of current.

areola The pigmented area around the nipple of the breast; also, any small space within a body tissue.

arithmetic and logical unit Portion of computer hardware in which arithmetic and logical operations are performed.

arithmetic check A verification of arithmetic computation; e.g., multiplying 4 by 2 to check against the product obtained by multiplying 2 by 4.

armature 1. The movable part of a magnetic circuit. 2. The portion of a relay or buzzer which is attracted by the electromagnet. 3. The rotating portion of a direct current generator or motor, or of an alternating current motor. (In an alternating current generator it is the field which rotates, and the assembly of windings which are fixed is known as the stator.)

arrhythmia Any variation from the normal rhythm of the heart beat (i.e., extra-systoles, tachycardia, bradycardia, etc.). Patient-monitoring systems are usually designed to indicate arrhythmias because changes in cardiac rhythm are relatively important signs in terms of diagnosis and therapy.

arrhythmia monitor A focal piece of equipment in almost any coronary care set-up; it expands the conventional ECG monitoring into a *predictive function* that detects the early presence of extra or unusual heart impulses (such as ventricular extra-systoles), which often precede ventricular tachycardia, fibrillation, and other kinds of potentially fatal arrhythmia. The arrhythmia monitor thus relieves those in attendance of continual patient monitoring so that they may apply the time saved to constructive activity. The device is primarily dynamic, performing beat-by-beat monitoring plus transient detection and display of abnormal ectopic beats. Secondarily, it selects and initiates recordings of the abnormal excursions so that observations, conclusions, and treatment can be instituted, particularly if the premonitory indications call for it.

arterial Pertaining to an artery. Thus, arterial tension means the pressure of the blood circulating in a given artery.

arterial hemorrhage See *hemorrhage*.

arterial occlusion A blockage of an artery, preventing the flow of adequate blood to an organ or tissue.

arteriography The photographing of blood vessels by special technique after they have been injected with a substance opaque to X-rays.

arterioles Small arteries with contractile muscular walls which control the supply of blood to the capillaries.

arteriosclerosis A condition marked by loss of elasticity, by thickening and hardening of the arteries. Arteriosclerosis of arter-

ies supplying blood to the myocardium eventually leads to infarction of at least a portion of the heart musculature.

arteriovenous aneurysm A communication between an artery and a vein, usually the result of injury.

artery A vessel carrying oxygenated blood from the heart to other portions of the body. The aorta is the principal artery, from which smaller arteries branch outward to deliver nourishing blood flow, through arteries diminishing in size to the porous capillaries.

arthrography Radiography of joint after the injection of radiopaque fluid to outline the joint space.

articulation 1. Speech spoken by a talker which is understood by a listener. Generally applied to unrelated words, as in code messages, as distinguished from intelligibility, which refers to related words. 2. In physiology, the act of enunciating speech. 3. A joint between rigid parts of an animal body.

artifact Any component of a signal that is extraneous to the variable represented by the signal. The term artifact refers to the presence of an unwanted signal such as power line frequency interference or noise. When recording, for example, the EEG, the presence of ECG or EMG information on the recording also constitutes artifact. By far the most common source of artifact is power line interference; however, power line leakage considerations are of paramount importance from a safety viewpoint and it is not uncommon for safety considerations to seemingly contradict artifact-elimination considerations.

artificial ear A specially constructed device which simulates the acoustic impedances of the human ear. It is equipped with a microphone and is used for testing earphones and hearing aids.

artificial heart A mechanism which simulates or aids the natural pumping action of the heart. Extracorporeal mechanisms are used in cardiac surgery, where the patient's heartbeat must be arrested to allow surgical precision. Blood flow is then maintained by the artificial heart. Intracorporeal mechanisms have been devised, for temporarily assisting the recuperating heart. See also *heart*.

artificial hypothermia Technique used in conjunction with major heart surgery, etc., in which the blood is cooled by passing it through a heat exchanger. The body temperature is lowered to approximately 85°F (29.44°C), at which level the oxygen requirements of tissues, especially the brain cells, are greatly reduced. This enables the circulation to be stopped for a time.

artificial intelligence 1. The design of computer and other data processing machinery to perform increasingly higher-level cybernetic functions. 2. The capability of a device to perform functions that are normally associated with human intelligence, such as reasoning, learning, and self-improvement. Related to machine learning.

artificial kidney An extracorporeal mechanism through which blood is passed for the removal of water and soluble wastes. The mechanism provides a function comparable to that of the natural kidneys, and is used when the natural kidneys are incapable of effective waste removal. Blood passes through a coiled tube of membranous material bathed in warm saline. Waste products selectively diffuse through the membrane and are expelled. The process is called dialysis.

artificial language In computer terminology, a language specifically designed for ease of communication in a particular area of endeavor, but one that is not yet natural to that area. This is contrasted with a natural language which has evolved through long usage.

artificial larynx An electromechanical device that simulates that portion of the larynx concerned with speech and permits conversation by a person whose larynx has been surgically removed. Normally, the instrument is held against the neck when speech is desired. This device is not always required, however, since many such patients learn to speak quite well by special methods involving the regurgitation of air from the stomach.

artificial valve See *heart valve*.

artificial voice Small loudspeaker mounted in a shaped baffle which is proportioned to simulate the acoustical constants of the human head; used for calibrating and testing close-talking microphones.

-ary Suffix meaning one who or that which.

ASCII code American Standard Code for Information Interchange. An 8-bit code (one bit is for parity check) which gives 128 combinations. Used for both synchronous and nonsynchronous teletypewriter and data transmission. Developed by the American Standards Association.

ascites An accumulation of serous fluid within the abdominal or peritoneal cavity.

aseptic Free from bacteria. In aseptic surgery all instruments, dressings, etc., are sterilized before use.

-asia Suffix meaning condition or state of.

-asis Suffix meaning condition or state of.

asphyxia Loss of consciousness due to inadequate oxygen supply; suffocation.

aspiration The operation of drawing off fluids from the body.

aspirator A device which produces negative pressure, for withdrawing fluids and gases from a cavity.

assemble 1. To gather, interpret, and coordinate data required for a computer program, translate the data into computer language and project it into the final, master routine, or program for the computer to follow. 2. To translate from a symbolic program to a binary program by substituting binary operation codes for symbolic operation codes and absolute or relocatable addresses for symbolic addresses in a computer.

assembler A computer program which operates upon symbolic input to produce machine instructions, carrying out the following functions: Translate symbolic operation codes, allocating storage, at least to the extent of assigning storage locations to successive instructions, and utilizing symbolic addresses so defined), computing absolute or relocatable addresses from symbolic addresses, generating sequences of symbolic instructions by the insertion of parameters supplied for each case into macro definitions, inserting library routines. An assembler differs from a compiler chiefly in that it does not make use of information of the overall logical structure of the program, but evaluates each symbolic instruction as though it stood alone or in the immediate context of a few preceding instructions produced by the assembler.

assembly 1. Process whereby instructions written in a symbolic form by the programmer are changed by a computer to machine language. 2. The act of fitting together components to make a mechanism or aggregation.

assembly language A computer language, having one-to-one correspondence with an assembly program, but is easier to use than numerics. It directs a computer to operate on a symbolic language program (using mnemonics like: SUB, ADD, TST, etc.) to produce a machine language program. See also *higher order language, machine language* and *source language*.

assembly language programming See *symbolic-language programming*.

assembly system An automatic programming software system which includes a programming language and a number of machine language programs. These programs aid the programmer by performing different programming functions such as checkout and updating.

associated corpuscular emission The full complement of secondary charged particles (usually limited to electrons) associated with an X-ray or gamma-ray beam in its passage through air. The full complement of electrons is obtained after the radiation has traversed sufficient air to bring about equilibrium between the primary photons and secondary electrons. Electronic equilibrium with the secondary photons is intentionally excluded.

associative memory A computer memory in which the storage locations are identified by their contents rather than their addresses. Enables faster interrogation to retrieve a particular data element. Also called *content-addressable memory*.

astable The condition of a device which has two temporary states. The device oscillates between the two states with a period and duty cycle predetermined by time constants.

astable circuit A circuit that alternates automatically and continuously between two unstable states at a frequency dependent on circuit constants. It can readily be synchronized at the frequency of any repetitive input signal. Examples are blocking oscillators and certain multivibrators.

astable multivibrator Multivibrator that can function in either of two semistable states, switching rapidly from one to the other, without external triggering. Also called *free-running multivibrator*.

asterognosis Loss of the ability to recognize the form of an object by touch.

asthenia Weakness of the body or a body part.

asthma Recurrent, acute, reversible episodes of difficult breathing due to bronchospasm. Allergy is the usual cause. The chief difficulty is in expiration.

astigmatism Inequality in the curature of the cornea or lens, of the eye, with consequent blurring and distortion of the images thrown upon the retina.

astringent A substance applied to produce local contraction of blood vessels and inhibit secretion, e.g., tannin, adrenalin.

asymmetrical cell Cell, such as a photoelectric cell, in which the impedance to the flow of current in one direction is greater than in the other direction.

asynchronous Without regular time relationship. Hence, as applied to program execution, unexpected or unpredictable in respect to instruction sequence.

asynchronous computer A computer in which each event or the performance of each operation starts as a result of a signal generated by the completion of the previous event or operation, or by the availability of the parts of the computer required for the next event or operations. Contrasted with synchronous computer.

asynchronous data Information which is sampled at irregular intervals with respect to another operation.

asynchronous device A device in which the speed of operation is not related to any frequency in the system to which it is connected.

asynchronous input/output The ability to receive input data while simultaneously putting out data.

asynchronous inputs The terminals in a flip-flop which affect the output state of the flip-flop independently of the clock. Called set, preset, reset, or clear. Sometimes these are referred to as *dc inputs*.

asynchronous logic 1. Logic networks in which the speed of operation depends only on the signal propagation through the network contrasted with synchronous logic where a clock pulse controls the speed of propagation. 2. A form of logic in which computer operations occur independently of time, in a sequential manner.

asynchronous operation 1. Generally, an operation that is started by a completion signal from a previous operation. It then proceeds at the maximum speed of the circuits until finished and generates its own completion signal. 2. (of flip-flops) In this mode, entry of data into a flip-flop does not require a gating or clock pulse. 3. Operation of a switching network by a free-running signal that triggers successive instructions. The completion of one instruction triggers the next.

asynchronous transmission 1. A transmission process such that between any two significant instants in the same group, there is always an integral number of unit intervals. Between two significant instants located in different groups, there is not always an integral number of unit intervals. In data transmission, this group is a block or a character. In telegraphy, this group is a character. 2. Transmission in which each information character is individually synchronized (usually by the use of start and stop elements).

asynergia Poor or absent coordination of movements of parts which normally act together. Typically this affliction results from disease or damage in the brain.

asystole A cessation of cardiac contraction with a silent ECG tracing.

ataxia Literally, disorder; applied to any defective control of muscles and consequent irregularity of movements.

ataxiagraph Instrument that determines defective muscular coordination by measuring body sway while patient stands erect with eyes closed.

-ate Suffix meaning possessing or characterized by.

atelectasis 1. Collapse or incomplete expansion of alveoli (or a larger lung zone, such as a segment or lobe). 2. Incomplete expansion of the lungs at birth or collapse of the adult lung.

atherosclerosis Narrowing of blood vessels resulting from atheromatous deposits.

athetosis A disease of the motor centers of the brain which manifests as a slow, worm-like movement of the hands and feet.

-ation Suffix meaning act or state of.

atmospheric pressure The barometric pressure of air at a particular place on the earth's surface. The nominal, or standard, value of atmospheric pressure is 760mm Hg (14.7 pounds per square inch) at sea level. Atmospheric pressure decreases with increasing altitude.

atom The smallest portion of an element which exhibits all the properties of the element. It consists of a positively charged nucleus (having almost the entire mass of the atom) surrounded by one or more negatively charged electrons. In a *neutral atom,* the number of electrons is such that their total charge equals that of the positive charge of the nucleus.

atomic charge The electric charge of an ion, equal to the number of ionizations multiplied by the charge on one electron.

atony Lack of normal tone or strength. Loss of motor neurons originating in the spinal cord often results in atony or the flaccid paralysis of one or more muscles. Stimulation of flaccid muscles has been attempted by means of electrical stimulators.

atresia Absence or closure of a normal body opening.

-atresia Suffix meaning without opening.

atria 1. The two thin-walled chambers of the heart into which the veins drain. 2. Plural of atrium.

atrial Pertaining to the atria of the heart, or to either the left atrium or right atrium.

atrial fibrillation Cardiac arrhythmia caused by the independent contraction of muscle bundles in the atrial walls. There is no coordinated atrial contraction and the ventricular contractions are stimulated irregularly. Fibrillation of an atrium, while a serious condition, is not immediately life-threatening, since the circulation is still maintained by the ventricles.

atrial flutter Cardiac arrhythmia caused by rapid atrial contractions, 200 to 300 per minute, stimulated by an excitable focus in the atrial wall. The ventricles are unable to contract at this rate and respond only to every second or third atrial contraction.

atrial septal defect Defect in the development of the heart leaving a hole in the wall separating the right and left atrium.

atrioventricular Located between an atrium and ventricle of the heart.

atrioventricular node A specialized cluster of cells located in the septum of the heart. Depolarization potentials arriving from the sinoatrial node are believed to initiate a delayed depolarization of these cells, which provides a stimulus potential to the neuromuscular tissue of the specialized conduction system. Conduction of those impulses to the myocardial musculature of the heart causes depolarization and contraction of those muscles and gives rise to the electrocardiographic potentials observable at the body surface.

atrium A chamber of a heart which receives venous blood. In the natural heart, the right atrium receives blood from the systemic venous circulation and the left atrium receives oxygenated blood from the venous circulation of the lungs.

atrophy A wasting away or diminution in the size of a cell, tissue, organ or other part.

atropine An alkaloid drug that inhibits the action of the parasympathetic division of the autonomic nervous system.

attenuate 1. To weaken. 2. To obtain a fractional part. 3. To reduce in amplitude an action or signal.

attenuation 1. A general term used to denote a decrease in magnitude in transmission from one point to another. It may be expressed as a ratio or, by extension of the term, in decibels. 2. The action of weakening a pathogen for use in a vaccine. 3. Decrease in magnitude in transmission from one point to another. May be expressed as a ratio or, by extension of the term, in decibels, or nepers. 4. Of a quantity associated with a traveling wave in a homogenous medium, the decrease with distance in the direction of propagation. NOTE: In a diverging wave, attenuation includes the effect of divergency. 5. Reduction or division of signal amplitude, retaining the characteristic wave-form; implies deliberately throwing away or discarding a part of the signal energy for the sake of reduced amplitude.

attenuator 1. Network of resistors which is included in an electrical circuit to introduce a reduction of signal strength. 2. A device for reducing the energy of a wave without introducing distortion. Attenuators are commonly combinations of fixed or adjustable resistances. By properly proportioning the series and shunt elements, the impedance of an attenuator, as viewed from either or both ends, may be made to have almost any desired value independent of the value of attenuation. In its many different forms and applications, the attenuator becomes known as a pad, gain control, level adjuster, volume control, etc.

atto Abbreviated a. Prefix meaning 10^{-18}.

attovolt 10^{-18} volt.

audio analgesia The use of music and/or white noise (in a headset worn by the patient) to reduce pain and discomfort during dental work.

audio-frequency choke An inductor which presents considerable impedance to audio-frequency currents but little to direct currents. Generally wound on a ferromagnetic core.

audiogram Chart showing the responsiveness of ear to sounds of differing pitch.

audiometer An electronic instrument used to measure an individual's hearing acuity and threshold. The simplest units perform this function by providing to the listener (usually through earphones) an audio signal (commonly a pure tone) of known intensity and frequency. More sophisticated instruments offer the listener a variety of signals, pure tones, and speech, through a variety of output transducers (earphones, bone vibrators, or loudspeakers).

auditory nerves See *cranial nerves.*

audiometry Measurement of hearing ability. The results are usually plotted as an audiogram.

auditory aphasia Inability to comprehend words or sentences, usually due to injury or lesion of the brain.

auditory sensation The perception of sounds by the ear. It is limited on the low-level end by the threshold of audibility, and on the high-level end by the threshold of feeling.

auricle 1. The external ear. 2. One of the upper cavities of the heart now usually termed atrium.

auricular Pertaining to the ear or to the auricles of the heart. Auricular fibrillation. See *atrial fibrillation.*

auricular flutter See *atrial flutter.*

auriculoventricular bundle Also called *atrioventricular (A.V.) bundle.* Normally the contraction of the heart is initiated at the sinoatrial node, the impulse then causing atrial contraction and reaching the atrioventricular node which it stimulates. The A.V. node is composed of specialized tissue continuous with the A.V. bundle through which the impulse is conducted to the ventricles to initiate their contraction. Defects in the A.V. bundle which impair conduction of the impulse result in heart block.

auscultation The practice of listening to the sounds produced by functional organs of the body and inferring their condition from the nature of the sound perceived. The stethoscope, invented by Laennec, is the principal clinical instrument for auscultation, although electronic listening aids

have also come into wide use, owing to the fact that precise filtering can be used to sharpen listening by suppressing unwanted information outside the frequency range of interest.

autistic child In child psychiatry, a child who responds chiefly to inner thoughts, who does not relate to his environment, and whose overall functioning is immature and often appears retarded.

autoantibody Antibody directed against a component normally present in the body.

autoclaving Sterilization under conditions of high-pressure steam heating in an enclosed chamber. The device used is an autoclave.

autocondensation A method of introducing high frequency alternating current into living tissue for therapeutic purposes. The patient is connected as one plate of a capacitor to which the current is applied.

autoconduction A method of introducing high frequency alternating currents into living tissues for therapeutic purposes. The patient is placed inside a coil and acts essentially as the secondary of a transformer.

automatic check Check performed by equipment built into a computer specifically for that purpose, and automatically accomplished each time a pertinent operation is performed. Sometimes referred to as a built-in check. Machine check can refer to an automatic check, or to a programmed check of machine functions.

automatic coding Any technique in which a computer is used to describe the steps to be followed in solving a given problem before the final coding of this problem for a given computer. See also *unwind.*

automatic data processing (ADP) 1. The

processing (classifying, sorting, calculating, summarizing, recording, printing) of data through the use of electronic digital computers, communications channels, devices used with such computers, and associated peripheral equipment. Includes preparation of source data in a form appropriate for such processing. Also called integrated data processing. 2. The automatic processing of digital information by obtaining input information in machine language, operating on the information by computer and other machines and producing processed output information.

automatic data processing equipment (ADPE) Electronic digital computers, communications equipment, devices used with such computers, and associated peripheral equipment. This also includes transcription and transmission devices that are designed especially for producing media for mass data processing such as punched cards, paper or magnetic tapes.

automatic data processing system Electronic system that includes electronic data processing, plus auxiliary and connecting communications equipment.

automatic programming A term for all techniques which are designed to simplify the writing and execution of programs. Assembly programs, which usually translate from the programmer's symbolic language to machine language, assign absolute addresses to instruction and data words, and integrate subroutines into the main routine, are examples of automatic programming aids.

automatic routine A computer routine that is executed independently of manual operation, but only if certain conditions occur within a program or record, or during some other process.

automation 1. The entire field of investi-

gation, design, development, applications and methods of rendering or making processes or machines self-acting or self-moving; rendering automatic; theory, art of technique of making a device, machine, process or procedure more fully automatic; the implementation of a self-acting or self-moving, machine. 2. A machine or control that automatically follows predetermined operations or responds to encoded instructions. 3. Automatically controlled operation of an apparatus, process or system by mechanical or electronic devices that take the place of human observation, effort and decision.

automaton A device that automatically follows preset operations or responds to encoded instructions. A machine which exhibits living properties.

autonomic Acting independently of volition; relating to, affecting, or controlled by the autonomic nervous system.

autonomic nervous system The functional division of the nervous system which supplies the glands, heart, and smooth muscle with both involuntary control nerves and sensory nerves. For example, the secretion of the salivary glands is largely under the control of this system. Autonomic nerves can further be classified as sympathetic or parasympathetic, depending upon their origin in relation to the vertebrae.

autoradiograph A photographic record of radiation from radioactive material in an object, made by placing the object very close to a photographic film or emulsion. This process is called autoradiography. It is used, for instance, to locate radioactive atoms, or tracers in metallic or biological samples.

autoradiography Photography   showing

localization of radioactive substance in a tissue section.

autotransformer 1. Tapped transformer with a single winding in which the whole winding acts as the primary winding, and only part of the winding acts as the secondary (stepdown); or part of the winding acts as the primary and the whole winding acts as the secondary (stepup). 2. Voltage, current, or impedance transforming device in which parts of one winding are common to both primary and secondary circuits.

auxiliary equipment The peripheral equipment or devices not in direct communication with, and not under direct control of, the central processing unit of a computer.

auxiliary operation An operation performed by equipment not under continuous control of the central processing unit.

auxiliary storage A storage device in addition to the main storage of a computer; e.g., magnetic tape, disc or magnetic drum. Auxiliary storage usually holds much larger amounts of information than the main storage, and the information is usually less rapidly accessible.

auxiliary ventricle An artificial ventricle added to the intact, natural heart to decrease the work of the left ventricle. Also known as a *booster heart*.

available power The power which a linear source of energy is capable of delivering into its conjugate impedance.

available time 1. The number of hours a computer is available for use. 2. The time during which a computer has the power turned on, is not under maintenance, and is known or believed to be operating correctly. Synonymous with available machine time. 3. The period that elapses from completion of corrective action or preventive maintenance to the next critical failure or preventive maintenance action. (Reliability term.)

avalanche The effect obtained when reverse bias applied to a semiconductor junction (or an electric field applied to a gas) starts a chain reaction in which the original electron flow causes a much larger number of electrons to join it. The result is regenerative, and can be destructive, unless power is externally limited.

avalanche breakdown A nondestructive breakdown in a semiconductor diode caused by the cumulative multiplication of carriers through field-induced impact ionization.

avalanche conduction A form of conduction in a semiconductor which is similar to the breakdown of a gas tube. Charged-particle collisions create additional hole-electron pairs, thus causing a transition from high impedance to a state of very low impedance.

avalanche diode An improved diode that is less susceptible to damage from the high reverse voltages. It will block in the reverse direction up to a predetermined breakdown voltage or avalanche point, after which the voltage drop will remain relatively constant independent of current. It can breakdown repeatedly at the predescribed reverse voltage without damage to the PN junction.

avalanche effect The cumulative multiplication of carriers in a semiconductor caused by an electric field across the barrier region strong enough to cause electrons to collide with valence electrons, releasing new electrons which experience more collisions, which release more electrons, etc.

avalanche noise 1. A junction phenomenon in a semiconductor. Carriers in a high voltage gradient develop sufficient energy to dislodge additional carriers through physical impact. This agitation creates ragged current flows which manifest themselves as noise. 2. The noise produced when a PN junction diode is operated at the onset of avalanche breakdown.

avalanche transistor   A transistor, which when operated at a high reverse bias voltage, will supply a chain generation of hole-electron pairs.

average acoustic output   Vibratory energy output of a transducer, as measured by a radiation pressure balance; expressed in terms of watts per unit area of the transducer face.

average current   The arithmetic mean of the instantaneous currents of a complex wave, averaged over one-half cycle. See *average value.*

average value   Also *average voltage.* Of a complex or sinusoidal alternating wave, an average of many instantaneous amplitudes taken periodically during one-half cycle of the wave. If the wave is sinusoidal, the average value is 0.637 times the peak value, or 0.9 times the effective voltage.

average voltage   See *average value.*

AWG   Abbreviation for American Wire Gauge.

axial lead   A connecting lead from a resistor diode or capacitor which comes out from an end, along the axis of the component.

axis   1. The second cervical vertebra on which the atlas rotates. 2. Line passing through the center of a body.

axis of pelvis   A curved line which is everywhere at right angles to the planes of the pelvic cavity.

axis traction   A force so applied to the fetus by forceps that its effect is always exerted along the axis of the pelvis.

axon   A fine cylindrical outgrowth from a neuron which transmits messages between different parts of the nervous system by means of impulses.

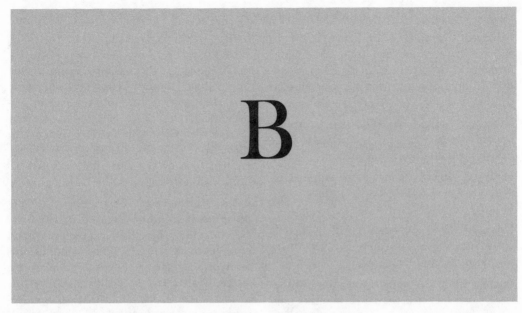

B

back bias 1. Degenerative or regenerative voltage which is fed back to circuits before its originating point. 2. Voltage applied to an input of stage to restore a condition which has been upset by some external cause. 3. See also *reverse bias*.

back electromotive force A voltage developed in an inductive circuit when the current through the circuit changes. The polarity of the back voltage is at each instant opposite to and less than that of the voltage which produces the current.

background count The evidence or effect on a detector of radiation, other than that which it is desired to detect, caused by any agency. In connection with health protection, the background count usually includes the radiation produced by naturally occurring radioactivity and cosmic rays.

background radiation The radiation in man's natural environment, including cosmic rays and radiation from the naturally radioactive elements, both outside and inside the bodies of men and animals. It is also called natural radiation. The term may

also mean radiation that is unrelated to a specific experiment.

backup 1. A redundant system or organic structure which can assume important functions if failure occurs in its counterpart. 2. An item kept available to replace an item which fails to perform satisfactorily. 3. An item under development, intended to perform the same general function performed by another item also under development.

backward diode A highly doped, alloyed germanium junction that operates on the principle of quantum-mechanical tunneling. The diode is backward because its easy-current direction is in the negative voltage rather than the positive-voltage region of the I-V curve. The backward diode has a negative resistance region, but the resultant *valley* of its I-V curve is much less pronounced than in tunnel diodes.

balanced 1. Electrically alike and symmetrical with respect to reference. 2. Arranged to provide balance between certain sets of terminals.

31

balanced amplifier Amplifier circuit in which there are two identical signal branches connected to operate in phase opposition and with input and output connections each balanced to reference.

balanced bridge A Wheatstone bridge circuit which when in a quiescent state has an output voltage of zero.

balanced circuit A circuit terminated by a network whose impedance losses are infinite.

balanced input A symmetrical input circuit having equal impedance from both input terminals to reference.

balanced line A transmission line consisting of two identical signal conductors having equal resistance and equal capacities with respect to the cable shield or with respect to ground.

balanced network 1. Hybrid network in which the impedances of opposite branches are equal. 2. A network in which the corresponding series impedance elements are identical, and symmetrical with respect to reference.

balanced oscillator An oscillator having the electrical center of its tank circuit at ground potential. The voltages between either end of the tank circuit and its center are equal in magnitude and opposite in phase.

balance output A three-conductor output (as from an amplifier) in which the signal voltage alternates above and below a third neutral wire. This symmetrical arrangement tends to cancel any hum picked up by long lengths of interconnecting cable.

balanced transmission line Transmission line having equal conductor resistances per unit length and equal impedances from each conductor to earth and to other electrical circuits. See also *twin line*.

ballast resistor Special type of thermally variable resistor used to compensate for fluctuations in alternating current power line voltage. Usually connected in series with the power supply to a receiver or amplifier. The resistance of a ballast resistor increases rapidly with increase in current through it, thereby tending to maintain essentially constant current despite variations in line voltage.

ballistocardiogram A waveform of the impulse imparted to or incurred by the body as a result of the displacement of blood upon each heartbeat. The period of this waveform's cycle is the time interval between heartbeats. Typically taken in one of two ways: by measuring the seismic disturbance imparted to the table upon which the patient is lying supine or by measuring the deflection of a heavy metal bar placed across the patient's ankles while he is lying supine.

ballistocardiograph A device for recording of the heart beat by detecting small movements of a bed or table on which the patient lies. Since each systolic ejection of blood from the heart produces a force in the direction opposite to that of the contractile force, the ballistocardiograph is able to register readings more or less related to the stroke volume.

ballistocardiography The measurement of the total body motions produced in response to the pumping actions of the circulatory system. The measurement of this parameter has evolved along two basic lines. In the first technique, ultra-low frequency devices attempt to support the body in the direction of interest with a spring-damping system which has a natural frequency much less than the exciting frequency of the cardiovascular activity. A second technique uses a high frequency suspension system and attempts to measure forces without motion. This technique allows the natural resilience of the body to play an important role.

ball-type valve (Star-Edwards)   See *heart valve.*

banana jack   Jack that fits a banana plug. Generally designed for panel mounting.

banana plug   Single-conductor plug with a spring-metal tip which somewhat resembles a banana in shape.

band-elimination filter   A network which is designed to pass freely signals of all frequencies except those within a definite band.

band-pass   A range, expressed in hertz, or kilohertz, expressing the difference between the limiting frequencies at which a desired fraction (usually half power) of the maximum output is obtained.

band-pass filter   A filter which passes, without appreciable attenuation or distortion, frequencies within a specified band while attenuating frequencies outside of that band.

B and S gauge   Brown and Sharpe wire gauge in which the conductor sizes rise in geometrical progression. Adopted as the American Wire Gauge standard.

band–stop filter   A filter having characteristics opposite to those of a band–pass filter. The band–stop filter attenuates frequencies within a defined band, and offers low attenuation of frequencies outside this band.

bandwidth   1. The range of frequencies over which a given device is designed to operate within specified limits. 2. A range of frequencies between upper and lower limits. Two methods of specifying are in common use: (a) Between the high and low frequencies where the power level is 3 dB below that at midband, and (b) 10 dB below.

barometric pressure   The weight of the atmosphere on a unit surface. The standard barometer reading at sea level and 59°F is 29.92 inches of mercury absolute.

baroreceptors   Aortic nerve endings of afferent branches of vagus and glossopharyngeal nerves. Receptors of pressure stimuli in reflex control of blood pressure.

barretter   The temperature sensitive element of a bolometer, used for radio-frequency power measurements. Usually a fine platinum wire sealed in a vacuum. See also *bolometer.*

barrier layer   Surface of contact between a metal and semiconductor. Acts as a rectifier of alternating currents. NOTE: Some barrier layers, when illuminated, generate a voltage through photovoltaic action. The junction between the copper and cuprous oxide in the photoelectric cell is a barrier layer.

barrier-layer cell   Type of photovoltaic cell in which light acting on the surface of contact between layers of copper and cuprous oxide causes an electromotive force to be produced.

barrier voltage   The voltage necessary to cause electrical conduction in a junction of two dissimilar materials, such as a PN junction diode.

basal metabolic rate   The rate of consumption of oxygen by the patient after an overnight fast and at least an hour's complete rest. This figure is expressed as a percentage of the normal average. Normal range ± 15 per cent.

basal metabolism   The rate of energy expenditure of the body at rest.

base   1. In a transistor, the center region between the emitter and the collector. The difference in potential between the base and the emitter determines the current delivered to the collector. The base is the control element in a transistor, as the grid is in a vacuum tube. 2. In a printed circuit board, the portion upon which the conductive pattern is placed. 3. In a vacuum tube, the insulated portion through which the electrodes are connected to the pins. 4. The quantity of characters for use in each of the digital positions of a numbering system. In the more common numbering systems the characters are some or all of the Arabic numerals as follows:

System Name	Characters	Radix
Binary	(0,1)	2
Octal	(0, 1, 2, 3, 4, 5, 6, 7)	8
Decimal	(0, 1, 2, 3, 4, 5, 6, 7, 8, 9)	10

Unless otherwise indicated, the radix of any number is assumed to be 10. For positive identification of a radix 10 number, the radix is written in parentheses as a subscript to the expressed number, e.g., $126_{(10)}$. The radix of any nondecimal number is expressed in similar fashion, e.g., $11_{(2)}$ and $5_{(8)}$.

base address 1. A number which appears as an address in a computer instruction, but which serves as the base, index, initial or starting point for subsequent addresses to be modified. 2. A number used in symbolic coding in conjunction with a relative address.

baseline A reference line, usually horizontal, in a scalar cathode-ray screen waveform display, or in a scalar physiologic record. The baseline sets the minimum value (usually zero), from which wave deflection heights can be measured and converted to a corresponding standard unit. In cardiovascular pressures, a wave height, measured in millimeters of deflection, could thus be interpreted in the standard pressure unit: millimeters of mercury.

basic See *alkali*.

basic need A need that is necessary for survival, i.e., the need for oxygen.

basilic vein Superficial vein of the arm.

basophils Cells of the anterior pituitary gland which secrete adrenocorticotrophic (ACTH) hormone.

batch 1. A group of records or documents considered as a single unit for the purpose of processing. 2. A number of records or documents grouped together for the purpose of processing as a single unit. 3. A yield of products (typically solid-state devices or printed circuits) resulting from a single processing run.

batch processing 1. In a computer, a systems approach to processing where a number of similar input items are grouped for processing during the same machine run. 2. The yield of identical products from a single production run.

battery 1. Device for converting chemical energy into electrical energy. 2. Series of several galvanic cells which, when assembled, produce electricity. 3. Term used to denote power source in telephone circuits regardless of whether a battery or generator supply is used. 4. Two or more cells coupled together in series or parallel, to furnish electric current by conversion of chemical, thermal, solar, or nuclear energy. In the series connection, the arrangement gives a greater voltage (two cells give twice the voltage, three cells give three times the voltage, and n cells give n times the voltage). The parallel connection gives the same voltage as an individual cell, but a current capability equal to the product of the number of cells times the current capability of one cell.

battery clip A metal clip having spring toothed jaws which can be snapped on a battery post, and also having a screw terminal to which a wire can be connected.

baud A unit of transmission speed or signalling elements per second of digital signals. It is the reciprocal of the length in seconds of the shortest element of the digital code. For example, if a teletypewriter is operating at 60 wpm, the length of the shortest element of the 7.42-unit code is 22 milliseconds. The baud rate is therefore $1/0.22 = 45.45$ bauds.

bayonet base A smooth cylindrical base having two horizontal projecting pins on opposite sides, to engage in corresponding slots in a bayonet socket and hold the base firmly in the socket.

BCD A system of representing numerical, alphabetic and special characters by means of binary notation. Abbreviation for Binary Coded Decimal when individual decimal digits are represented by some binary code. For example: 16 might be represented in an 8-4-2-1 BCD notation as 0001 (for 1) and 0110 (for 6). In pure binary, 16 is 10000.

B.E. Abbreviation for barium enema.

bead thermistor A thermistor consisting of a small bead of semiconducting material such as germanium, placed between two wire leads. Used for microwave power measurement, temperature measurement, and as a protective device. The resistance decreases as the temperature of the device increases.

beat 1. A single contraction of the heart. 2. Periodic variations that result from the mixing of waves of different frequencies. The term is applied both to the linear addition of two waves, resulting in a periodic variation of amplitude, and the nonlinear addition of two waves, resulting in new fre-

quencies, of which the most important usually are the sum and difference of the original frequencies.

beat frequency A new wave which is created when two different frequencies are combined in a nonlinear circuit. Its frequency is the difference between the two combined frequencies.

beat-frequency oscillator 1. A generator of test signals or tone whose amplified output is the difference frequency between a fixed frequency oscillator and a variable frequency oscillator both produced in the instrument. 2. A low-power variable-frequency oscillator used in a communication receiver to produce an audible signal when beat against a continuous-wave radio signal. See also *heterodyne*.

beat note 1. The audible effect created when the waves of two sound-producing devices of slightly different frequency are combined in air. 2. The accoustical frequency created when two radio-frequency waves of slightly different frequencies are supplied to and mixed by a nonlinear device.

beats 1. The periodic variations of amplitude whch result when two periodic waves having different frequencies are mixed. 2. The signal formed when two signals of different frequencies are simultaneously present in a nonlinear device. The frequency of the beat is equal to the difference in frequency between the two primary signals. For example: beats are produced in a superheterodyne receiver between the incoming signal and the signal generated by the receiver local oscillator.

beat tone Musical tone due to beats, produced by the heterodyning of two high-frequency wave trains.

bel A ratio of two signal levels, such as at the output and input, in logarithmic terms. It is too large a unit to use for precise measurement and we use tenths or decibels. These are defined in power terms as $10 \log_{10} \dfrac{\text{output}}{\text{input}}$ or in voltage or current terms as $20 \log_{10} \dfrac{\text{output}}{\text{input}}$.

benign Not a threat to life; nonmalignant.

beryllium window Apparatus designed to allow minimum penetration of the beam in X-ray therapy.

beta 1. A band of electroencephalographic waves whose frequency range is approximately 13 to 22 Hz. The associated mental state is irritation, anger, jitter, frustration, worry, or tension. 2. The current gain of a transistor when connected as a grounded-emitter amplifier, expressed as a ratio of a small change in base current to the resulting change in collector current, collector voltage being constant. 3. The numerical ratio of ac output current to input current (gain) in a transistor in a common emitter circuit.

beta cells Insulin-producing cells in the islets of Langerhans in the pancreas.

beta ray A stream of high-speed electrons or positrons of nuclear origin, more penetrating (will penetrate up to one centimeter of tissue) but less ionizing than alpha rays.

betatron Doughnut-shaped accelerator for accelerating electrons by magnetic induction to produce beta rays. See *accelerator*.

Bevatron High energy, proton accelerator. See *accelerator*.

bezel The flange or cover used for holding an external graticule or CRT cover in front of the CRT in an oscilloscope. May also be used for counting a trace recording camera or other accessory item.

bfo Abbreviation for beat frequency oscillator. See also *heterodyne*.

bi Prefix meaning two.

bias 1. Electrical, mechanical, or magnetic force which is applied to a device to establish a reference level for operation of the device. 2. A fixed voltage applied to an element in a circuit. The bias applied to the grid of a vacuum tube or to the base of a transistor determines the operating range and normal current which will flow through the device. 3. The magnetic flux applied to reed relay contacts by a permanent magnet, so as to aid or oppose the flux produced by the electromagnetic coil surrounding the reed contacts. 4. The alternating current applied to an erase head in a tape or disc recorder. 5. The mechanical force applied by a spring to the armature of a relay, so as to require production of a specified magnetic flux by the relay coil to secure contact closure.

bias distortion 1. Distortion resulting from operation on a nonlinear portion of the characteristic curve of a vacuum tube or solid-state amplifying device, due to improper biasing. 2. Distortion in magnetically recorded data resulting from beat frequencies between signal and erase bias oscillator frequencies.

bicuspid 1. Having two points or cusps. 2. The two teeth immediately behind the canines in each jaw are bicuspids. 3. Bicuspid or mitral valve, the valve between the left atrium and left ventricle of the heart. Rarely the aortic valve is bicuspid.

bidirectional transducer A transducer capable of measuring stimulus in both a positive and a negative direction from a reference zero or rest position. Also called a *differential transducer*.

bifilar resistor A resistor wound with a

wire doubled back on itself to reduce the inductance.

bifilar transformer A transformer in which the primary and secondary windings are wound together in the same direction side by side. Near unity coupling results from this type of winding, so that there is a very efficient transfer of energy from primary to secondary.

bifurcation 1. Branching, as in blood vessels. 2. Providing separate, parallel outputs from a common input cable. Typically, a bifurcated ECG cable is used to allow two separate ECG amplifiers to receive and process the cardiac potentials provided from a single set of ECG electrodes. In this way, two different ECG leads can be visualized from common electrodes.

bigeminy Name applied to the pulse when a double impulse is produced by *coupled heartbeats*. An extra heartbeat occurs just after the normal beat.

bilateral 1. Having a voltage current characteristic curve that is symmetrical with respect to the origin; that is, being such that if a positive voltage produces a positive current magnitude, an equal negative voltage produces a negative current of the same magnitude. 2. Having, pertaining to, or occurring on, the two sides of the body.

bilateral transducer A transducer capable of transmission in either direction between its termination. Also called *reversible transducer*.

bile Gall. The secretion of the liver; greenish, bitter and viscid. Alkaline. Specific gravity: 1.010 to 1.040. It consists of water, inorganic salts, bile salts, bile pigments. About 570 to 850 milliliters secreted daily.

bile duct Duct transporting the bile from the liver to the duodenum.

biligraphin Iodine-containing radiopaque material used to visualize the gall bladder and biliary apparatus by intravenous cholangiography.

bimetallic strip A composite strip of two metals welded or alloyed together, used to open and close contacts in thermostats and thermal delay switches.

bimetal thermometer Thermometer that uses two metals of dissimilar temperature coefficients of expansion bonded together. Change in temperature causes bimetal to warp by a known amount, activating an indicator calibrated in corresponding degrees of temperature.

bimetal thermostat A device which responds to temperature changes in which the sensing element is a strip formed of two dissimilar metals. The two pieces are joined together and curl when the temperature is changed, due to the unequal thermal coefficients of expansion, thus opening or closing electrical contacts.

binary 1. Characteristic or property involving a selection, choice or condition in which there are two possibilities—that is, on or off, 1 or 0. 2. A number representation system using two digits: zero and one. In the normal decimal-based system, ten digits (0 to 9) are used to express any number. In the binary system, numerical values double in progression from right to left. For example:

$$1 = 1$$
$$10 = 2$$
$$100 = 4$$
$$1000 = 8$$

Therefore:

$$11 = 3$$
$$101 = 5$$
$$110 = 6$$

$$111 = 7$$
$$1001 = 9$$
$$1010 = 10$$
$$1011 = 11$$
$$1100 = 12$$
$$1101 = 13$$
$$1110 = 14$$
$$1111 = 15$$

Then by adding additional places:

$$10000 = 16$$
$$100000 = 32$$
$$1000000 = 64$$
$$1000001 = 65$$
$$1000011 = 67$$

Digital computers use the binary number system in their computations expressing one and zeroes by the presence or absence of an electrical voltage. 3. Neurological signals are also spoken of being binary in nature, since the nerve fiber is either polarized or completely depolarized, and it passes rapidly from one state to the other. See *T flip-flop, number system.*

binary cell 1. A storage cell of one binary digit capacity, e.g., a single-bit register. 2. Any device or circuit that can be placed in either of two stable states to store a bit of binary information.

binary chain A cascaded series of binary cells, each of which can be in one of two possible states (on or off). They are so arranged that each circuit can affect or modify the condition of the circuit following it.

binary code 1. A coding system in which the encoding of any data is done through the use of bits, i.e., 0 or 1. 2. A method of representing numbers in a scale of two (on or off, high level, or low level, one or zero, presence or absence of a signal) rather than the more familiar scale of ten used in normal arithmetic. Electronic circuits designed to work in two defined states are

much simpler and more reliable than those working in ten such states.

binary coded character One element of a notion system representing alphanumeric characters such as decimal digits, alphabetic letters, and punctuation marks by a predetermined configuration of consecutive binary digits.

binary coded decimal (BCD) A binary numbering system for coding decimal numbers in groups of 4 bits. The binary value of these 4-bit groups ranges from 0000 to 1001 and codes the decimal digits 0 through 9. To count to 9 takes 4 bits; to count to 99 takes two groups of 4 bits; to count to 999 takes three groups of 4 bits.

Binary Code		Decimal Code
0000	=	0
0001	=	1
0010	=	2
0011	=	3
0100	=	4
0101	=	5
0110	=	6
0111	=	7
1000	=	8
1001	=	9
0001 0000	=	10

binary coded decimal notation A system of number representation in which each decimal digit of the number is expressed by a different code written in binary digits.

binary coded decimal number A number usually consisting of successive groups of figures, in which each group of four figures is a binary number that represents, but does not necessarily equal arithmetically, a particular figure in an associated decimal number; e.g., if the three right-most figures of a decimal number are 262, the three right-most figure groups of the binary coded decimal number might be 0010, 0110, 0010.

binary coded decimal system System of number representation in which each decimal digit is represented by a group of binary digits.

binary coded digit One element of a notation system used for representing a decimal digit by a fixed number of binary positions.

binary coded octal system Octal numbering system in which each octal digit is represented by a three-place binary number.

binary counter An interconnection of flip-flops having a single input so arranged to enable binary counting. Each time a pulse appears at the input, the counter changes state and tabulates the number of input pulses for readout in binary form. It has a 2^n possible counts where n is the number of flip-flops.

binary digit (bits) Since all modern computers are composed of two-state logic and memory, they perform all their functions in binary arithmetic. A single binary digit, which is represented by a single two-state logical or memory element, is called a bit. The largest decimal numeral that can be represented by an N bit binary numeral is $2^N - 1$. For convenience, some binary/decimal equivalents are listed below:

Binary		Decimal
0	=	0
1	=	1
10	=	2
11	=	3
100	=	4
101	=	5
110	=	6
111	=	7
1000	=	8
1001	=	9
1010	=	10

1011	=	11
1100	=	12
.		.
.		.
.		.
11111	=	31
100000	=	32
.		.
.		.
.		.
111111	=	63
1000000	=	64

2. Character used to represent one of the integers smaller than the radix 2, that is, either a zero or one.

binary logic Digital logic elements which operate with two distinct states. The two states are variously called true and false, high and low, on and off, or 1 and 0. In computers, they are represented by two different voltage levels. The level which is more positive (or less negative) than the other is called the high level, the other, the low level. If the true (1) level is the most positive voltage, such logic is referred to as *positive true* or *positive logic*.

binary notation The writing of numbers to the base two. For example, numbers zero to 15 are written as 0, 01, 10, 11, 100, 101, 110, 111, 1000, 1001, 1010, 1011, 1100, 1101, 1110, 1111. The position of the digits designates various powers of two; thus, 1011 means $8 + 0 + 2 + 1$ which equals 11.

binary number A number, usually consisting of more than one figure, representing a sum in which the individual quantity represented by each figure is based on a radix of two. The figures used are 0 and 1.

binary numeral The binary representation of a number; e.g., 101 is the binary numeral and V is the Roman numeral of the number of fingers on one hand.

binary row Pertaining to the binary representation of data on punched cards in which adjacent positions in a row correspond to adjacent bits of the data. For example, each row in an 80-row card may be used to represent 80 consecutive bits of two 40-bit words.

binary scale (or numbering system) A numbering system having rules much simpler than those of the familar decimal numbering system. This simplicity makes the binary numbering system ideal for computers which are highly illiterate devices.

Where the decimal numbering system uses ten marks (0 through 9) thus having a radix (or base) of 10, the binary system uses only two marks (0 and 1) thus having a radix of 2.

In the decimal system, we think in tens. For example, the number 35 means: $10 + 10 + 10 + 5 = 35$. Or it can be written as: $3(10) + 5(1) = 35$. Or we can write it as: $3(10^1) + 5(10^0) = 35$.

In the binary system, we deal with powers of 2 rather than powers of 10. To convert the decimal number 35 to a binary number, first line up the various powers of 2 since 2 is the radix or base of the binary system. Above each power of 2, show its value in a decimal number.

We then get: 32 16 8 4 2 1 .

$$2^5 \quad 2^4 \quad 2^3 \quad 2^2 \quad 2^1 \quad 2^0$$

Remembering that in the binary system we have only two marks, 0 and 1, we then convert the decimal number 35 to a binary number as follows:

$1(2^5) + 0(2^4) + 0(2^3) + 0(2^2) + 1(2^1) + 1(2^0)$

The resulting binary number is 100011.

binary scaler A flip-flop having only a single input (called a T flip-flop or toggle). Each time a pulse appears at the input, the flip-flop changes state, generating uniform 1 and 0 levels for successive processing stages.

binary search A search which finds an element in an ordered table by successively halving a search interval and looking at the remaining half where that element is known to exist.

binary signal A voltage or current which carries information by varying between two possible values, corresponding to 0 and 1 in the binary system.

binary signaling Communications mode in which information is passed by the presence and absence, or plus and minus variations, of one parameter of the signaling medium.

binary system A system of numeric representation and mathematical computation based on powers of two.

binding energy The minimum energy required to dissociate a nucleus into its component neutrons and protons. Neutron or proton binding energies are those required to remove a neutron or a proton, respectively, from a nucleus. Electron binding energy is that required to remove an electron from an atom or a molecule.

binding post A threaded stud equipped with binding nuts under which wires can be clamped to make an electrical connection. Part of a terminal. The stud commonly used has #8 or #10 fine-pitch screw threads.

biochemical fuel cell An electrochemical generator of electrical power in which bio-organic matter serves as the fuel source. An electrochemical reaction usually takes place with air as the oxidant at the cathode and employs microorganisms to catalyze the oxidation of the bio-organic matter at the anode.

biochemistry Branch of science which investigates the chemical processes in organic matter and its by-products. Biochemistry studies all phyisological questions, such as the ways and means by which food and other materials are used in the body (metabolism). In this connection, a special study has to be made of hormones and enzymes. Chemical studies are made of foodstuffs and especially of vitamins. A further important biochemical field is the secretion of the ductless glands. Through biochemical methods, protoplasm and other living materials have been analyzed, and many of these complex substances have been synthetically produced.

bioelectric See *bioelectricity*.

bioelectric discharge A self-propagating electric current generated on the surface of nerve and muscle cells by potential differences across excitable cell membranes.

bioelectricity 1. Collectively, the electrical phenomena which appear in living tissues. 2. Electric currents and potential differences in living tissues. Muscle and nerve tissue, for example, are generators of bioelectricity, although the potential registered may be less than one millivolt in some cases.

bioelectric potential See *bioelectricity*.

bioelectrogenesis The practical application of electricity drawn directly from the bodies of animals, including humans, to power electronic devices and appliances.

bioelectronics 1. The application of electronic theories, hardware, and techniques to the problems of medicine and the life sciences. 2. The integrated, long-term electronic control of various, impaired, physiologic systems by means of low-power electronic and electromechanical devices. The cardiac pacemaker is, thus, a bioelectronic instrument.

biofeedback The formation of a man-machine information loop, through which the mind gradually gains some control over bodily functions once thought to be autonomous and involuntary. Awareness of existence of a bodily function (e.g., blood pressure, heart rate, alpha rhythm of the EEG), through use of detection instrumentation, can open new pathways of control on the part of the subject. Instrumentation makes apparent the success of this control, thus feeding back and reinforcing the subject's powers of control. With adequate learning experience aided by instrumentation, it is believed that the information and control loop can eventually be closed within the individual, making external instrumentation unnecessary thereafter.

biogalvanic battery A device that uses metals and the body's oxygen and fluids to generate electricity.

bio-instrumentation Instruments and apparatus intended for measurement and study in the fields of medicine and the life sciences.

biological battery See *biogalvanic battery*.

biological dose The radiation dose absorbed in biological material. Measured in rems.

biological half-life The time required for a biological system, such as a man or an animal, to eliminate, by natural processes, half the amount of a substance (such as a radioactive material) that has entered it.

biological shield A mass of absorbing material placed around a reactor or radioactive source, to reduce the radiation to a level that is safe for human beings.

biological warfare Employment of living organisms, toxic biological products, and plant growth regulators to produce death or casualties in man, animals, or plants; or defense against such action.

biologic energy Energy that is produced by bodily processes and that can be used to supply electrical energy for implanted devices such as electronic cardiac pacemakers or bladder stimulators. The biologic energy can result from muscle movement (such as that of the diaphragm), temperature differences, pressure differences, expansion of the aorta, oxidation of materials within the gastro-intestinal tract, and other processes.

bioluminescence The production of light by living organisms, as glowworms, some deep-sea fish, some bacteria, and some fungi.

biomechanics The application of mechanical laws to living animals. In common usage, this term is applied only to the locomotor system. Analysis of the motion of limbs permits the design of electronic devices that allow paralyzed individuals to perform useful tasks. Actual motions are accomplished either by stimulation of muscles or by programmed movement of braces.

biomedical engineering That branch of engineering science which applies the methods and hardware of technology to the creation of instrumentation and systems to serve the fields of medicine and the life sciences.

biomedical oscilloscope Any oscilloscope designed or modified for use in medical applications. Many such oscilloscopes are characterized by slow sweep rates and long-persistence screens. (These characteristics are necessary because of the low frequencies of biological signals.) Newer designs employ electronic shift register storage of biological signals and scanning techniques to refresh a display screen, thus producing a nonfading waveform display. Such oscilloscopes are capable of presenting many successive waves, and can freeze the data on screen, indefinitely, for careful study.

biomedical telemetry The process of measuring or recording at a distance such biological variables as pulse rate, electrocardiogram, temperature, etc. Typically, a radio link is used between the patient and the receiving equipment. The biological signal to be telemetered modulates the transmitter (usually a miniaturized FM oscillator). At the receiver, the radio signal is demodulated to recover the original biological signal.

Biomedical Telemetry Systems are used in progressive care wards and in similar applications where patients are ambulatory but must still be monitored. The freedom obtained in wireless ECG monitoring is, however, gained only by acceptance of a somewhat lower overall level of dependability than is usual in hard-wire ECG monitoring. The low-power radio signal emitted by a patient-borne transmitter is subject to reflection and interruption by walls, metal surfaces, and objects and people in hallways. Careful planning and siting of antennas, knowledge of the range and technical factors affecting telemetry signals, and study of its suitability for an application should precede the decision to use telemetry in place of hard-wired ECG monitoring.

biometer An instrument for measuring the amount of life by assessing the respiration.

biometrics Science of statistics applied to biological observations.

biometry Application of mathematics to biological problems.

bionics 1. The study of living systems, for the purpose of relating their characteristics and functions to the development of mechanical and electronic hardware. 2. The reduction of various life processes of nature to mathematical terms, so that they can be duplicated or simulated with systems hardware. 3. The art concerned with electronic simulation of biological phenomena. 4. The study of systems which function after the manner of living organisms. 5. The utilization of information gained from the study of biological systems in the design and development of man-made, engineering systems. For example, a radiant-energy detector has been designed on the basis of the snake infrared detector, and some sonar systems have used the bat sounding mechanism as a model.

biophysical Pertaining to that branch of knowledge concerned with the application of physical principles and methods to biological problems.

biopotential A voltage difference measured between two points in living cells, tissues, and organisms. The ECG, EEG, EMG, and EOG are gross examples of biopotentials. Similarly, a potential difference measured along neuromuscular tissue can be termed a biopotential. Also, the potential difference across a cell membrane can be called a biopotential. The term is nonspecific and should be used with care to avoid confusion.

biopsy Removal of a small piece of tissue from the living body for microscopic or chemical examination to assist in disease diagnosis.

bioseries In evolution, a historical sequence formed by the changes in any one single heritable character.

biota The animal and plant life of a region; fauna and flora, collectively.

biotelemetry Technique of measuring or monitoring vital processes and transmitting data without wires to a point (or points) remote from the subject. See *biomedical telemetry*.

bipolar 1. Having two poles, polarities, or directions. 2. Applied to amplifiers or power supplies having an output which can vary in either polarity from zero. 3. The semiconductor technology employing two-junction transistors. 4. A transistor structure whose electrical properties are determined within the silicon material. Memories using this technology are characteristically high-speed devices. 5. General name for NPN and PNP transistors since working current passes through semiconductor material of both polarities (P and N). Also applied to integrated circuits that use bipolar transistors.

bipolar electrode Electrode, without metallic connection with the current supply, one face of which acts as anode surface and the opposite face as a cathode surface when an electric current is passed through a cell.

bipolar transistor Transistor that uses both negative and positive charge carriers.

bistable 1. A circuit element with two operating states. For example, a flip-flop in which one transistor is saturated while the other is turned off. It changes state for each input pulse or trigger. 2. A device capable of assuming either one of two stable states. See also *multivibrator*.

bistable element Another name for flip-flop. A circuit in which the output has two stable states (output levels 0 or 1) and can be caused to go to either of these states by input signals, but remains in that state permanently after the input signals are removed. This differentiates the bistable element from a gate also having two output states but which requires the retention of the input signals to stay in a given state. The characteristic of two stable states also differentiates it from a monostable element which keeps returning to a specific state, and an astable element which keeps changing from one state to the other.

bistable multivibrator Type of relaxation oscillator having two stable states and requiring two input pulses to complete the cycle.

bit 1. Abbreviation of binary digit. 2. Unit of information content. It equals one binary decision, or the designation of one of two possible and equally likely values or states of anything used to store or convey information A binary digit may be conveyed by one binary code element. 3. Unit of storage capacity. The capacity, in bits, of a storage device is the logarithm to the base two of the number of possible states of the device.

bit density Number of bits which can be placed, per unit area or volume, on a storage medium. For example, bits per inch of magnetic tape. Also called *record density*.

bit location Storage position on a record capable of storing one bit.

bit-parallel Describes a method of information transfer in which all of the bits (binary digits) constituting one alphanumeric character are transmitted simultaneously on parallel transmission paths.

bit rate 1. The speed at which bits are transmitted. 2. The rate at which binary digits, or pulses representing them, pass a given point on a communications line or channel. Clarified by band and channel capacity.

bit-serial Describes a method of information transfer in which each bit (binary digit) is transmitted sequentially, one after another. The teletypewriter uses bit-serial transmission.

bits per minute Words per minute times characters per word times bits per character.

bit stream 1. A term commonly used in conjunction with a transmission method in which character separation is accomplished by the terminal equipment and the bits are transmitted over the circuit in a consecutive line of bits. 2. A binary signal without regard to grouping by character.

black body An object that theoretically absorbs all infrared radiation incident upon it. At the same time, it is an object that, when held at a uniform temperature, emits the maximum amount of radiation in any part of the spectrum obtainable from any temperature radiator at the same temperature.

black box 1. A generic term which is used to describe an unspecified device which does, however, perform a special function, or in which known inputs produce known outputs in a fixed relationship. 2. A unit of interface equipment, specified only in terms of its performance.

black light Invisible light rays outside the violet end of the light spectrum. Exposure to this light causes some hydrocarbons to fluoresce or give off visible light.

blanking The process of cutting off the electron beam in a cathode-ray tube, so as to prevent appearance of unwanted trace lines on the phosphor screen. Typically, the beam is blanked while it is being shifted to another position on the screen, in preparation for writing new data.

blanking time The length of time the electron beam of a CRT is shut off.

bleeder A high resistance connected across the dc output of a high-voltage power supply which serves to discharge the filter capacitors after the power supply has been turned off, and to provide a stabilizing load.

bleeder current Current drawn continuously from a power supply to improve its voltage regulation or to increase the value of the voltage drop across a particular resistor.

blind spot Point where the optic nerve enters the retina, where no light receptors are located.

blivet See *land*.

block A conduction defect in the A.V. node, such that electrical activity from the atria does not pass through to the ventricular area. This is known as a complete block. The conduction defect may be such that every other electrical impulse from the atria passes through to the ventricles. This is known as a 2-to-1 or second degree block. In the condition known as 1-to-1 or first degree block, every impulse passes through but is excessively delayed in the A.V. node. 2. Group of computer words considered as a unit by virtue of their being stored in successive storage locations. 3. The set of locations or tape positions in which a block of words is stored or recorded. 4. Circuit assemblage that functions as a unit, e.g., a circuit building block of standard design, and the logic block in a sequential circuit.

block diagram 1. Diagram in which the essential units of any system are drawn in the form of rectangular blocks, and their relation to each other indicated by appropriate connecting lines. 2. Diagram in which the principal divisions or sections of a circuit are indicated by geometric figures and the path of the signal or energy by lines and/or arrows. 3. In computer programming, the graphic representation of the procedures by which data is processed within the system. Used by programmers as an aid to program development.

blocking Difficulty in recollection, or interruption of a train of thought or speech, caused by unconscious emotional factors.

blocking antibody An antibody inhibiting effect of another antibody.

blocking capacitor Capacitor which introduces a comparatively high series impedance for limiting the current flow of low-frequency alternating current or direct current without materially affecting the flow of higher frequency alternating current.

blocking oscillator Relaxation oscillator consisting of an amplifier (usually single-stage) with its output coupled back to its input by means which include capacitance, resistance, and mutual inductance.

blood The fluid connective tissue which circulates throughout the body's blood vessels and major organs, conveying oxygen and nutrients, and removing wastes. Blood constituents can be divided into cellular and noncellular components. The cellular component consists of the red blood cells (RBCs or erythrocytes) and a smaller number of white blood cells (WBCs or leucocytes). The noncellular component (plasma) is a complex solution containing many proteins. One of these, fibrinogen, is important in the blood-clotting mechanism and can be removed by allowing the plasma to clot. This leaves the serum which contains albumin and globulins.

blood cell counter Device that counts the number of *leucocytes* (white blood cells) and *erythrocytes* (red blood cells) in blood.

blood chemistry Determination of the content of various blood chemicals; the most usual are: sugar, for diabetes; urea nitrogen (BUN), for kidney or liver disease; uric acid, for gout; and cholesterol, for vascular and liver disease.

blood corpuscles Blood cells; usually divided into red blood cells (RBCs or erythrocytes) and white blood cells (WBCs or leucocytes). The red blood cells are biconcave in shape and measure about seven microns in diameter. They have no nuclei and contain hemoglobin. The white blood cells are divided into three groups: (1) Cells which contain granules in the cytoplasm; these are classed as basophil, neutrophil, or eosinophil (acidophil) granulocytes according to the staining of the granules. (2) Lymphocytes, which have an even nucleus and relatively little clear cytoplasm. (3) Monocytes, relatively large cells with kidney-shaped nuclei. The function of the red blood cells is to transport oxygen. The function of the various white blood cells is in the defense of the body against infection by phagocytosis and the production of antibodies.

blood count Determination of the number and percentage of red and white blood cells from a blood sample that is obtained by puncturing a vein or the skin. It consists of a red blood cell count (RBC), white blood cell count (WBC), and platelet count.

blood culture Investigation to detect the presence of pathogenic germs by special culturing in artificial media.

blood grouping Blood typing for selecting and matching blood transfusion donors and for the diagnosis of various diseases. For a blood transfusion to be successful, it is essential that the blood of the donor be compatible with that of the patient. Blood grouping is decided according to the presence or absence of certain agglutinogens in the corpuscles, of which there are two: A and B. The international nomenclature of the different groups is as follows: AB, A,B,O. In Group AB are those who may receive blood from any other group and are called universal recipients. Group A may receive blood from Groups A and O. Group B may receive blood from Groups B and O. Group O may receive only from Group O. Thus, Group O can give blood to all other groups, and there-

fore is the universal donor. Before transfusion, a direct match is always made between the red cells of the donor and the serum of the recipient. Any clumping together or agglutination of the corpuscles (which can be seen even with the naked eye) means incompatibility. The Rhesus (Rh) Factor. In human beings of most races, 85 percent possess this agglutinogen in their red cells, and are termed Rh positive. The remaining 15 percent, Rh negative, are liable to form an antibody (agglutinin) against the agglutinogen, if it is introduced into their circulation. It may occur in an Rh negative woman if she becomes pregnant with a fetus whose blood cells are Rh positive or if an Rh negative person is transfused with Rh positive blood. Other blood groups include M, N, P, Lewis, Duffy, Kell, Lutheran, etc.

blood platelets Small fragmentary bodies produced by the breakdown of special cells, megakaryocytes, in the bone morrow. Platelets are important in initiation of clotting in blood vessels.

blood poisoning See *septicemia*.

blood pressure The result of the pumping action of the heart which empties blood into a closed system of elastic blood vessels. The volume capacity of this system changes as a result of the stretching of the vessels and by caliber changes in response to nervous and chemical stimuli. Pressure is measured in terms of mm of mercury (Hg) above atmospheric pressure during contraction (systolic) and relaxation (diastolic) states of the cardiac cycle, thus giving rise to two numeric readings. The blood pressure is maintained at a normal level by a combination of five factors: cardiac output, peripheral resistance, blood volume, elasticity of the arterial walls, and blood viscosity.

blood sedimentation rate   Also called erythrocyte sedimentation rate (E.S.R.). It is a measure of the rate at which red cells clump together. Clumping of RBCs, and therefore a raised sedimentation rate, is increased by the presence of certain proteins in the blood. The test is a nonspecific index of disease.

blood volume   The calculated amount of blood in the whole body. About 8 pints or 4.5 liters in the average adult.

blooming   Increase in the size of a cathode-ray tube scanning spot caused by defocusing when the brightness control is set too high.

B-mode (brightness mode)   A method of echographic data presentation in which the horizontal of the oscilloscope trace represents time while the amplitude of the echo is represented by the intensity of corresponding portions of the trace. Strong echoes yield high intensity.

body burden   The amount of radioactive material present in the body of a man or an animal.

body capacitance   Capacitance introduced into an electrical circuit by the proximity of the human body. The body is electrically conductive and large in surface area, thus making it one plate of a capacitor, the other plate usually being the earth. This capacitance can become part of an electrical circuit through the contact, upsetting a delicate adjustment, causing static discharge, and producing other unpredictable results. Well-designed equipment and good grounding practices eliminate any effect of this phenomenon.

body electrodes   Electrodes placed on, or in, the body to couple electrical impulses from the body to an ECG, EEG, or other measuring/recording device.

boil   A staphylococcal infection of the skin, causing inflammation around a hair follicle.

bolometer   A radiation detector that converts incident radiation into heat. The conversion causes a temperature change in the material used in the detector. The temperature change is, in turn, measured to give an indication of the amount of incident radiant energy.

bolus   1. A large round mass such as that of food before it is swallowed. 2. A premeasured dose of medication.

bond   A low-resistance electrical connection between two lead cable sheaths, between two ground connections, or between similar parts of two circuits.

bonded strain gauge   Strain-sensitive elements arranged to facilitate bonding to a surface in order to measure applied stresses. Other forms of stimuli may also be measured with bonded strain gauges by collecting the applied force in a column or other suitable force-summing member and measuring the resultant stresses. (Pressure transducers typically employ bonded strain gauge elements in a Wheatstone bridge configuration.)

bonded transducer   A transducer which employs the bonded strain gauge principle of transduction.

bone conduction vibrator   An electromechanical transducer that is applied to the head and that produces the sensation of hearing by vibrating the bone with which it is in contact.

bone marrow   Fatty substance contained within the marrow cavity of bones. In the flat bones, and with children in the long bones as well, the fat is replaced by active blood-forming tissue, which is responsible for production of the granular leucocytes, the red cells, and platelets.

bone marrow puncture Method by which specimen of blood-forming marrow tissue is obtained. The bone is punctured and a specimen of marrow cells withdrawn through a needle.

bone seeker A radioisotope that tends to accumulate in the bones when it is introduced into the body. An example is strontium–90, which behaves chemically like calcium.

boolean 1. Pertaining to the processes used in the algebra formulated by George Boole. 2. Pertaining to the operations of formal logic.

Boolean algebra The mathematics of logic which uses alphabetic symbols to represent logical variables and 1 and 0 to represent states. There are three basic logic operations in this algebra: AND, OR, and NOT. See also *NAND, NOR, Invert,* which are combinations of the three basic operations.

booster heart See *auxiliary ventricle.*

bootstrap 1. A technique or device designed to bring itself into a desired state by means of its own action. For example, a machine routine whose first few instructions are sufficient to bring the rest of itself into the computer from an input device. 2. A name given to a circuit which increases output voltage by placing its developed potential in series with the supply voltage, thus making substantially higher voltage available to supply other stages.

bootstrap loader A very short program loading routine, used for loading other loaders in a computer. Often implemented in a read-only memory (ROM).

boss See *land.*

boundary In a solid-state device, an interface between P and N material at which

donor and acceptor concentrations are equal.

bourdon tube A tubular configuration employed in certain transducers as the force-summing member. The tube is sealed at one end and may be twisted, circular, helical, or spiral. A pressure differential between the inside and outside tends to straighten the tube, thus producing a displacement for transduction.

bps Bits per second. In serial transmission, the instantaneous bit speed within one character as transmitted by a machine or a channel.

brachial Relating to the arm or a comparable process.

brachytherapy Radiation treatment using a solid or enclosed radioisotopic source on the surface of the body, or at a short distance from the area to be treated.

brady Prefix meaning slow.

bradycardia Abnormal slowness of cardiac contraction, usually with a heart rate of less than 60 beats per minute. See *Stokes-Adams syndrome.*

bradypnea A very slow respiratory rate.

brain The main integrating mass of nervous tissue situated in the skull. The brain plays a central role in all complex activities: learning, thinking, perception, motor functions, communication, respiration, and circulation, to name a few. The principal parts of the brain are the *hindbrain,* the *midbrain,* and the *forebrain.*

Within the *hindbrain* are the cerebellum and the medulla, the latter containing vital centers for respiration and heart rate, but also including centers that relay sensory impulses upward to the midbrain and forebrain. The cerebellum is a principal center for motor coordination. It helps to smooth and give precision to our movements, and makes use of vestibular and kinesthetic impulses to control posture and balance.

The *midbrain* is a bridge between the hindbrain and forebrain, providing ascending and descending tracts for conveying impulses between them. Here, also, are centers for vision and hearing.

The *forebrain* is the seat of intellect, and has these major components: the cerebral cortex, the thalamus, the hypothalamus, septum, and amygdala. Outside the forebrain, but running through the midbrain and hindbrain, is the reticular activating system. Chief feature of the forebrain in man is the cerebral cortex. Within its four lobes (frontal, parietal, temporal, and occipital) are expressive and receptive sensory centers controlling movements and actions. Here, also, are the centers of intelligence, personality, and memory.

brain-stern The portion of the brain remaining after the cerebellum and cerebral hemispheres have been removed.

brain waves The electrical waveforms generated by the brain and picked up by means of metal (usually silver/silver chloride) electrodes on the surface of the scalp. See *electroencephalography*.

branch 1. A set of program instructions executed between two successive decision instructions. 2. To select a branch as in 1.

branching In a computer, a method of selecting the next operation to execute while the program is in progress, based on the computer results.

branch instruction A form of instruction that enables a programmer to instruct the computer to choose between alternate subprograms, depending on conditions determined by the computer during execution of the program. Also called transfer instruction, or conditional jump instruction.

branchpoint A point in a computer's routine where one, two, or more choices is selected under control of a routine, e.g., a conditional transfer or conditional jump.

breadboard An experimental feasibility model of a circuit in which the components are fastened temporarily to a chassis or board without regard to final layout. Also used as a verb—to breadboard a circuit.

breadboarding The construction of a system or portion of a system for the purpose of studying it characteristics, without any regard for layout or appearance.

breadboard model Assembly in rough form to prove the feasibility of a circuit, device, system, or principle.

break contact See *normally closed contact*.

breakdown 1. Disruptive electric discharge through insulation on wires, insulators, or other materials separating circuits, or between electrodes in a vacuum tube or gas-filled tubes. 2. Of a semiconductor, a phenomenon whose initiation is observed as a transition from a region of high dynamic resistance to a region of substantially lower dynamic resistance, for increasing magnitude of bias.

breakdown diode A semiconductor designed to conduct in the reverse direction at some specific value of reverse bias. It may operate in either the avalanche or zener mode.

breakdown impedance Of a semiconductor, the small-signal impedance at a specified direct current in the breakdown region.

breakdown region Of a semiconductor diode, the entire region of the volt-ampere characteristic beyond the initiation of breakdown for increasing magnitude of bias.

breakdown voltage 1. The voltage at which there is a disruptive current discharge through insulation or a dielectric. 2. Avalanche breakdown in a semiconductor.

breakover In a silicon controlled rectifier or related device, a transition into forward conduction caused by the application of an excessively high anode voltage. In some cases this is destructive to the device.

breakover voltage The value of positive anode voltage at which an SCR switches into the conductive state with gate circuit open.

break-point instruction In digital computer programming, an instruction which together with a manually operated control, causes the computer to stop.

breeder reactor A reactor that produces fissionable fuel as well as consuming it, especially one that creates more than it consumes. The new fissionable material is created by capture in fertile materials of neutrons from fission. The process by which this occurs is known as *breeding*.

breeding See *breeder reactor*.

breeding gain See *breeding ratio*.

breeding ratio The ratio of the number of fissionable atoms produced in a breeder reactor to the number of fissionable atoms consumed in the reactor. Breeding gain is the breeding ratio minus one.

bremsstrahlung (braking radiation) Electromagnetic radiation emitted by a fast-moving charged particle (usually an electron) when it is slowed down (or accelerated) and deflected by the electric field surrounding a positively charged atomic nucleus. X-rays produced in ordinary X-ray machines are bremsstrahlung.

bridge 1. A device in which voltage measuring instrument is bridged or connected across two branches of a circuit. Bridge devices are used for measuring resistance changes in variable resistance transducers and, therefore, have an important application in the detection of changes in body temperature, respiration, and blood pressure. See Wheatstone bridge. 2. An arrangement of four diodes for full-wave rectification of alternating current. See *bridge rectifier*.

bridge circuit A measuring network of four elements connected in a diamond configuration. A potential is connected across a pair of opposite nodes and an indicator across the other pair of nodes. If the bridge circuit is balanced, no potential appears across the indicator.

bridge hybrid See *hybrid junction*.

bridge rectifier A full-wave rectifier with four elements connected in series as in a bridge circuit. Alternating current is applied to one pair of opposite junctions, and direct current is obtained from the other pair of junctions.

bridge transformer See *bridging transformer*.

bridging amplifier An amplifier with an input impedance that is sufficiently high so that its input may be bridged across a circuit without substantially affecting the signal level at that point.

bridging connection Parallel connection by means of which some of the signal energy in a circuit may be withdrawn frequently, with imperceptible effect on the normal operation of the circuit.

bridging contacts A contact form in which the moving contact touches two stationary contacts simultaneously during transfer.

bridging loss Loss resulting from bridging the bridging point, and measured before the bridging, to the signal power delivered to that part of the system following the bridging point, and measured before the bridging, to the signal power delivered to the same part after the bridging.

bridging transformer A transformer designed to couple two circuits having at least nominal ohmic isolation, and operating at different impedance levels, without introducing significant frequency or phase distortion, or significant phase shift. Also called *bridge transformer* and *hybrid coil*.

British Thermal Unit (BTU) The quantity of heat required to raise the temperature of one pound of water one degree Fahrenheit.

broadband amplifier An amplifier having essentially flat response over a wide range of frequencies.

bronchi (singular bronchus) Tubes into which the trachea divides, leading into the lungs.

bronchial breathing Abnormal breath sounds heard on auscultation over diseased lung.

bronchial carcinoma Cancer arising from the lining of a bronchus.

bronchiectasis Pathological dilatation of bronchi.

bronchography Instillation of radioopaque dye in the bronchi, so that they are apparent on X-ray.

bronchophony Voice resonance heard over bronchi.

bronchoscope An instrument for seeing into the main bronchi.

bronchoscopy Inspection of the bronchi with a lighted instrument (bronchoscope).

bronchospirometry Method of assessing the function of each lung separately by passing a catheter down the trachea into the right or left main bronchus.

bronchus Either of two primary divisions of the trachea that lead into the right and the left lung.

bronzed diabetes See *hemochromatosis*.

Brown & Sharpe (B&S) gauge A system for measuring and specifying the diameter of nonferrous wires. Now called the American Wire Gauge (AWG).

brush A piece of conductive material, usually carbon or graphite, which makes continuous contact with a moving surface, such as the commutator of an electric motor or generator armature, and forms the connection between it and the remainder of the circuit.

brute-force filter Type of power supply filter which relies upon large values of capacitance and inductance to smooth out pulsations, rather than on resonant effects of a tuned filter.

brute supply A type of power supply that is completely unregulated. It employs no circuitry to maintain output voltage constant with changing input line or load variations. The output voltage varies in the same percentage and in the same direction as the input line voltage varies. If the load current changes, the output voltage changes inversely; an increase in load current drops the output voltage.

bryozoans Minute animals, usually forming plant-like colonies from the tidal zone to great depths. Also called *moss animals*.

BTU Abbreviation for British Thermal Unit.

bubble chamber A device used for detection and study of elementary particles and nuclear reactions. Charged particles from an accelerator are introduced into a super-heated liquid, each forming a trail of bubbles along its path. The trails are photographed, and by studying the photograph, scientists can identify the particles and analyze the nuclear events in which they originate.

buccal Pertaining to the mouth, but especially the cheeks.

bucket 1. In a computer, a general term for a specific reference in storage, e.g., as a section of storge. The location of a word, a storage cell, etc. 2. A colloquial term for an integrator circuit.

bucking voltage A voltage having opposite polarity to that of another voltage which it opposes.

buffer 1. An isolating circuit used to avoid any reaction of a driven circuit upon the corresponding driving circuit. 2. Generally, a device used as an interface between two circuits or equipments to reconcile their incompatibilities or to prevent variations in one from affecting the other. 3. In a digital computer, a circuit used for transferring data from one unit to another when temporary storage is required because of different operating speeds, or times of occurrence of events. 4. Any substance which tends to resist a change in pH (acidity or alkalinity) when acid or base is added. Buffers in the blood are of particular importance, as they help to prevent acidosis or alkalosis.

buffered computer A counting system with a storage device which permits input and output data to be stored temporarily in order to match the slow speed of input-output devices with the higher speeds of the computer. Thus, simultaneous input-output-computer operations are possible.

A data transmission trap is essential for effective use of buffering since it obviates frequent testing for the availability of a data channel.

buffered terminal A terminal which contains storage equipment so that the rate at which it sends or receives data over its line does not need to agree exactly with the rate at which the data is entered or printed.

buffer element A low-impedance inverting driver circuit. Because of its very low source impedance, the element can supply substantially more output current than the basic circuit. As a result, the buffer element is valuable in driving heavily loaded circuits or minimizing rise-time deterioration due to capacitive loading.

buffer storage 1. Synchronizing element between two different forms of storage, usually between internal and external. 2. Input device in which information is assembled from external or secondary storage and stored ready for transfer to internal storage. 3. Output device into which information is copied from internal storage and held for transfer to secondary or external storage. Computation continues while transfers between buffer storage and secondary or internal storage, or vice versa, take place. 4. Device which stores information temporarily during data transfers.

buffer storage unit A unit in a computer used to store data temporarily before it is sent to another destination. It often has the capability to accept and give back data at widely varying rates so that data transfer takes place efficiently for each device connected to it; that is, a high speed device is not slowed down by the presence of a slow-speed device in the same system unless the data transfer is directly between the two devices.

bundle branch block A defect of the specialized conduction of the heart, resulting in occasional, frequent, or continual missed beats, in which the natural activating impulse experiences difficulty in passing through the bundle branch. This condition may occur in either the left or right bundle branch; hence, the terminology left bundle branch block (LBBB) and right bundle branch block (RBBB).

bundle of His 1. A portion of the heart's specialized conduction system through which natural pacing impulses pass en route to the muscles surrounding the ventricles. The structure is located in the interventricular septum, near the septal leaflet of the tricuspid valve. 2. A small band of cardiac muscle fibers transmitting the wave of depolarization from the atria to the ventricles during cardiac contraction.

burn-in The operation of an item to stabilize its failure rate.

bursa A small sac interposed between movable body parts.

bus 1. Term used to specify an uninsulated conductor (a bar or wire); may be solid, hollow, square, or round. 2. Sometimes used to specify a bus bar. 3. Circuit over which data or power is transmitted. 4. Communications path between two switching points. 5. A common pathway for a number of signals in multiplexed communications, or for routing data between various elements in a computing system.

bus bar A heavy conductor, frequently a rigid bar or strap, used to make common connections between circuits. Sometimes called simply *bus*.

bypass 1. An electrical device which shunts out or away from the main circuit a selected current or frequency, such as separating direct from alternating current.

2. In gas pressurizing of cables, a pipe which permits gas to flow around a plug or stricture in the cable. 3. To cause to be shunted or passed around. 4. A revascularization procedure, in which an occluded artery segment is shunted by an added segment, tied into the artery before and after the occlusion. The new segment permits unimpeded flow of oxygenated blood around the occlusion.

bypass capacitor Capacitor for providing an alternating current path of comparatively low impedance around some circuit or element. Frequently used to prevent unwanted signal coupling between stages, through the impedance of a common power supply. Also used to provide an ac signal return to reference from a line also carrying direct current.

bypass filter Filter which provides a low-attenuation path around some other equipment as, for instance, a carrier-frequency filter is used to bypass a telephone repeater station.

by-product material Any radioactive material, except source material or fissionable material, obtained during the production or use of source material or fissionable material. It includes fission products, and many other radioisotopes produced in nuclear reactors.

byte 1. An 8-bit grouping of binary digits (bits). May contain two 4-bit characters, or some other combinations of bit subgrouping to facilitate information handling. 2. The number of bits that a computer processes as a unit. This may be equal to, or less than, the number of bits in a word. For example, both an 8-bit and a 16-bit length computer may process data in 8-bit bytes. 3. The smallest addressable unit of main storage in a computer system. The byte typically consists of eight data bits and one parity bit.

C

cable 1. An insulated conductor (single conductor cable) or a combination of conductors insulated from one another (multiple-conductor cable). 2. A small number of large conductors or a large number of small conductors, bundled together, usually color-coded, and with a protective overall jacket.

cable shield A metallic layer applied over insulation covering a cable, composed of woven or braided wires, foil wrap, or metal tube, which acts to prevent electromagnetic or electrostatic interference from affecting conductors within.

cachexia A profound and marked state of general ill health and malnutrition.

cadmium cell In all scientific research, it is essential that certain standards accepted by all investigators always be available for checking measuring instruments and experimental equipment. The Weston cadmium cell is a primary cell which gives a constant voltage of 1.0186 volts at a temperature of 20°C. It is known as a standard cell because others are compared with it. The assembly is fitted in a case for protec-

tion and temperature stabilization. This cell has a very high internal resistance and only very minute currents in the order of microamps should be taken from it.

cadmium selenide photoconductive cell A photoconductive cell that uses cadmium selenide as the semiconductor material. It has a fast response time and high sensitivity to longer light wavelengths, such as those emitted by high-wattage incandescent lamps and some infrared light sources.

cadmium sulfide photoconductive cell A photoconductive cell in which a small wafer of cadmium sulfide is used to provide an extremely high dark-light resistance ratio. Some cells can be used directly as light-controlled switches, operated directly from the 120-volt ac power line.

caecum The blind intestine, a cul-de-sac at the commencement of the large intestine.

caffeine An alkaloid substance obtained from tea leaves or coffee beans. It is a cardiac stimulant.

calculator 1. A device capable of performing at least the basic arithmetic. Computations: addition, subtraction, multiplication, and division. 2. A calculator as in 1. which requires frequent manual intervention. 3. Generally and historically, a device for carrying out logical and arithmetical digital operations of any kind.

calculus A stone formed in the gallbladder or urinary system. It may consist of cholesterol or calcareous substances.

calibration 1. Process of comparing an instrument or device with a standard to determine its accuracy or to devise a corrected scale. 2. Taking measurements of various parts of electronic equipment to determine the equipment's performance level and whether it conforms to technical specifications. 3. Relating a measurement made in one system of units (e.g., millimeters of deflection) to another system of units, more usual for a particular measurement (e.g., millimeters of mercury).

calomel electrode An electrode used in pH meters for determining the difference in potential of a solution.

calorie Metric term for the standard unit of heat. One kilocalorie (C) is the amount of heat required to raise 1 liter of water by 1 degree Centigrade. The amount of heat produced in the body of the combustion of food can be estimated. One gram carbohydrate, 1 gram protein each yield 4.1 C. One gram fat yields 9.3 C. A diet should yield an adequate number of calories per day. One calorie (c) is the amount of heat necessary to raise 1 gram of water by 1 degree Centigrade.

calorimeter An apparatus for determining the amount of heat yielded by combustion of a substance, usually a food item.

camera tube Electron beam tube used in a television camera to convert an optical image into an electrical signal. An electron-current or charge-density image is formed

from an optical image and scanned in a predetermined sequence to provide a representative electric signal.

Canadian Standards Association (CSA) Canadian board which issues standards and specifications prepared by various governmental and industrial committees. Many of its standards have the force of law in Canada by reason of statutes.

cancer (neoplasm) A cellular tumor (swelling) resulting from uncontrolled tissue growth. Its natural evolution is to spread locally or to other body locations through the blood and lymph stream.

candela Unit of luminous intensity, which is of such value that the luminous intensity of a full radiator at a freezing temperature of platinum is 60 candela per centimeter squared (formerly candle).

candle Unit of luminous intensity; luminous intensity of 1/60th of one square centimeter of projected area of a blackbody radiator operating at the temperature of solidification of platinum. See *candela*.

candlepower Unit for the measure of the intensity of a source of white light. Initially the standard for this unit was a spermaceti candle weighing six to the pound and burning at the rate of 120 grams per hour.

candoluminescence A light-producing phenomenon which produces white light without need for the very high temperatures normally required in the generation of bright white light.

cannibalize To remove parts from one piece of equipment and use them to replace like, defective parts in a similar piece of equipment in order to keep the latter operational.

cannula 1. A hollow tube, often fitted with a trocar, or sharpened at one end, for insertion into a cavity, organ, or vessel of the body. Numerous cannulae of varied sizes and shapes have been devised to fulfill specific clinical needs. 2. Term applied to the hollow shaft of a hypodermic syringe, which is sharpened at one end and shaped to puncture the skin for purposes of injection.

canthus Either of the two angles formed by the junction of the eyelids, designated outer or lateral, and inner or medial.

cap Abbreviation for capacitor, or capacitance.

capacitance 1. Property that exists whenever two conductors are separated by an insulator, permitting storage of electricity. 2. The property of an electric device comprised of conductors and associated dielectrics which determines, for a given rate of change of potential difference between the conductors, the displacement currents in the device. Also the property which determines how much electrical charge will be stored in the dielectric for a given potential difference between the conductors.

capacitive coupling 1. The coupling between two portions of a circuit which exists by virtue of mutual capacitance. 2. Interstage coupling in a multistage amplifier which employs a capacitor connected from the output of one stage to the input of the next stage.

capacitive divider Two or more capacitors placed in series across a source, making available a portion of the source voltage across each capacitor. The voltage across each capacitor will be inversely proportional to its capacitance.

capacitive feedback Process of returning part of the energy in the output circuit of an amplifier to the input circuit by means of a capacitance common to both circuits.

capacitivity See *dielectric constant.*

capacitor A device for storing electric energy, blocking the flow of direct current, and permitting the flow of alternating current to a degree permitted by the frequency and its capacitance. Consists of one or several pairs of conducting plates separated by a dielectric, which may be air, waxed or oiled paper, mica, glass, mylar, etc.

capacitor bank A number of capacitors connected together in series, parallel, or in series-parallel.

capacitor-input filter A power supply filter in which the first element encountered by the unfiltered wave from the rectifier is a capacitor shunted across the circuit.

capacity 1. Capacitance. 2. The rated load-carrying capability of a device or machine. 3. The ability to carry current. 4. The ability to perform work. 5. The output capability of a cell or battery over a period of time expressed in ampere-hours (amp-hr). 6. In a calculator, the maximum number of digits that can be entered as one factor or obtained in a result. In most machines, the capacity is equivalent to the number of digits in the display.

capillaries 1. The extensive network of microscopic vessels which communicate with the arterioles and the venules. The walls are formed of a single layer of endothelium. 2. Any of the smallest vessels of the cardiovascular system, connecting arterioles with venules and forming networks throughout the body.

capillary hemorrhage See *hemorrhage.*

capstan The drive spindle in a tape recorder which, in conjunction with the pinch-wheel, pulls the tape across the heads at constant speed.

capture A process in which an atomic or nuclear system acquires an additional particle; for example, the capture of electrons by positive ions, or capture of electrons or neutrons by nuclei.

capture effect Effect characteristic of the FM system whereby the stronger of two signals, even when in the same channel, captures the tuner so that it responds mostly to the stronger, pushing the weaker unwanted signal well into the background.

carbohydrate Compound of the general formula $C_x(H_2O)$, such as sugar and starch. Carbohydrates are of central importance in cell metabolism.

carbon film resistor A resistor formed by vacuum depositing a thin carbon film on a ceramic form.

carbon resistor Small rod of carbon composition having axial leads, and suitably protected by an outer covering. Obtainable in graded resistance values from 10 ohms to 22 megohms and in 5%, 10%, and 20% accuracy tolerance. Power dissipation rating is determined by physical size and ranges from $1/8$ watt to 2 watts.

carboxyhemoglobin Hemoglobin which has combined with carbon monoxide to form a stable compound. This union prevents the hemoglobin from reacting with oxygen, making carbon monoxide a poisonous gas.

carcinogenic Term applied to substances producing or predisposing to cancer.

carcinoma in situ Early stage of carcinoma in which the growing cells have not invaded surrounding tissues.

carcinomatosis The spread of carcinomatous metastases.

card field In data processing, a set of card columns, either fixed as to number and position or, if variable, then identifiable by position relative to other fields. Corresponding fields on successive cards are normally used to store similar information.

card hopper See *card stacker*.

cardia 1. Of the heart. 2. The aperture between the esophagus and the stomach.

cardiac Relating to the heart.

cardiac apex See *apex*.

cardiac arrest Stopping of the heart.

cardiac arrhythmia Any irregularity in the rhythm of the heartbeat. Tachycardia (fast rate) and bradycardia (slow rate) are well-known arrhythmias.

cardiac arrhythmia monitor Electronic instrument that detects, identifies, and warns of deviation from normal heart wave.

cardiac catheter A thin flexible tube designed to be inserted into the cardiovascular system, usually by cutdown of a vein. The tube is guided to the heart under fluoroscopy and while continuously monitoring intracardiac pressure through a sterile saline fluid column which couples the pressure wave to an external transducer. Many specialized catheters have evolved for measurement of flow, volume, cardiac output and valvular gradients.

cardiac catheterization Investigation carried out to diagnose valvular abnormalities, circulatory deficiency and similar problems of the heart. A catheter is introduced through a vessel in the arm into a chamber of the heart, from which pressure recordings can be obtained.

cardiac catherization system Instrumentation that permits simultaneous and sequential determinations of pressure in areas of patient's heart, ECG and cardiac output. Computerized sytsem presents on-line evaluation of data, protocol.

cardiac cycle The sequence of events

which results in a contraction of the heart. Electrophysiologically, it is reported by the ECG, which shows atrial contraction (P wave), ventricular contraction (QRS complex), and repolarization (T wave).

cardiac failure A contraction of the heart insufficient to expel a volume of blood equal to that which fills it. There is a damming back of blood on the venous side of the circulation with consequent clinical signs of congestive cardiac failure.

cardiac massage Direct cardiac massage consists of squeezing the heart rhythmically to stimulate the normal heart beat in an attempt to restart the circulation, having first gained access to the heart through an incision in the chest. More recently external cardiac massage has been widely used to restart the heart. This consists of pressing on the chest with the patient lying flat on his back so that the heart is rhythmically compressed between the front and back of the thoracic cage.

cardiac monitor An instrument employed in hospitals for monitoring the heart action of cardiac and other critically ill patients. Typically, the cardiac monitor includes a cathode-ray tube for displaying the electrocardiogram, and a cardiotachometer for indicating heart rate. Alarm circuits are usually included in the monitor to alert hospital personnel in the event the patient's heart rate rises above or drops below preset limits.

cardiac output The volume of blood pumped by the heart each minute. In the human adult, this is about 5 liters per minute. Artificial hearts must maintain this flow rate.

cardiac output (Thermodilution Method) The blood pumped into the aorta by the left ventricle, each minute (expressed in liters per minute). This can be determined by injecting a known amount of cold, sterile solution into the right atrium or superior vena cava, and detecting the resulting change in blood temperature in the pulmonary artery. (A thermistor accomplishes this temperature measurement and provides a corresponding electrical signal.) Cardiac output is inversely proportional to the integral of the temperature change.

cardiac pacemaker 1. The natural pacemaker of the heart. A cluster of cells in the sinoatrial node which sequences depolarization of cardiac musculature. 2. A device that controls the frequency of cardiac contractions, usually implanted when the heart's natural pacemaker fails. 3. A device that stimulates the heart and controls its rhythm by means of electrodes placed on the chest wall or implanted under the skin.

cardiac resuscitation The emergency treatment for cardiac arrest or ventricular fibrillation. Blood flow and respirations are artificially maintained by an individual or team, until full resuscitative facilities are available. A defibrillator may be used to apply countershock to restore normal heart rhythm.

cardiac tamponade A condition where the action of the heart is impeded by the accumulation of fluid in the pericardium.

cardiology That branch of medicine dedicated to the care of patients suffering from heart and circulatory diseases, and to the acquisition of new knowledge about the cardiovascular system.

cardiomyopathy Disease of heart muscle (myocardium) not caused by a specific infection.

cardiomyotomy Operation to relieve muscular spasm at the lower end of the esophagus.

cardio-omentopexy A surgical procedure

consisting of grafting part of the omentum on to the surface of the heart. The omentum is taken through the diaphragm. The object is to provide a better blood supply in cases where the coronary circulation is deficient.

cardiopathy Disease of the heart.

cardiopulmonary Descriptive term which unites the closely interrelated function of the heart and lungs.

cardiopulmonary analyzer Equipment for evaluation of heart and lung performance.

cardiopulmonary bypass Directing the flow of blood from the entrance of the right atrium, avoiding both the heart and the lungs, conditioning it, and reentering at the aorta. Circulation is mechanically sustained usually via a pump oxygenator.

cardiopulmonary resuscitation (CPR) Simultaneous forced ventilation of the lungs and squeezing of the heart ventricles (usually through periodic, externally applied compression of the sternum) to sustain the flow of oxygenated blood throughout the system. The technique is often applied in cases of cardiac arrest, to prevent tissue damage, until other measures can be taken to restore normal respiration and circulation.

cardiopulmonary simulator A device which simulates the physiological functions of a sudden-death victim; normally used as a training aid.

cardiostimulator A device used to stimulate the heart, and/or regulate its beat. See *pacemaker* and *defibrillator*.

cardiotachometer A heart-rate measuring instrument whose reading is proportional to the minute rate of input pulses. The meter is ordinarily calibrated to read directly in beats per minute (BPM).

cardiotachoscope 1. An early electronic

patient-monitoring instrument, devised by M. L. Scheiner and A. Himmelstein in 1950. The instrument provided cathode-ray tube display of a patient's ECG and employed a second cathode ray tube to display heart rate, by controlling the sweep length of a line trace behind a calibrated transparent overlay. Alarm devices were provided to detect out-of-limit heart rate. 2. A type of patient-monitoring instrument dedicated to display of ECG waveform and heart rate.

cardiovascular An all-inclusive term pertaining to the heart and circulatory system.

cardioversion An elective technique for reducing a patient's cardiac rate through the application of a synchronized electrical (direct current) discharge to the precardium. A defibrillator is used, with a low energy setting. Discharge is synchronized to the patient's ECG by detecting the R-wave of the QRS complex and supplying a triggering pulse to the defibrillator at that instant. The object of synchronizing is to ensure that discharge will not occur during the vulnerable period of the T-wave (the refractory period during which the myocardium is repolarizing). Failure to synchronize presents the risk of inducing ventricular fibrillation.

carditis Inflammation of the heart muscle.

card punch A machine having a typewriter keyboard used to punch information into data cards. Also called a *keypunch*.

card punching The basic method for converting source data into punched cards. The operator reads the source document and by depressing keys, converts the information into punched holes.

card reader 1. A mechanism that senses information punched into cards. 2. An input device consisting of a mechanical punch card reader and related electronic circuitry which transcribes data from punch cards to working storage or magnetic tape.

card stacker A machine which stacks punch cards into a pocket or bin after they have been passed through a computer. Also called *card hopper*.

caries Decay of a tooth or a bone.

carotid Name given to one of the two great arteries of the neck which supply oxygenated blood to the brain, and to structures connected with them.

carotid body Specialized tissue found at the bifurcation of the carotid artery (into internal and external carotids) which is sensitive to chemical changes in the blood.

carotid sinus Region of the carotid artery just below its bifurcation which is sensitive to pressure and acts as one of the blood pressure regulating mechanisms. Carotid sinus syncope may occur in individuals who have hypersensitive carotid sinuses; loss of consciousness may follow pressure on the sinus by an abrupt movement of the neck, etc.

carrier 1. High-frequency sinusoidal current which can be modulated with voice or digital signals for bulk transmission via cable or radio circuits. 2. In a semiconductor, the mobile electron or mobile hole which carries the current. 3. A stable isotope, or a normal element, to which radioactive atoms of the same element can be added to obtain a quantity of radioactive mixture sufficient for handling, or to produce a radioactive mixture that will undergo the same chemical or biological reaction as the stable isotope. A substance in weightable amount which, when associated with a trace of another substance, will carry the trace through a chemical, physical, or biological process. 4. An individual who harbors disease organisms in his body and does not manifest symptoms, but who can pass on the infection to others.

cartridge Endless loop tape in a packaged, standardized container.

cartridge fuse A short tube of fiber containing a fusible link or wire which is connected to metallic ferrules at the ends of the tube. Serves to interrupt excessive currents by melting of the fusible link.

castration Removal of the primary sex organs (gonads).

castration anxiety Anxiety due to danger (fantasied) of loss of the genitals or injuries to them. May be precipitated by everyday events that have symbolic significance and appear to be threatening, such as loss of job, loss of a tooth, or an experience of ridicule or humiliation.

catabolism The phase of metabolism associated with the breaking down (lysis) of material by biochemical reaction in living tissue.

catadioptric Optical system combining refractive and reflective elements.

catalepsy A condition usually characterized by trance-like states. May occur in organic or psychological disorders or under hypnosis.

catalyst A substance which initiates and/or accelerates a chemical reaction, but normally does not enter into the reaction.

cataphoresis The introduction of positively charged ions (cations) through the unbroken skin, by means of an electric current.

cataract Opacity of the normally transparent eye lens; this condition leads to impaired vision and stems from hereditary, nutritional, inflammatory, toxic, traumatic, or degenerative causes.

catastrophic failure A sudden failure without warning, as opposed to degradation failure; or a failure whose occurrence can prevent the satisfactory performance of an entire assembly or system.

catatonic state (catatonia) A state characterized by immobility with muscular rigidity or inflexibility and at times by excitability. Virtually always a symptom of schizophrenia.

catelectrotonus The increased sensitivity produced in a nerve or muscle in the region of contact with the cathode when an electric current is passed through it.

cathartic A medicine which hastens bowel evacuation.

catheter Tubular instrument for exploration and for withdrawing or introducing fluids, usually in communication with a body cavity or vessel. Urinary catheters are often inserted through the urethra to withdraw urine from the bladder. Electrodes can also be inserted into catheters for purposes of electrical stimulation of, for example, the heart. See *cardiac pacemakers*. A principal application is in cardiac catheterization, where the catheter may be saline filled and connected to an external pressure transducer. This catheterization procedure allows precise recording of intracardiac and intravascular pressures, by fluid transmission of the pulsatile wave to the transducer. Other cardiac catheterization procedures are conducted to assess cardiac output (by dye or thermodilution methods), and for angiographic study of the heart's arteries.

catheterization Introduction of a tubular instrument called a catheter into a body cavity to inject or withdraw liquids (usually via the bladder for urine withdrawal). Also, a major diagnostic procedure for assessment of heart performance by measure-ment of intracardiac pressures, valve gradients, cardiac output, conduction of activating impulses, and blood sampling. See also *catheter*.

cathodoluminescence Luminescence produced when high-velocity electrons bombard a metal in a vacuum. Small amounts of the metal are vaporized in an excited state and emit radiation characteristic of the metal.

cathode 1. A source of electrons. 2. The negatively charged pole or element from which the current (a flow of electrons) leaves. 3. The negative pole of an electric battery or cell. 4. The output element of a rectifying diode. 5. The terminal of a polarized electrolyitc capacitor which must be connected to the negative, or less positive, terminal of a voltage source. 6. The more noble, lower-potential electrode of a corrosion cell where the action of the corrosion current may reduce or eliminate corrosion, or the negatively charged metallic parts of an impressed-current system. 7. That terminal of a semiconductor diode which is negative with respect to the other terminal, when the diode is biased in the forward direction.

cathode bias Bias obtained by placing a resistor in the return circuit, thereby making the cathode of a vacuum tube positive with respect to ground.

cathode coupling Use of an input or output element in the cathode circuit for coupling energy to another stage.

cathode follower An electron tube circuit having a very low output impedance, a high input impedance, and a negative gain. The input signal is applied between control grid and ground, and the output signal is taken from cathode to ground. See also *emitter follower*.

cathode-ray oscilloscope Test instrument

which makes possible the visual inspection of alternating current waveforms. Consists of an amplifier, time base generating circuits, and cathode-ray tube for translation of electrical energy into light energy.

cathode-ray storage An electrostatic storage device that utilizes a cathode-ray beam for access to the data.

cathode-ray tube (CRT) A vacuum tube in which a beam of electrons is focused on a phosphor screen. The screen glows wherever the beam strikes it. The beam is electronically deflected and its intensity varied in accordance with signals supplied to the CRT, thereby providing a visual display of an electronic signal. The familiar television picture tube is a cathode-ray tube, as are the screen display tubes of patient monitors, physiological recorders, and computer video terminals.

cation Positive ion that moves toward the cathode in a discharge tube, electrolytic cell, or similiar equipment. The corresponding negative ion is called *anion*.

catwhisker A small, sharp-pointed wire used to make contact with a sensitive point on the surface of a semiconductor. The early semiconductor diodes and the earlier crystal detectors used catwhiskers.

caudad Toward the tail, or cauda. In man, it means downward.

cauterize To produce tissue destruction by chemical, electrical, thermal, cryogenic, or mechanical methods.

cavitation 1. The turbulent formation including growth and collapse of bubbles in a fluid. It occurs when the static pressure at any point in fluid flow is less than fluid vapor pressure. 2. The production of gas filled cavities in a liquid when the pressure is reduced below a certain critical value with no change in the temperature. Ordi-

narily this is a destructive effect, as the high pressures produced when these cavities collapse often damage mechanical components of hydraulic systems, but the effect is turned to advantages in ultrasonic cleaning.

cavitation noise The noise produced in a liquid by the collapse of bubbles that have been created by cavitation.

CCD Abbreviation for charge-coupled device. A device in which circuits made by MOS technologies allow charges to be transferred from point to point in a sequential manner.

-cele Suffix meaning hernia, protrusion, swelling, or tumor.

cell 1. A living protoplasmic mass containing a nucleus. The protoplasm of the nucleus is the karyoplasm and that of the remainder of the cell is the cytoplasm. Many different cellular structures make up the body organs and tissues, all sharing this common arrangement. 2. Single unit, such as a battery, that produces a direct voltage by converting chemical energy into electrical energy. 3. Single unit that produces a direct voltage by converting radiant energy into electrical energy. 4. Single unit that produces a varying voltage drop because its resistance varies with illumination. 5. Elementary unit of storage in a computer. 6. A microscopic mass of protoplasm bounded by a semipermeable membrane, usually including one or more nuclei and various nonliving products, capable alone, or interacting with other cells, of performing all the fundamental functions of life.

cell counter An electronic instrument used to count the white or red blood cells and other very small particles.

cell membrane The cell boundary, about 100 angstroms thick, consisting of a bimolecular, lipid (fat) layer with an outer, protein coat. The excitability of nerve and muscle cells depends upon the existence of a resting potential across the cell membrane.

celsius temperature scale The metric scale of temperature, on which water freezes at 0 degrees C and boils at 100 degrees C. Absolute zero is −273 degrees C. To convert a Centigrade temperature to Fahrenheit, multiply by 9/5 and add 32.

-centeses Suffix meaning puncture.

centigrade temperature scale The older name for the Celsius temperature scale. Although officially abandoned by international agreement in 1948, the term is still in common usage.

central nervous system General term incorporating the brain and spinal cord, as opposed to the peripheral nervous system which includes the nerves and sensory receptors outside the brain and spinal cord. Disease of the neurons and supporting cells that constitute the brain and spinal cord is always serious since it is believed by many authorities that nerve cells, once destroyed, can never be replaced by the division of other neurons.

central processing unit That component of a computing system which contains the arithmetic, logical, and control circuits of the system, and which performs the interpretation and execution of instructions. Synonymous with *main frame*.

cephalad Toward the head; the opposite of caudad.

cephalic Directed towards or situated on, in, or near the head.

ceramic A clay-like material, consisting mainly of aluminum and magnesium oxides which after molding and firing is used as insulation. It will withstand high temperatures and is less fragile than glass. Glazed ceramic is porcelain.

ceramic-based microcircuit A microminiature circuit printed on a ceramic substrate. Usually consists of combinations of resistive, capacitive, or conductive elements fired on a wafer-like piece of ceramic.

ceramic capacitor A capacitor whose dielectric is a ceramic material such as steatite or barium titanate, the composition of which can be varied to give a wide range of temperature coefficients. The electrodes are usually silver coatings, fired on opposite sides of the ceramic disc or slab, or fired on the inside and outside of a ceramic tube. After connecting leads are soldered to the electrodes, the unit is usually given a protective insulating coating.

cerebellum 1. Outgrowth from the hindbrain overlying the medulla oblongata. Concerned with the coordination of movement. 2. A large dorsally projecting part of the brain, especially concerned with the coordination of muscles and the maintenance of bodily equilibrium.

cerebral cortex See *brain*.

cerebral hemorrhage Rupture of an artery of the brain (usually in the region of the internal capsule), due to either high blood pressure or disease of the arteries. Escape of blood causes destruction of brain tissue, and paralysis occurs of that side of the body which is opposite to the injured side of the brain. If the hemorrhage has occurred on the left side of the brain, then speech is affected in right-handed subjects. Also called *cerebral vascular accident*.

cerebral palsy A condition in which the control of the motor system is affected due to a lesion in the brain resulting from a birth injury or prenatal defect. The popular term is *spastic*.

cerebral thrombosis Thrombus formation

in vessels supplying the cerebral cortex or its major connections.

cerebral vascular accident (CVA) See *cerebral hemorrhage.*

cerebrospinal fluid The fluid which surrounds the brain and spinal cord.

cerebrum The larger part of the brain occupying the cranium. See *brain.*

cermet A metal-dielectric mixture used in making thin film resistive elements. The first half of the term is derived from *cer*amic and the second half from *met*al.

cermet films Film materials with two or more phases which are combinations of metals and dielectrics. They are used to circumvent the resistivity temperature-coefficient relationship for monophasic materials, in order to obtain large sheet resistances with low temperature coefficients.

cerumen Wax found in the ears.

CGS (centimeter-gram-second) system The metric system of units for expressing the magnitude of space, mass, and time. The fundamental units of these quantities are the centimeter, gram, and second.

chain reaction A reaction that stimulates its own repetition. In a fission chain reaction, a fissionable nucleus absorbs a neutron and fissions, releasing additional neutrons. These, in turn, are absorbed by other fissionable nuclei, releasing still more neutrons. A fission chain reaction is self-sustaining when the number of neutrons released in a given time equals or exceeds the number of neutrons lost by absorption in nonfissioning material, or by escape from the system.

chance failure See *random failure.*

channel 1. Portion of the spectrum assigned for the operation of a specific carrier and the minimum number of sidebands necessary to convey intelligence. 2. Single path for transmitting electric signals. NOTE: The word path includes separation by frequency division or time division. Channel may signify either a one-way or two-way path, providing communication in either one direction only or in two directions. See also *alternate channel.* 3. That portion in electric computers of a storage medium which is accessible to a given reading station. 4. Path along which information, particularly a series of digits or characters, may flow. 5. In computer circulating storage, one recirculating path containing a fixed number of words stored serially by word. 6. The main current path between the source and the drain in a field effect transistor or other semiconductor. 7. The smallest subdivision of a circuit (or of a trunk route), by means of which a single type of communication service is provided, i.e., a voice channel, telemetry channel, or a data channel. 8. An instrumentation term referring to an amplifying input of a recorder having many such inputs. The signal applied to a recorder channel is amplified and outputted onto paper, film, or tape, for subsequent study.

channel capacity 1. Maximum number of binary digits or elementary digits to other bases which can be handled in a particular channel, per unit time. 2. Maximum possible information transmission rate through a channel at a specified error rate. Channel capacity may be measured in bits per second or bauds. 3. Also used to denote total number of individual channels within a telemetry, recording, video, or communication system.

channel designator Number associated with each channel, tributary, or trunk for reference purposes. Also called *channel sequence number.*

channel pulse Telemetering pulse representing intelligence on a channel by virtue of its time or modulation characteristics.

character 1. One symbol of a set of elementary marks or events that may be combined to express information. A group of characters, ·in one context, may be considered as a single character in another, as in the binary coded decimal system. 2. One of a set of symbols used to present information on a display device. 3. The electrical, magnetic, or mechanical profile used to represent a character in a computer, and its various storage and peripheral devices. A character may be represented by a group of other elementary marks, such as bits or pulses.

character density The number of characters that can be stored per unit of length; e.g., on some makes of magnetic tape drives, 200, 556, or 1100 bits can be stored serially, linearly, and axially to the inch.

character display tube A form of cathode-ray tube in which the cathode-ray beam can be shaped by either 1. electrostatic or electromagnetic deflection, or 2. passing the beam through a mask, into symbols or letters.

character generation technique 1. The manner in which characters are formed on the face of a cathode-ray tube; point placement, stroke, monoscope or extrusion is given. 2. The method of computing the path of the electron beam while tracing the character; analog, digital, mask, fixed-scan, or by software.

character generator 1. A unit that accepts input in the form of one of the alphanumeric codes and prepares the electrical signals necessary for its display in the proper position on a Dot Matrix, TV system, or CRT. 2. That part of the display controller which draws alphanumeric char-

acters and special symbols for the screen. A character is automatically drawn and spaced every time a character code is interpreted. 3. A hardware or software device which provides the means for formulating a character font and which also may provide some controlling function during printing.

characteristic curve A curve plotted on graph paper which shows the relationship between two changing values.

characteristic impedance 1. The impedance that a transmission line would have if it were infinitely long. 2. Ratio of voltage to current at every point along a transmission line on which there are no standing waves. 3. Square root of the product of the open and short-circuit impedance of the line.

character printer Device capable of producing hard copy in which printing is accomplished a character at a time.

character reader An input device of a computer which can directly recognize printed or written characters, there being no need to first convert them into punched holes or polarized magnetic spots.

charge 1. To replenish the electrical charge in a battery. 2. To store electrical energy in a capacitor. 3. To accumulate an excess of electrons (negative charge) or scarcity of electrons (positive charge) on an object by means of friction or induction.

charge-coupled device (CCD) A high-density charge storage device finding use in high-speed serial memory applications. CCD processing is closely related to MOS technology.

charged particle An ion; an elementary particle that carries a positive or negative electric charge.

charger A device, rectifier, or DC generator which supplies direct current at a voltage suitable for charging a storage battery.

charger-eliminator A battery charger with a low-noise, low-impedance output which can either charge a storage battery or supply a dc load directly, without a storage battery in parallel.

charge transfer The process in which an ion takes an electron from a neutral atom with a resultant transfer of electronic charge. In order for this to be possible, however, the ion and the atom must be of the same species.

charging The process of supplying electrical energy for conversion to stored chemical energy.

charging current The current which flows to a capacitor when it is first connected to a source of electrical potential.

chassis The metal box or frame on and in which components of an electronic equipment are mounted.

check Of a computer, a process of partial or complete testing of the correctness of machine operations, the existence of certain prescribed conditions within the computer, or the correctness of the results produced by a program. A check may be made automatically by the equipment or may be programmed.

check bit A redundant bit in a computer carried with a group of bits in such a way that an inaccurate retrieval of that group of bits is detected if the number of errors is odd.

check character One or more characters carried in a computer in such a fashion that if a single error occurs, excluding compensating error will be reported.

check digit In computers, one or more redundant digits in a character or word, de-

pending upon the remaining digits, so that if a given digit changes from the desired state the malfunction is easily detected.

chelate Molecule in which a central inorganic ion is covalently bonded to one or more organic molecules, with at least two bonds to each molecule, used as a laser dopant.

chelating agent Substance which forms a complex with metals, thus rendering them chemically inactive.

chemical dosimeter A detector for indirect measurement or radiation by indicating the extent to which the radiation causes a definite chemical change to take place.

chemiluminescence Emission of light during a chemical reaction.

chemoreceptor Nerve-ending capable of detecting and differentiating substances according to their chemical structure by contact with the molecules of the substances, e.g., taste, smell.

chemotherapy The prevention or treatment of a disease by chemicals which act as antiseptics within the body, usually without serious toxic effects on the subject.

Cheyne-Stokes respirations Respirations with rhythmical variations in intensity occurring in cycles, often with periods of apnea.

chill Involuntary contractions of the voluntary muscles, with shivering and shaking.

chip (die) 1. A single piece of silicon which has been cut from a slice by scribing and breaking. It can contain one or more circuits, but is packaged as a unit. 2. In semiconductor usage, a single piece of material (germanium, silicon, etc.) containing one or more diodes, transistors, or other components which are interconnected or which may be connected externally to form a circuit. 3. A single substrate on which all the active and passive circuit elements have been fabricated using one or all of the semiconductor techniques of diffusion, passivation, masking, photoresist, and epitaxial growth. A chip is not ready for use until packaged and provided with external connectors. The term is also applied to discrete capacitors and resistors which are small enough to be bonded to substrates by hybrid techniques. 4. A tiny piece of semiconductor material scribed or etched from a semiconductor slice on which one or more electronic components are formed. The total number of usable chips obtained from a wafer is the yield.

chloramphenicol An important broad-spectrum antibiotic used to treat typhoid fever, certain types of food poisoning, and other infections caused by microorganisms that are resistant to other antibiotics.

chloroquine compounds Antimalarial drugs also used against certain forms of extraintestinal amebiasis. They are also used in a few unrelated diseases, such as arthritis.

chlorpromazine and other tranquilizers Over 30 compounds with mild depressant action used in certain mental illnesses and anxiety states, and for antinauseant and sedative effects.

choke An inductance used in a circuit to present a high impedance to higher frequencies, but not appreciably limiting the flow of direct current. Also called *choke coil* or *inductor*.

choke-input filter A power supply filter in which the first element encountered by current from the rectifier is a series inductance.

cholecystography Radiographic examination of the gall bladder and bile duct. The bile is rendered visible by the introduction of a radio-opaque substance which is usually ingested some time before the procedure is undertaken.

chopper 1. A device which interrupts a direct current signal at an audio rate so that it can be passed through a transformer. 2. An electromechanical or electronic device used to interrupt a dc or low-frequency ac signal at regular intervals to permit amplification of the signal by an ac amplifier. It may also be used as a demodulator to convert an ac signal to dc. 3. A rotating shutter for interrupting an otherwise continuous stream of particles. Choppers can release short bursts of neutrons with known energies; used to measure nuclear cross sections.

chordae tendineae Thin, musculotendinous bands extending between the walls of the ventricles of the heart and the tricuspid and mitral valves.

chromatograph An instrument used in analytical chemistry to determine contents of a chemical mixture through the use of preferential adsorption of gases or liquids in an ascending molecular weight order, onto a solid adsorptor such as activated charcoal, alumina, or silica gel.

chromatography Separation of components of a mixture by their physical properties.

chromosomes When a cell divides, the genetic material present in the nucleus becomes segregated into thread-shaped bodies which are visible under the microscope. These are known as chromosomes and consist of connected strands of DNA molecules known as genes. In man there are 46 chromosomes per cell: 22 pairs of autosomes and two sex chromosomes; females have two X chromosomes, males one X and one Y. The Y chromosome is shorter than the X chromosome, so that male offspring having an X-Y pair continue to be the determinants of sex in succeeding generations.

chronaxy 1. The time required for the excitation of a nervous element by a definite stimulus. 2. The minimum time at which a current just double the rheobase will excite contraction.

chronic bronchitis Lung disease characterized by productive cough for at least two successive months for two years.

chronic exposure Term used to denote radiation exposure of long duration, by fractionation or protraction.

chronic obstructive pulmonary disease (COPD) Lung disease characterized by cough, dyspnea, wheeze, and spitting. Thus, there are features of emphysema, chronic bronchitis, and asthma.

chronic renal failure This may be the end result of a large number of widely different kidney diseases.

chronistor A subminiature elapsed-time indicator which employs electroplating principles to totalize equipment operating time up to several thousand hours.

Ci Abbreviation for curie.

-cid(e) Suffix meaning cut or kill.

cilium A minute hairlike process attached to the free surface of a cell, as in the nose.

cinefluorographic study Use of a motion picture camera to record the visual images produced by a fluoroscopic instrument, so that a moving picture record of a procedure can be made, and reviewed as often as desired. Used in a variety of studies, including cinecoronary angiography.

circuit 1. Electronic path between two or more points, capable of providing a number of channels. 2. Number of conductors connected together for the purpose of carrying an electrical current. 3. Connected

assemblage of electrical components such as resistors, capacitors, and inductors having desired electrical characteristics. 4. A combination of electrical components that performs some function, such as amplification or detection.

circuit breaker 1. A device which interrupts the flow of current in a circuit when the current exceeds a predetermined value. 2. A device which interrupts the flow of current when the current reverses direction, or when there is a predetermined deviation of current, voltage, or impedance from a standard value. 3. An electromagnetic device that opens a circuit automatically when the current exceeds a predetermined value. It can be reset by operating a lever or by other means.

circuit component An element of a circuit, such as a resistor, capacitor, coil, transformer, or semiconductor.

circuit elements Discrete units of resistance, inductance, and capacitance which, when interconnected, form an electrical circuit.

circulating currents Electric currents through the axon and surrounding solution by which an impulse at one region of an axon membrane stimulates an adjacent region, resulting in conduction of the impulse.

circulating memory A type of memory in which a data stream circulates in a loop. One example is a string of shift-register stages with the last output connected to the first input. At every clock pulse, a particular bit is accessed as it passes a certain point in the circuit. Circulating memories also use other delay techniques, including electrical and acoustical delay lines.

circulating storage A device, using a delay line or shift register, which stores information in a train or pattern of pulses, in which the pattern of pulses issuing at the final end is sensed, amplified, reshaped, and reinserted into the delay line at the beginning end.

circulation Arterial blood received into the left atrium passes through the mitral valve to the left ventricle. It is then discharged into the aorta, and through its smaller arterial branches, to the capillaries, into small veins, then larger, until on reaching the superior and inferior venae cavae it passes into the right atrium. The venous blood which is received into the right atrium passes through the tricuspid valve into the right ventricle, and from there into the pulmonary artery, which divides into two branches, one going to each lung. The artery divides in the lung into capillaries, and here, the blood, by means of the hemoglobin in the red corpuscles, takes up oxygen from the inspired air. Oxygenated blood returns to the heart by the four pulmonary veins, two from each lung, entering the left atrium.

Veins from the pancreas, spleen, stomach, and intestines unite behind the pancreas and form the portal tube or vein. This takes blood, rich in the products of digestion, to the liver where it divides into smaller vessels and capillaries. Blood leaves the liver by the hepatic veins which enter the inferior vena cava.

circulatory failure Blood flow inadequate for metabolic needs.

circumcision Removal of all or part of the foreskin of the penis.

cirrhosis Chronic liver ailment, characterized by an increase in its fibrous support tissue that results in a progressive destruction of liver cells and impairment of the organ's functions.

-cis Suffix meaning cut or kill.

clamping Preventing a pulsed signal from

drifting with respect to a reference voltage by tying some repetitive portion of the waveform to that reference voltage. Also called *dc restoration*.

clamping diode A diode used to fix a voltage level at some point in a circuit.

clamp-on-ammeter An alternating current ammeter which has a built-in current transformer whose core can be clamped around the conductor in which current is to be measured. The magnetic circuit thus formed allows measurement of current flow in the conductor, without breaking the conductor.

-clasia Suffix meaning remedy.

class A amplifier Amplifier in which the input bias value of a device is chosen for operation in the linear portion of its characteristic curve. The output signal is an amplified duplicate of the input signal.

class AB amplifier Amplifier in which the input bias and alternating input signal are such that the output current flows for appreciably more than half, but less than the entire input signal cycle.

class B amplifier Amplifier in which the input bias is approximately equal to the cut-off value so that the output current is approximately zero when no input signal is applied, and flows for approximately half of each cycle when an alternating input signal is applied.

class C amplifier Amplifier in which the input bias is appreciably greater than the cutoff value, so that the output current is zero when no alternating input signal is applied, and flows for appreciably less than half of each cycle when an alternating input signal is applied.

clavicle The collar bone. It articulates with the sternum and the scapula.

clearing time That interval required, for

chemical action of fixer or an X-ray or graphic record emulsion, to dissolve unexposed salts.

clear terminal See *reset terminal.*

-cleisis Suffix meaning closure.

clinic 1. A place for the care of the sick and the acquisition of knowledge about disease. 2. A building, or part of a building, in which medical instruction can be given to students at the bedside, or in the presence of the patient, based upon the instructing physician's observation and examination of the patient's condition throughout the course of a disease. 3. A place where ambulatory patients can be given medical treatment.

clinical 1. The findings of a physician regarding a patient's disease through personal observation and examination at the bedside, in contrast with the findings obtained through laboratory studies. 2. Bedside treatment, as well as the care of ambulatory patients, conducted for instructional and therapeutic purposes.

clinical scalar electrodardiogram See *electrocardiogram.*

clipping circuit A circuit which removes, by electronic means, one or both peak excursions of a waveform, at a predetermined level.

clock 1. The primary source of synchronizing signals for a computer data system or data transmission system. In high-speed systems, the clock may be an oscillator, the output of which is used as a reference or timing frequency by the system. 2. A device that measures and indicates time of day or specific operations. 3. Equipment providing a time base used in a transmission system to control the timing of certain functions such as the duration of signal elements, the sampling, etc.

clock circuit 1. A circuit which provides accurately timed pulses of uniform length which can be used to control and synchronize other circuits. 2. An electronic circuit employing analog or digital techniques to measure time intervals.

clocked flip-flop A flip-flop circuit that is set and reset at specific times by adding clock pulses to the input so that the circuit is triggered only if both trigger and clock pulses are present simultaneously.

clock frequency In digital computers, the master frequency of periodic pulses which are used to schedule the operation of the computer.

clock input That terminal on a flip-flop whose condition or change of conditions conrtols the admission of data into a flip-flop through the synchronous inputs and thereby controls the output state of the flip-flop. The clock signal performs two functions: 1. It permits data signals to enter the flip-flop. 2. After entry, it directs the flip-flop to change state, accordingly.

clock pulse A pulse which is used to gate information into a flip-flop when it is used in the synchronous mode. (In J-K connected flip-flops it will cause counting if the data inputs are both held at a logic 1.)

clock rate The rate at which a word or characters of a word (bits) are transferred from one internal computer element to another. Clock rate is expressed in cycles (if a parallel-operation machine—words; if a serial operation machine—bits) per second.

clofibrate A drug used in reducing elevated cholesterol and triglyceride concentrations in the blood.

clomiphene citrate A drug that improves fertility, sometimes causing multiple births.

closed circuit An electrical circuit

through which current can flow, such as when a power switch is turned on. The opposite of an open circuit, through which current cannot flow.

closed circuit jack A jack which has its through circuits normally closed. Circuits are opened by inserting a mating plug.

closed loop 1. An automatic control system in which feedback is used to link a controlled process back to the original command signal. The feedback mechanism compares the actual controlled value with the desired value, and if there is any difference, an error signal is created that helps correct the variation. In automation, feedback is said to close the loop. 2. A system employing a man-machine interface, which places the intelligence and/or coordination of the human into the loop.

closed loop system A family of automatic control units in which the output of a system is fed back for comparison with the input, the purpose being to reduce any difference between input command and output functional response.

closed loop voltage gain The voltage gain of an amplifier with feedback.

clotting time Time taken for blood to clot when bleeding has occurred. Normal time for human blood is 4 to 13 minutes at 37°C.

cloud chamber A device in which the tracks of charged atomic particles such as cosmic rays or accelerator beams are displayed. It consists of a glass-walled chamber filled with a supersaturated vapor, such as wet air. When charged particles pass through the chamber, they trigger a process of condensation, and so produce a track of tiny liquid droplets, much like the vapor trail of a jet plane, to permit study of the particle's motions and interactions.

-clysis Suffix meaning injection.

cm Abbreviation for centimeter. One-hundredth of a meter.

CMOS Abbreviation for complementary metal-oxide semiconductor. Family of devices implementing MOS technology. Digital logic is performed by both N-channel and P-channel field-effect transistors, also amplification. Favored for low power disipation, good noise immunity, faster speeds than ordinary MOS devices, adaptability to wide range of logic levels (swings), wide operating-temperature range, adaptability to power-supply variations, and excellent fanout DTL/TTL compatability. Disadvantages include more complicated processing, limited system interface, and larger chip size.

coagulation 1. This term has three relevant interpretations: In the hematological sense, it refers to the clotting of blood. In the surgical sense, electric coagulation is the breaking down of tissue into a necrotic mass by means of the application of bipolar current. On the other hand, a laser photocoagulator is a device that focuses intense light on the posterior aspect of a patient's eye for purposes of welding a detached retina by means of small burns. 2. Thickening of a fluid into curds or clots (almost always applied to blood).

coarctation 1. A compression of the wall of a vessel, narrowing its lumen and reducing the volume (or flow). 2. A stricture or occlusion resulting from outside force deforming a vessel.

coaxial cable Two electrical conductors sharing the same axis, constructed with one conductor as a wire, passing through the exact center of a tube, which is the second conductor and acts as a shield. The conductors are insulated from each other by a solid nonconductor running the length of the cable, or by spacers placed at intervals. This system is most desirable in circuits transmitting high frequencies, for it minimizes transmission line losses.

coaxial line A transmission line consisting of a central wire or tube completely surrounded by an outer tubular conductor. The two conductors have a common axis, and are separated by a solid dielectric or by dielectric spacers, sometimes augmented by an insulating gas. A coaxial line has no external field, and is not disturbed by external fields.

cobalt Trace element. Its absence from the diet of the young may lead to anaemia.

cobalt bomb Source of irradiation in deep X-ray therapy.

COBOL Abbreviation for common business–oriented language.

cochlea A division of the labyrinth of the ear of higher vertebrates that is usually coiled like a small shell and is the seat of the hearing organ.

Cockcroft-Walton accelerator A device for accelerating charged particles by the action of a high direct current voltage on a stream of gas ions in a straight insulated tube. The voltage is generated by a voltage multiplier system consisting essentially of a number of condenser pairs connected through electronic switching devices. The particles (which are nuclei of an ionized gas, such as protons from hydrogen) gain energies of up to several million electron volts from the single acceleration so produced. Named for the British physicists, J. D. Cockcroft and E. T. S. Walton, who developed this machine in the 1930s.

code 1. A system of characters and rules for representing information in a language that can be understood and handled by the computer. See also *ASCII*. 2. A system of symbols used to represent information in a form suitable for transmission or processing. 3. To prepare a routine in machine language for a specific computer. 4. Encode, to express given information by means of a code.

coded decimal Describing a form of notation by which each decimal digit separately is expressed in another number system; e.g., in the 8-4-2-1 coded decimal notation, the number twelve is represented as 0001 0010, for 1 and 2; whereas in pure or straight binary notation it is represented as 1100. Other coded decimal notations used are the 5-4-2-1, the excess three, and the 2-3-2-1 codes.

code line A single instruction usually written on one line, in a code for a specific computer to solve a problem. The instruction is usually stored as a whole in the program register of the computer while it is executed, and it may contain one or more addresses of registers or storage locations in the computer where numbers or machine words are to be obtained or sent, and one or more operations to be executed.

coding A system of symbols and rules that tell the computer how to handle information—where to get it, what to do with it, where to put it, where to go for the next step, etc.

coercive force The value of magnetomotive force required to reduce residual magnetic flux to zero.

coherent electroluminescence devices See *diode laser.*

coherent light 1. A single frequency of light. 2. Light having characteristic similar to a radiated radio wave that has a single frequency.

coherent radiation 1. Radiation whose waves are in phase in time and space.

2. Radiation in which the phase between any two points in the radiation field has a constant difference, or is exactly the same throughout the duration of the radiation.

cohesion A force which unites particles.

coil One or more turns of wire, in a cylindrical or doughnut shape (torus), used to create a magnetic flux or to add inductance to a circuit.

coincidence circuit A circuit consisting of several gates in parallel, from which an output signal is produced only when all inputs are present simultaneously.

coincidence counting The use of electronic devices to detect when two or more pulses from separate counters occur within a given time interval. This is done to determine whether the pulses were produced by the same particle, for example in scintillation counting, or whether they correspond to the same event.

Col Abbreviation for collimator.

cold Idiomatic expression applied, in general, to electrical circuits that are disconnected from voltage supplies and at ground potential. Opposed to hot, which means carrying an electrical charge.

cold cathode A cathode in an electron tube which releases electrons at room temperature, aided by the presence of gases within the tube envelope. (Contrasted with thermionic emission occurring when a cathodic material is raised in temperature by an electrical heater wire, as in most electron tubes.)

cold joint A soldered connection which was inadequately heated with the result that the wire is held in place by rosin flux, not solder.

collagen An albuminoid supportive protein found in connective tissue.

collector 1. In a transistor, the region into which majority carriers flow from the base under the influence of a reverse bias across the two regions. The collector is analogous to the plate of a vacuum tube. 2. The external terminals of a transistor that is connected to this region. 3. In certain electron tubes, such as the klystron, an electrode to which electrons or ions flow after they have completed their function.

collector cutoff The operating condition of a transistor when collector current is reduced to the leakage current of the collector-base junction.

collector junction Of a transistor, a junction normally biased in the high-resistance direction, through which the current can be controlled by the introduction of minority carriers.

colloid A suspension of submicroscopic particles (typically, between 1 and 100 millimicrons in diameter), uniformly distributed throughout a fluid suspending medium.

color code 1. A system of colors applied to the insulation of cable conductors to permit their easy identification. The traditional color code is based on repetitive use of the five colors blue, orange, green, brown, and slate. 2. A code specified by the EIA to identify the electrical value and accuracy of a resistor or capacitor.

colorimeter Optical or photoelectric instrument for measuring color differences or colors; basic analytical instrument for analysis of chemical compounds and solutions by colorimetric technique; also used for measuring color of blood for hemoglobin analysis.

color video display equipment Electronic means for presenting diagnostic scan in color tones rather than shades of gray on television monitor.

colostomy Surgical creation of an opening into the colon, thereby forming a new anal aperture.

colpitts oscillator Oscillator in which a parallel-tuned circuit is connected between a vacuum tube grid and plate, or junction transistor base and collector, and in which the tank capacitance contains two voltage-dividing capacitors in series, with their common connection at cathode or emitter potential. When these two capacitances are the plate-to-cathode and grid-to-cathode capacitances of a vacuum tube, then the circuit is known as an *ultra-audion oscillator*. Other configurations are possible, but the tapped capacitance is the distinguishing circuit feature.

column-binary code Code used with punched cards, in which successive bits are represented by the presence or absence of punches on contiguous positions in successive columns as opposed to rows. Used in connection with 36-bit-word computers, where each group of three columns is used to represent a single word.

column values A binary word is read from right to left, with each digit representing an ascending power of 2. Thus, a five-bit word would have the column values: $2^4 = 16$, $2^3 = 8$, $2^2 = 4$, $2^1 = 2$, and $2^0 = 1$. See *binary coded decimal*.

coma 1. Insensibility, stupor, or sleep. 2. State of unconsciousness.

comatose State of coma.

command A group of signals or pulses initiating one step in the execution of a computer program.

commissure Bundle of nerve fibers connecting the right and left sides of the brain and spinal cord.

common 1. Shared by two or more circuits. Used to designate the terminal of a three-terminal device that is shared by the input and output circuits. (Thus, a transistor may be operated in a common-base configuration, a common-collector configuration, or a common-emitter configuration.) Vaccum tube connections may be characterized in a similar way, but grounded is normally used instead of common. 2. A point that acts as the reference potential for several circuits; a ground.

common base Transistor circuit configuration in which the base terminal is common to the input circuit and to the output circuit and in which the input terminal is the emitter terminal and the output terminal is the collector terminal.

common-base feedback oscillator A common-base bipolar transistor amplifier with a positive feedback network between the collector (output) and the emitter (input). Loop gain is adjusted to produce oscillations at a desired frequency, usually determined by selected values of resistance and capacitance, a tuned circuit, or a piezoelectric crystal.

common business oriented language (COBOL) A specific language by which business data processing procedures may be precisely described in standard form. The language is intended not only as a means for directly presenting any business program to any suitable computer, for which a compiler exists, but also as a means of communicating such procedures among individuals.

common collector Transistor circuit configuration in which the collector is common to the input circuit and to the output circuit and in which the input terminal is the base terminal and the output terminal is the emitter.

common-collector connection See *grounded-collector amplifier*.

common emitter Transistor circuit configuration in which the emitter is common to the input circuit and to the output circuit and in which the input terminal is the base and the output terminal is the collector.

common-emitter connection See *grounded-emitter amplifier.*

common language A method of computer representation which is common to a related group of digital computers; that is, fortran is a common language on many makes of computers.

common machine language A machine sensible information representation which is common to a related group of data processing machines. A distinguishing characteristic is that, generally, one line of coding specifies one instruction to the computer.

common mode characteristics A principal characteristic of differential and operational amplifiers, descriptive of performance. Both inverting and noninverting inputs receive a common signal, and so, no difference should exist between the two; hence, a perfect amplifier would have no output. In biomedical applications such as ECG, EEG, and EMG measurement, artifact rejection is closely related to the common mode characteristic of the input amplifier stage.

common mode error (CME) (Referred to the input.) A generally small offset voltage appearing between the input terminals of a differential operational amplifier as a function of the common-mode voltage.

common-mode gain The ratio of the output voltage of a differential amplifier to the common-mode input voltage. The common-mode gain of a theoretically ideal differential amplifier is zero.

common-mode input That signal applied in phase (i.e., common mode) equally to both inputs of a differential amplifier.

common-mode input capacitance The equivalent capacitance of both inverting and noninverting inputs of an operational amplifier with respect to ground.

common-mode input impedance The open-loop input impedance of both inverting and noninverting inputs of an operational amplifier with respect to ground.

common-mode input resistance The equivalent resistance of both inverting and noninverting inputs of an operational amplifier with respect to ground or reference.

common-mode input voltage The maximum voltage that can be applied between the two inputs of a differential amplifier, together, and ground (or reference), without causing damage.

common-mode output voltage The output voltage of an operational amplifier resulting from the application of a specified voltage common to both inputs.

communicable disease A disease capable of being transmitted from one person to another.

commutation 1. Mechanical process of converting the alternating current which flows in the armature of direct current generators into the direct current generator output. 2. Sampling of various quantities in a repetitive manner, for transmission over a single channel in telemetering. 3. Repetitive sampling of two or more information channels for transmission over a single data link. 4. The process by which a silicon controlled rectifier (SCR), in the conducting state, is made to revert to the nonconducting state. This generally requires shunting away current from the conducting junctions or opening the circuit.

commutation duty cycle 1. In a multichannel telemetry system, the actual data link *on time* for a particular telemetry channel, expressed as a percentage of the total time allocated for that channel. 2. Channel dwell period expressed as percent of channel interval.

commutation rate The number of commutator inputs sampled per second. See *channel sampling rate.*

commutator The part of a moving electromechanical device which makes contact with a fixed conductor in order to transmit electrical energy to or from the electromechanical device. In a motor or generator, the commutator is the part of the revolving armature against which the stationary brushes rest.

comparator 1. A circuit or device which is designed to reval the quantitative or qualitative difference between two inputs. 2. An analog comparator is capable of comparing two allegedly identical inputs and can reveal differences in polarity, phase, or amplitude. 3. A digital comparator (typically an exclusive NOR or coincidence gate) is capable of comparing logic states and duration. 4. A circuit which compares two inputs and normally provides zero output while these two are in agreement. The output can be made proportional to degree of difference (analog) or can rise to a logic level (digital) when a *miscompare* occurs. 5. A circuit intended to ensure that noise or artifact are not accepted by a data system as *signal,* by comparing two versions of the signal and rejecting an input when there is a difference.

compatibility 1. The ability of one unit to be used with another without detrimental effect on the signal through mismatch. 2. The congruity of a medicine, or substance in a medicine, with respect to another medicine, as demonstrated by their ability to mix without deleterious effect or loss of therapeutic power. 3. The ability of two tissues from different donors to merge without rejection.

compatible monolithic integration circuit Device in which passive components are deposited by thin-film techniques on top of a basic silicon-substrate circuit containing the active components and some passive parts. See also *all-diffused monolithic integrated circuit.*

compensated amplifier An amplifier which is made wideband by the addition of low-frequency compensation and high-frequency compensation.

compensation 1. A defense mechanism, operating unconsciously, by which the individual attempts to make up for (i.e., to compensate for) real or fancied deficiencies. 2. A conscious process in which the individual strives to make up for real or imagined defects in such areas as physique, performance, skills, or psychological attributes.

compile 1. To prepare a machine language program from a computer program written in another programming language by performing the usual functions of an assembler, and also making use of the overall logical structure of the program, or generating more than one machine instruction for each symbolic statement, or both. 2. To produce a machine language routine from a routine written in source language by selecting appropriate subroutines from a subroutine library, as directed by the instructions or other symbols of the original routine supplying the linkage, which combines the subroutines into a workable routine and translating the subroutines and linkage into machine language. The compiled routine is then ready to be loaded into storage and run; for example, the compiler does not usually run the routine it produces.

compiler 1. A computer program which translates a higher level language into either assembly language or machine language. 2. A computer program more powerful than an assembler. In addition to its translating function, which is generally the same process as that used in an assembler, it is able to replace certain items of input with series of instructions, usually called subroutines. Thus, where an assembler translates item for item, and produces as output the same number of instructions or constants which were put into it, a compiler will do more than this. The program which results from compiling is a translated and expanded version of the original. See also *assembler*.

complement 1. A quantity expressed to the base N, which is derived from a given quantity by a particular rule; frequently used to represent the negative of the given quantity. 2. A complement of N, obtained by subtracting each digit of the given quantity from N-1, adding unity to the least significant digit, and performing all resultant carrys; e.g., the twos complement of binary 11010 is 00110; the tens complement of decimal 456 is 544. 3. A complement of N-1, obtained by subtracting each digit of the given quantity from N-1; e.g., the ones complement of binary 11010 is 00101; the nines complement of decimal 456 is 543.

complementary MOS (CMOS) An MOS or IC device involving both P-channel and N-channel MOS-FETS. This technique increases logic speed but requires additional processing steps that reduce circuit density and raise cost per function.

complex A group of associated ideas that have a common emotional tie. These are largely unconscious and significantly influence attitudes and associations. Examples are: Inferiority complex—feelings of inferiority stemming from real or imagined physical or social inadequacies that may cause anxiety or other adverse reactions. The individual may overcompensate by excessive ambition or by the development of special skills, often in the very field in which he was originally handicapped. Oedipus complex—attachment of the child for the parent of the opposite sex, accompanied by envious and aggressive feelings toward the parent of the same sex. These feelings are largely repressed (i.e., made unconscious) because of the fear of displeasure or punishment by the parent of the same sex. In its original use, the term applied only to the male child.

complex wave A waveform which varies from instant to instant, but can be resolved into a number of sine wave components, each of a different frequency and probably of a different amplitude.

compliance A ratio measure of resistance to distention. Compliance relates change of volume to the associated change in pressure. An easily distended lung has high compliance. A stiff lung has low compliance.

component A functional part of a subsystem or equipment which is essential to operational completeness of the subsystem or equipment. It may consist of a combination of parts, assemblies, accessories, and attachments, but may also be any self-contained element with a specific function.

component part The physical realization of an individual electrical element in an independent body which cannot be further reduced or divided without loss of its stated function. This term is commonly applied to active devices. Examples are transistors, thyristors, and magnetic cores as well as resistors, capacitors, and inductors.

component stress From a reliability viewpoint, those factors of usage (or test) which tend to affect the failure rate of component parts. This includes voltage, power, temperature, frequency, rise time, etc. However, the principal stress, other than electrical, is usually the thermal-environment stress.

Compton effect When X-rays strike the surface of a material of low atomic weight, some of the scattered X-rays are of longer wavelength than the striking rays. Compton, who first observed this phenomenon, explained that when high-energy photons (X-rays) struck an electron contained within material of low atomic weight, the wavelength was reduced by the quantity of energy imparted to the electron; in mechanical terms, it is similar to the elastic collision when two billiard balls meet.

compulsion An insistent, repetitive, and unwanted urge to perform an act that is contrary to the person's ordinary conscious wishes or standards. Failure to perform the compulsive act results in overt anxiety.

computer A device that can accept information, process it, and supply the results of the processing. Generally, the processing involves performing arithmetical and logical operations on the information. In recent years, the mammalian brain has been spoken of as being analogous to the computer in that the latter possesses artificial intelligence. Both analog and digital computers have been used in numerous biomedical, data-processing applications in essentially all experimental and clinical fields. The digital is the larger, more expensive type used in business and industry. It handles figures (digits) somewhat the

way an adding machine does. It produces a special numerical result by breaking down a problem into a succession of the four arithmetical operations of addition, subtraction, multiplication, and division, by any of the procedures known in numerical analysis.

The analog computer uses physical quantities as analogs to the variables being solved. It handles continuous data such as charts and curves. Each point on a curve is the analog of the information plotted, just as distances along a slide rule are analogs of numbers. The analog computer produces other charts or curves which are usually less precise than the digital results. This type of computer is used mostly in engineering studies.

computer-aided monitoring Application of a digital computer to the tedious, rate analysis of repetitive information, most particularly in the evaluation of ECG data from many beds in a coronary or intensive care unit. The computer typically plays a twofold role: detection and accumulation. It detects arrhythmic episodes, usually classifying them and bringing them to the attention of the duty nurse. And, it stores classified data, so that the nurse can periodically review the progress of a patient, or the trends in his condition, so that therapy can be adjusted. The effect of computer-aided monitoring is the elimination of time-consuming scope-watching from the nursing routine, freeing the nurse for constructive care of the patients.

computer control Parts of a digital computer which affect the carrying out of instructions in proper sequence, the interpretation of each instruction, and the application of the proper signals to the arithmetic unit and other parts according to this interpretation.

computer diagnosis The use of data-processing systems for sifting and refining raw data, physiologic parameters, and the results of laboratory analyses, in order to arrive at one or a number of possible clinical diagnoses. The object is to place a logical, statistical summary before the human practitioner, so that his judgment can be used to the full, rather than wasted on the tedious reduction of data.

computer interface Peripheral equipment for attaching a computer to scientific or medical instruments.

computer language The communication method used to instruct a computer to perform various operations. See *machine language* and *high-level language*.

computer terminal Peripheral computer equipment for entering and retrieving data. Sometimes incorporates keyboard and cathode-ray tube for display.

computer turn-around time The total time required for the submission of the program to the machine room, the processing of the program through the computer, and the delivery of the processed program back to the programmer.

computer word 1. A series of 1's and 0's which are grouped into units. These words are intelligible to the computer and represent alphabetic, numeric, and special characters. Many computer systems, at the programmer's discretion, can act on part of a word while leaving the rest of the word intact for future use. 2. A sequence of bits or characters treated as a unit and capable of being stored in one computer location. Synonymous with *machine word*.

computing machine An automatic device designed to carry out well-defined operations in mathematics, data reduction, and automation.

condensation 1. A chemical reaction in which two or more molecules combine with a resulting separation of water or some other simple substance. If a polymer is formed, the process is called polycondensation. Also see *polymerization*. 2. Term applied to the change of a fluid from the gaseous phase to the liquid phase.

condenser 1. Outmoded term for capacitor. 2. That portion of distillation or refrigeration equipment in which vapors give up heat and assume the liquid phase.

conditional jump Computer instruction which causes the proper one of two (or more) addresses to be used in obtaining the next instruction, depending upon some property of one or more numerical expressions or other conditions.

conditioned reflex A reflex which is modified by experience in such a way that the original sensory stimulus is replaced by a different learned stimulus. Pavlov's dog salivated at the sound of a bell, associating this with the introduction of food into its mouth.

conductance Ability of a material to conduct or carry an electric current. It is the reciprocal of the material's resistance and is expressed in mhos. (The word *ohm*, spelled in reverse, gives *mho,* the unit of conductance.)

conductimeter See *conductivity meter*.

conduction The flow of a neurological impulse, electricity, heat, or sound along or through a substance. Electricity flows in metals through the movement of electrons, and in semiconductors through the movement of electrons or holes. In a nerve, transient physiochemical changes occur in the nerve membrane, and these sweep along the fiber to its termination, effecting excitation of other nerves, muscle, or gland cells, depending upon the nerve's function.

conduction band A partially filled energy band of a material in which electrons can move freely, allowing the material to carry an electric current.

conductive paste 1. A substance applied to the skin to lower its electrical resistance in an area to which electrodes will be applied (ECG, defibrillator, etc.) Ordinarily, the paste contains an abrasive substance and an electrolyte, suspended in a mild base. Pastes are usually formulated for specific applications. 2. A gel applied to the skin to lower its acoustical impedance, and to accommodate an ultrasonic probe. Typically used in echocardiography, where it enables ultrasonic energy to penetrate to underlying tissues without severe attenuation at the skin interface.

conductive pattern The alternate to electrical wiring formed on the surface of an insulating base. A printed circuit board comprises areas of copper conductors, separated by the insulating base, from which the copper has been etched away by photochemical processing.

conductivity Ability of a material to allow electrons to flow, measured by the current per unit of voltage applied. It is also the reciprocal of resistivity.

conductivity cell Typically, two metal plates or electrodes firmly spaced within an insulating chamber which serves to isolate a portion of the liquid under test. This arrangement makes the measured resistance independent of both sample volume and proximity to surfaces like tank walls and metal piping. The electrodes are generally coated with a deposit of spongy black platinum which increases greatly the effective surface and thereby reduces the polarizing effect of the passage of current between the electrodes.

conductivity meter An instrument that measures and/or records electrical conductivity. Also called *conductimeter*.

conductor 1. Any substance (typically, a wire or cable) which is suitable for carrying an electric current. 2. One wire of a pair of wires. 3. That part of an electrical circuit which carries the current, as opposed to the dielectric 4. The opposite of an insulator; hence, any solid, liquid, or gaseous material which permits the flow of electrical current can be termed a conductor. 5. An instrument used to direct surgical knives (also called a director). 6. A material which promotes the transfer of heat.

congenital Existing at or previous to birth.

congenital heart disease Heart disease present from birth. Due to developmental abnormalities of the cardiovascular system (e.g., atrial-septal defect).

congestion An abnormal accumulation of blood in a part.

conjunctivitis Acute or chronic inflammation of the conjunctiva—the delicate transparent membrane lining the eyelids and covering the exposed surface of the eyeball. It results from the action of bacteria, allergens, and physical or chemical irritants.

connector 1. An electrical connecting device consisting of a mating receptacle and plug. 2. An automatic switch which closes a number of contacts simultaneously. 3. A two-digit automatic step-by-step switch which connects a trunk to a subscriber's telephone line. It tests for busy before making the connection, and may be arranged to search (hunt) for an idle terminal within a group of terminals. 4. Any device for making a temporary or semipermanent electrical connection.

consciousness The normal state of awareness.

console 1. That part of a computer which is used for communication between the operator or service engineer and the computer. 2. A centralized equipment grouping, usually arranged for convenient use and observation of equipment by a human operator.

constant current power supply A regulated power supply, which acts to keep its output current constant in spite of changes in load, line voltage, or temperature. Thus, for a change in load resistance, the output current remains constant to a first approximation, while the output voltage changes by whatever amount necessary to accomplish this.

constant-current transformer A transformer that automatically maintains a constant current in its secondary circuit under varying conditions of load impedance, when supplied from a constant-potential source.

constant field See *stationary field.*

constant-k filter A filter in which the series and shunt impedances are inverse elements (one inductive, one capacitive), and the product of the series and shunt impedances is a constant, independent of frequency.

constant voltage charger A battery charger which maintains a constant output voltage, thus allowing the charging current to taper off as the battery becomes charged, and finally allowing the battery to float, supplying any minor variations in the load current. See also *floating battery.*

constant voltage power supply A regulated power supply which acts to keep its output voltage constant in spite of changes in load, line voltage, or temperature. Thus,

for a change in load resistance, the output voltage of this type of supply remains constant to a first approximation, while the output current changes by whatever amount is necessary to accomplish this.

constant voltage transformer A power transformer which will supply a constant voltage to an unvarying load, even with changes in the primary voltage.

constipation Difficult passage of hard feces.

contact 1. The disc or bar of precious metal on a jack, key, or relay spring which touches another similar contact, thus making a temporary, low-resistance connection through which current can flow. 2. The act of effecting physical closure between two conductive surfaces, providing a path for current flow. 3. The element in a connector which makes the actual electrical contact between two halves of a connector. Also the point of joining in an electrical connection.

contact coupling A method of coupling ultrasonic energy from the echographic transducer to the patient where the anatomical structure to be studied by echographic methods is in direct contact with the transducer.

contact EMF A small voltage (electromotive force), established whenever two conductors of different materials are brought into contact. Typically encountered in electrical connections, where copper wire is connected by physical means to another metal. Also experienced in applying electrodes to the human body, by virtue of polarization effects at the skin/electrode interface. A principal cause of artifact and drift in physiological monitoring.

contact potential The potential difference between the contacting surfaces of two dissimilar electrically conductive substances, or between two materials joined to a common reference and related by an electron current flow (e.g., the potential between grid and cathode of an electron tube).

contact resistance   1. The ohmic resistance between the contacts of a switch, connector, or relay. It may be an extremely small value, typically in the milliohm range. Contact resistance is normally measured from terminal to terminal. 2. The resistance between an electrode and the underlying skin, which can be reduced by abrasion and conductive paste, but cannot be eliminated.

contacts   Conducting elements which act together to complete or to interrupt a circuit.

contaminate   To render unsterile.

continuous rating   The rating of a component or equipment which defines the substantially constant conditions which can be tolerated for an indefinite time without significant reduction of service life.

continuous scan thermograph   Equipment for presenting a continuous, visible thermal pattern (thermogram) of a patient or an object on a cathode-ray tube.

contour   A curve drawn on a two-dimensional chart through points satisfying f (x, y) = c, where c is a constant and f is same function, such as the field strength of a telemetry transmitter.

contourograph   1. A device capable of plotting a two-dimensional chart through points satisfying f (x, y) = c, for a given function. 2. A recording device, having triggered sweep, which can record repetitive physiological information (such as a single-lead ECG) in y-axis sweeps, while paper is moving at a controlled rate in the x-axis. The result is an overlaying series of

cardiac cycles, in which any abberation of rate, rhythm, or focus is immediately evident, by reason of its distortion of the usual pattern.

contracture   A shortening or distortion of muscle tissue.

control data   One or more items of data that control the identification, selection, execution or modification of another routine, operation, or record file in a computer. CONTROL DATA: a trademark of Control Data Corporation in respect of data processing equipment and related products.

control electrode   Electrode used to initiate or vary the current between two or more electrodes. The gate terminal of a silicon controlled rectifier is an example, and its analog is the grid of a thyratron electron tube. The trigger wire of a xenon strobe lamp is another example.

control grid   The electrode of a vacuum tube upon which a signal voltage is impressed to regulate the flow of current between an emitting cathode and collecting plate.

control grid bias   Average dc voltage measurable between the control grid and cathode of a vacuum tube. Bias sets the operating point of the device, thus determining what signal levels will be required to achieve amplification, and whether the output will be a true representation of the input.

controlled avalanche device   A semiconductor device that has a very specific maximum and minimum breakdown voltage characteristic and is also able to operate and indefinitely absorb momentary power surges in this avalanche region without damage, providing that there is sufficient current-limiting resistance.

controlled rectifier   A rectifier using a silicon controlled rectifier (SCR) as the rectifying element. The SCR can be triggered at any point above zero in the alternating current cycle, to control the load current. Will operate on either 60 Hz or 50 Hz power supply, and is usually triggered by a network which shifts the phase of the applied line voltage, so that voltage applied to the triggerable gate of the SCR leads voltage applied to its anode.

controlled thermonuclear reaction   Controlled fission, that is, fission produced under research conditions, or for production of useful power.

control register   1. In a digital computer, the register which stores the current instruction controlling the operation of the computer for a cycle. Also called *instruction register*. See also *program register*.

control unit   The section of a computer that supervises all information transfers and arithmetic operations in the computer. In most computers, it also controls the sequence of operations, and initiates the proper commands to the computer circuits after decoding an instruction.

conversational mode   Communication between a terminal and the computer in which each entry from the terminal elicits a response from the computer and vice versa.

conversational operation   Similar to interactive mode, but the computer user must wait for a question to be posed by the computer before responding.

conversion rate   1. The number of complete conversions an analog-to-digital converter can perform per unit time, usually specified in cycles (or conversions) per second. It must take into account not only conversion time, but recovery time, as well, and will usually be less than the reciprocal of conversion time. 2. A measure of the frequency at which conversions are made.

conversion time   1. In an analog-to-digital converter, the elapsed time between a command to perform a conversion and the appearance at the converter output of the complete digital representation of the analog input. Typical times in successive-approximation converters range from 0.8 microsecond to 400 microseconds. 2. The interval between the time an encoder input reaches a final value and the time the output settles to a within-tolerance representation. For programmed, or commanded, encoders, conversion is the interval between the time a command is received and the time the output settles to a within-tolerance value. For continuous or tracking encoders, conversion time is the interval between the time the significant change occurs at the input and the time the output settles.

converter   1. A device which can transform information or signals having one set of characteristics into an equivalent form having different characteristics. 2. A device capable of altering the frequency of a modulated carrier, changing it to one of higher or lower frequency, but having the same modulation characteristic. 3. A device that accepts an input that is a function of maximum voltage and time, and converts it to an output that is a function of maximum voltage only. 4. A device capable of converting impulses from one mode to another, such as analog to digital, or parallel to serial, or one code to another. 5. See specific listings under type of converter, i.e., BCD-to-seven segment, etc.

copper   A soft, yellowish-red metal which is very ductile and malleable. When pure, it is an excellent conductor of electricity.

copper loss   See I^2R *loss*.

copper-oxide photovoltaic cell A self-generating photovoltaic cell in which light acting on the surface of contact between layers of copper and cuprous oxide produces a voltage.

copper-oxide rectifier A metallic rectifier in which the rectifying barrier is the junction between metallic copper and cuprous oxide. A disc of copper is coated with cuprous oxide on one side, and a soft lead washer is used to make contact with the oxide layer.

copper sulfide rectifier A semiconductor rectifier in which the rectifying barrier is a junction between magnesium and copper sulfide elements.

core 1. A magnetizable portion of a device or component. 2. The portion of a solenoid which is moved by the magnetic field. 3. A movable portion of an inductor, by means of which its inductance can be changed. 4. The component of a transformer which confines the magnetic field for efficient energy transfer between primary and secondary. 5. Small toroids of magnetic material in a computer memory which can be polarized in either one of two directions by computer and whose direction of polarization can be read by the computer. The memory of most digital computers is primarily core.

core dump See *dump*.

core memory A ferromagnetic type of digital computer memory where each storage element is made up of individual miniature ferrite toroids, threaded onto select, sense read, and write wires, so that each core is addressable. Each addressed toroid can be magnetized in one direction or the other representing an 0 or a 1; that is, a single bit. This is a bistable permanent memory. If power is removed, the stored information remains. See also *internal storage*.

core plane One horizontal network of magnetic cores containing a core common to each storage position.

core storage A form of high-speed storage of information in a digital computer using a multiplicity of magnetic cores. See *core memory*.

cornea 1. Transparent epidermis and connective tissue which forms the front surface of the eye. 2. The transparent part of the coat of the eyeball that covers the iris and pupil and admits light to the interior.

corneal reflex Closing of the eyelids as a result of irritation of the cornea.

corner frequency The frequency at which the open loop plot of gain-versus-frequency changes slope. In a band-pass filter network, the corner frequency is usually specified at the 3 dB point of the filter response curve.

corona 1. A bluish static discharge which forms on a conductor suspended in dry air when the electrostatic stress in the air exceeds about 25,000 volts peak per inch. Corona is accompanied by a hissing sound and the production of ozone. 2. Any electrically detectable, field intensified ionization in an insulating system that does not result immediately in catastrophic breakdown. (Corona always precedes dielectric breakdown.) 3. A type of discharge, sometimes visible, in the dielectric of an insulation system, caused by an electric field and characterized by the rapid development of an ionized channel which does not completely bridge the electrode. May be continuous or intermittent. Not a material property but related to the system, including electrodes. Appears as pulse trains on voltage peaks when insulation is stressed above a critical value.

corona discharge A phenomenon that occurs when an electric field is sufficiently strong to ionize the gas between electrodes and cause conduction. The effect is usually associated with a sharply curved surface, which concentrates the electric field at the emitter electrode. The process is operable between an inception voltage and a spark breakdown voltage. These potentials and the current-voltage characteristics within the operating range are affected by the polarity of the corona electrodes as well as the composition and density of the gas in which the discharge occurs.

corona effect (of ac) The effect produced when two wires, or other conductors having a very great difference in voltage are placed near each other, so that the electric field between them stresses the air or other intervening insulators.

corona failure High-voltage failure initiated by corona degradation at areas of high-voltage stresss such as metal inserts or terminals.

corona loss A loss or discharge which occurs when two electrodes having a great difference of potential are placed near together. The corona loss takes place at the critical voltage and increases very rapidly with increasing voltage difference.

coronary care See *coronary care unit.*

coronary care unit An area dedicated to the intensive care of high-risk cardiac patients, staffed by specially trained nursing personnel, equipped with knowledge, training in cardiopulmonary resuscitation, antiarrhythmic drugs, and electronic patient monitoring systems, often including computers. An important task of the staff is to detect the premonitory events which may herald an acute episode in a patient's course and to administer prompt counteractive therapy. The objective is to shepherd the patient through the acute phase of the disease (where risk of further damage or death is greatest) and to promote a speedier recovery. Coronary care units have significantly reduced the mortality statistics associated with the incidence of heart attack.

coronary thrombosis Stage or complication of coronary artery disease which often proves fatal, particularly in adults over the age of fifty. It is caused by a thrombus (clot) forming in the diseased coronary artery.

coronary vessels Arteries and veins carrying the blood supply of the heart muscle.

corona shield A shield placed around a high potential point to redistribute electrostatic lines of force and prevent corona.

corona start voltage (CSV) The voltage difference at which corona discharge is initiated in a given system.

corpus The body; or, a body, as in corpus callosum, the connective transverse commissure between the cerebral hemispheres of the brain.

corrosion Electrochemical action which causes gradual destruction of the surface of a metal by oxidation or chemical combination. Also characterized by reduction of the electrical conduction efficiency between the metal and a contiguous substance or to the disintegrating effect of strong electrical currents or ground return currents in electrical systems. The latter is known as *electrolytic corrosion.*

cortex The outer or superficial part of an organ or body structure; especially, the outer layer of gray matter of the cerebrum and cerebellum.

cortical Of, relating to, or consisting of the cortex.

corticosteroids Generic term referring to

steroid hormones from the adrenal cortex, and to synthetic steroids with similar metabolic actions.

corticotrophin ACTH Hormone from the anterior pituitary gland stimulating the adrenal cortex.

costal Pertaining to the ribs.

coulomb The electrical unit of charge. One coulomb equals 6.25 billion billion (6.25×10^{18}) electrons. An electric current of one coulomb per second equals one ampere.

Coulomb's law The force of attraction or repulsion between two charges of electricity concentrated at two points in an isotropic medium is proportional to the product of their magnitudes and is inversely proportional to the square of the distance between them. Also called *law of electrostatic attraction*.

count 1. The external indication of a device designed to enumerate ionizing events. It may refer to a single detected event or to the total registered in a given period of time. The term often is erroneously used to designate a disintegration, ionizing event. 2. The cumulative total of a quantity of events registered by an indicator or stored within a memory, such as a shift register. 3. The output of a circuit which produces a pulse for a specific number of input pulses. 4. The quantity of bodies of interest (e.g., white or red blood cells, spermatozoa, or pollutant or allergenic particles in a specified sample of a substance).

counter 1. A device capable of changing states in a specified sequence upon receiving appropriate input signals. The output of the counter indicates the number of pulses which have been applied. See also *divider*. A counter is made from flip-flops and gates. The output of all flip-flops is accessible to indicate the exact count at all times. 2. Circuit which counts input pulses. One specific type is a circuit which produces one output pulse each time it receives some predetermined number of input pulses. The same term may also be applied to several such circuits connected in cascade to provide digital counting. 3. Less frequently, an accumulator. 4. Instruments used to detect ionizing radiation of very short wavelength (about one thousandth the wavelength of visible light). Natural sources of this radiation are radium, uranium isotopes, cosmic rays, and ores in which these elements are present; man-contrived sources are nuclear devices used for medical therapy and for generating electric power, high-voltage CRTs, X-ray machines, and atomic weapons.

counterelectromotive force A voltage generated in an inductive circuit by a change in current in the same circuit. The counter EMF is opposite in polarity to the voltage which produces the current in the circuit.

counterirritant An agent applied to the skin which causes dilatation of superficial vessels, used to relieve irritation.

counterpulsation The application of a pulsatile force acting against one that is already present. The term is usually applied with regard to blood flow from the left ventricle and involves the addition of an auxiliary ventricle to supply the counterpulsating force. By correctly adjusting the timing of the counterpulsation, the work of the heart is decreased.

counter tube An electron tube specifically designed for radiation detection that converts an incident particle or burst of incident radiation into a discrete electric pulse. This is generally done by utilizing the current flow through a gas that is ionized by the radiation. Also called *radiation counter tube*.

counting circuit A circuit which is actuated by successive pulses and can be arranged to count either a definite number of pulses or to give a distinct indication corresponding to each individual pulse. Flip-flops and gates are the typical elements which compose these circuits.

counts Clicking noises made by a radiation-detecting instrument in the presence of radiation. See *scintillation counter*.

couple 1. Two dissimilar metals in contact which form a galvanic couple or thermocouple. 2. A pair of opposing forces which tend to produce rotation. 3. To link two circuits together so that energy is transferred from one to the other through mutual capacitance or inductance.

coupling Abnormal heart beat which occurs in overdose of digitalis. The normal heart beat is followed by an extra ventricular contraction (ventricular extrasystole). The latter may be too weak to transmit a pulse, owing to the foreshortened ventricular filling time and the small volume of blood discharged by this secondary contraction.

coupling capacitor A capacitor which joins two circuits or amplifier stages by allowing alternating currents to pass but blocking direct currents. Careful selection of coupling capacitors and input-output impedances is essential in biomedical amplifiers to ensure that undesirable differentiation or integration of the phenomena signals do not result in severe distortion of what is being measured.

coupling medium In echographic studies, a gel-like sonic conductive material used to transfer the ultrasound energy from the transducer to the skin. It lessens the amount of ultrasound reflection at the transducer–skin interface, thus reducing near-field energy losses.

cpm Abbreviation for cycles per minute.

cps Obsolete abbreviation for cycles per second (of frequency). Now called hertz, abbreviated Hz.

cpu Central Processor Unit. See *general-purpose digital computer*.

CR Abbreviation for central ray.

cranial nerves Peripheral nerves emerging from the brain, as distinct from those emerging from the spinal cord. There are twelve pairs of cranial nerves: (1) Olfactory —the sensory nerves of the nose; (2) Optic —the sensory nerves from the eyes; (3) Oculomotor—motor nerves supplying the eye muscles; (4) Trochlear—supplying eye muscles; (5) Trigeminal—sensory from parts of face and tongue and motor to jaw muscles; (6) Abducens—supplying eye muscles; (7) Facial—sensory from face and motor to muscles of expression; (8) Auditory—sensory nerves from the ear and vestibular apparatus; (9) Glossopharyngeal—concerned with sensory and motor aspects of swallowing; (10) Vagus—a parasympathetic nerve with motor and sensory fibers distributed to esophagus, stomach, heart, and lungs; (11) Accessory—nerve to trapezius and sternomastoid muscles; (12) Hypoglossal—motor nerve to muscles below the pharynx.

cranium Hemispherical bone structure which surrounds and protects the brain. It consists of eight bones; the occiput behind, the frontal bone (forming the forehead) in front, two parietals at the sides with two temporals (containing the hearing apparatus) below, and a couple of smaller bones, the sphenoid and ethmoid.

creepage The conduction of electricity across the surface of a dielectric.

creepage path The path across the surface of a dielectric between two conductors.

Lengthening the creepage path reduces the possibility of arc damage or tracking.

cretinism See *hypothyroidism.*

critical 1. A stage in a chain of events beyond which lie serious or unpredictable consequences. 2. A stage in nuclear fission, wherein a chain reaction becomes sustainable.

critical angle The maximum angle of incidence for which light or ultrasonic energy will be transmitted from one medium to another. Approaching the interface at angles greater than the critical angle will cause energy to be reflected back into the first medium.

critical mass The smallest mass of fissionable material that will support a self-sustaining chain reaction under stated conditions.

critical voltage (of gas) The voltage at which a gas ionizes preliminary to dielectric breakdown of the gas.

crossover frequency Of a dividing network, the frequency at which equal powers are delivered to the higher band and the lower band when both are terminated in their normal load.

crosstalk 1. Unwanted sound or data reproduced by a system element associated with a given carrying channel, resulting from cross-coupling to another channel carrying sound, or other electric waves; or, by extension, the electric waves in the distributed channel which result in unwanted sound or garbled data. 2. Undesired power injected into a circuit (e.g., a communication circuit) from other circuits. May be intelligible or unintelligible. 3. Interference caused by energy being coupled from one circuit to another by stray electromagnetic or electrostatic coupling. See *artifact.*

crowbar A term describing the action in a power supply which effectively creates a high overload on the actuating member of the protective device. This anticipatory, protective action may be triggered by a slight increase in current or voltage to a circuit which is susceptible to damage.

crowbar voltage protector A separate circuit which monitors the output of a regulated power supply and instantaneously throws a short circuit (or crowbar) across the output terminals of the power supply whenever a preset voltage limit is exceeded. An SCR is usually used as the crowbar device.

CRT frame One full-screen scan of a CRT, of the many intensity-modulated trace lines that produce the dark and light pattern (raster) on the tube face.

CRT refresh rate Number of times per second (expressed in frames per second) that the trace passes a given point on the tube face.

CRT update Frequency, usually expressed in seconds, for updated display information (alphanumeric characters) to be presented to the CRT from an outside source.

CRT update rate Speed, expressed in characters per second, for displaying new alphanumeric characters on the screen. This is a function of the CRT electronics and operating mode.

cryogenic device A part, usually a semiconductor device, intended to function best at low temperatures near absolute zero.

cryogenics The science of producing and applying the effects of extremely low temperatures in the range usually below $-50°C$. In electronics, this covers special electrical effects available only at extremely low temperatures. These effects can be used in computers, memory systems, superconducting electromagnets, very low noise amplifiers, etc. Medical applications for lowered temperature include hypothermia, cryosurgery, and other techniques under active study.

cryolectronics Technology concerning the characteristic of electronic components at cryogenic temperatures.

cryosurgery Method of surgical removal of tissue by local freezing.

cryotronics The branch of electronics that is concerned with applications of cryogenics. A contraction of *cryo*genic elec*tronics*.

crystal A solid whose atoms are uniformly arranged in a repeating fashion. In electronics, this is frequently used to refer to a quartz or other piezoelectric crystal used to control the frequency of an oscillator. In semiconductor technology, it refers to a material such as silicon, which is the foundation for integrated circuits and the active and passive devices used in hybrid circuits.

crystal diode A two-electrode semiconductor device that utilizes the rectifying properties of a junction between P-type and N-type material in a semiconductor, as in a junction diode, or the rectifying properties of a sharp wire point in contact with a semiconductor material, as in a point contact diode. Also called *crystal rectifier, diode,* and *semiconductor diode*.

crystal filter A low-loss and extremely sharp filter which uses quartz crystal resonant elements instead of inductor-capacitor elements. The Q of a quartz crystal is about 20,000.

CSV Abbreviation for corona start voltage.

cue An item of behavior such as a gesture

or word which communicates how a situation is perceived.

cuff electrode An electrode in the shape of the letter C designed for application of potentials to small circular bodies, such as peripheral nerves.

cumulative dose (radiation) The total dose resulting from repeated exposures to radiation of the same region, or of the whole body.

cuprous oxide rectifier A dry-disc instrument rectifier in which rectification occurs at a layer of cuprous oxide on a copper plate. Current will flow from the cuprous oxide to the copper, but not in the reverse direction. The efficiency is in the range of 60% to 70%.

curare A dried aqueous extract especially of a vine used in medicine to produce muscular relaxation.

cure 1. To change the physical, chemical, or electrical properties of a material by chemical reaction, by the action of heat and catalysts, alone or in combination, with or without pressure; specifically, to convert a low molecular weight polymer or resin to an insoluble, infusible state. 2. To arrest the course of a disease through therapy or intervention, and to restore to health.

curie Unit used in nuclear physics and defined as the quantity of any material giving 3.70×10^{10} disintegrations per second, which is the rate of decay of 1 gram of elemental radium.

curie point 1. Temperature at which a ferromagnetic substance loses its magnetism. 2. The critical temperature at which piezoelectric materials lose their polarization, and therefore their piezoelectric properties.

curie temperature See *curie point*.

current The volume flow of electrons through a conductor, per unit of time. Current is measured in amperes, in milliamperes, and in microamperes. The opposition of a conductor to current flow is expressed in its resistance value; the higher its resistance, the lower the current for a specific, applied voltage. See *Ohm's law*.

current amplification Ratio of output signal current to input current for an electron tube, transistor, transducer, or magnetic amplifier, the multiplier section of a multiplier photo-tube, or any other amplifying device. Often expressed in decibels by multiplying the common logarithm of the ratio by 20.

current carrying capacity The maximum current that can be continuously carried without causing permanent change in or damage to the electrical or mechanical properties of a device or conductor.

current hogging 1. A condition where one of several parallel logic circuits has a lower resistance and takes most of the available current, resulting in unequal current sharing. 2. A similar condition experienced in charging two or more batteries in parallel.

current limiter A device that detects current leakage and prevents potential shock hazard by minimizing or interrupting current flow. See *ground fault interrupter*.

current-limiting resistor A resistor inserted in a circuit to limit the flow of current to a safe value in the event of a fault or short circuit.

current limit-sense voltage The voltage between the sense and limit terminals of a regulator which will cause current limiting.

current transformer A highly accurate instrument transformer having a high-current low-resistance primary, often of only one-half turn, which is placed in series in a power conductor carrying a large alternating current. The low-current secondary is connected to a low-current ammeter having a scale indicating the current in the transformer primary. See also *instrument transformer*.

curve tracer An instrument with which one current or voltage can be displayed, as a function of another voltage or current, with a third voltage or current as a parameter.

cutaneous Pertaining to the skin.

cutoff frequency The frequency response limitation, either upper or lower end, of a given component or piece of apparatus where the response falls off by 3dB, then is even lower beyond this frequency.

cuvette That chamber-like portion of a densitometry apparatus used in determination of cardiac output by the dye-dilution technique through which a mixture of blood and green dye is drawn by a constant withdrawal syringe, to ascertain the difference in light-transmissive properties of the blood.

cyanosis A dark, bluish appearance of the skin, lips, and nails due to inadequate oxygenation of the blood.

cybernetics 1. Comparative study of the control and intracommunication of information-handling machines and nervous systems of animals and man in order to understand and improve communications. 2. The science that is concerned with the principles of communication and control, particularly as applied to the operation of machines and the functioning of organisms.

cycle 1. An interval of space or time in which one set of events or phenomena is completed. 2. Any set of operations that is repeated regularly in the same sequence.

The operations may be subject to variations on each repetition. 3. In alternating current, the time for a change of state from a zero through a positive and negative maximum, and back to zero. 4. A set of operations performed in a predetermined manner. 5. In computer terminology, a sequence of operations that is repeated regularly, or the time it takes for one such sequence to occur.

cycles Incorrectly used as an expression of frequency, when cycles per second is meant. See also *hertz*.

cycles per second (CPS) The older unit of frequency, now replaced by the international standard hertz.

cycle time 1. Time interval between the appearance of corresponding parts of successive cycles. 2. The time required to complete one loop. It is the basic time unit of computer operation. 3. The measure of how long it takes to obtain information from a memory and then to write information back into the memory. Also called a read/write cycle time, as it is normally the addition of the write time and the read time.

cyclic binary code. See *gray code*.

cyclic code Positional notation, not necessarily binary, in which quantities differing by one unit are represented by expressions which are identical except for one place or column, and the digits in that place or column differ by only one unit. Cyclic codes are often used in mechanical devices because no ambiguity exists at the change-over point between adjacent quantities.

cyclic memory A memory that stores information continuously but provides access for reading or changing any piece of stored information only at multiples of a fixed time called the cycle time.

cycling A periodic oscillation in an automatically controlled system between the high limit and the low limit at which the controls operate.

cyclograph A device in which an electron beam moves in two directions at right angles.

cyclotron 1. Apparatus invented by Professor E. O. Lawrence for the acceleration of particles like protons and deuterons in order to give them high energies and thus provide very energetic projectiles for atomic bombardment. The idea of the device is that a stream of p particles (protons or deuterons) is made to circulate in a spiral and is given an acceleration by means of an alternating supply feeding two electrodes, called dees, the frequency of the alternation being such that the polarity of the dees is reversed just when the particles have reached half the way around their circle. Cyclotrons are an essential part of every establishment where atomic research is carried out. 2. Type of accelerator of nuclear particles (protons or deuterons) that uses an oscillating electric field and a fixed magnetic field to accelerate the particles. See *accelerator*.

cylindrical-film storage Computer storage device in which each storage element consists of a short length of glass tubing having a thin film of nickel-iron alloy on its outer surface. Wires threaded through the cylinders serve as bit and sense lines, while conducting straps at right angles to the cylinders serve as word lines.

cyst Any normal or abnormal sac in the body, especially one containing a liquid or semiliquid material.

cystic fibrosis An inherited disease of the glands of external secretion, affecting mostly the pancreas, respiratory tract, and sweat glands, and usually appearing in infancy.

cystitis Acute or chronic inflammation of the urinary bladder, caused by infection or irritation from foreign bodies (kidney stones) or chemicals. Its symptoms are frequent voiding accompanied by burning sensation.

cystogram X-ray study of urinary bladder.

cystometer Instrument that measures the condition of the bladder by measuring its pressure and capacity.

cystoscope Instrument for examining the urinary tract.

cytoglomerator Apparatus for processing blood before freezing it for storage, and after thawing.

cytology The study of cells.

cyton The body of a nerve cell exclusive of its various processes.

cytophotometer Instrument that measures and characterizes optical properties of cells.

cytoplasm The protoplasm content of a cell, not including the protoplasm within the nucleus of the cell.

-cytosis Suffix meaning condition of cell.

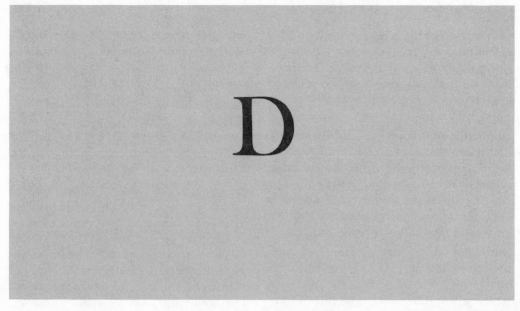

D

DAC Abbreviation for digital-to-analog converter.

D/A converter Abbreviation for digital-to-analog converter.

damped oscillation 1. An oscillation in which the amplitude of each succeeding wave becomes less. 2. An oscillation which dies out because the driving force has been removed and the oscillating circuit contains resistance, which dissipates oscillatory energy.

dampen 1. To diminish progressively in amplitude; usually said of waves or oscillations. 2. To deaden vibrations.

damping 1. Reduction of energy in a mechanical or electrical system by absorption or radiation. 2. Act of reducing the amplitude of the oscillations of an oscillatory system; hindering or preventing oscillation or vibration; diminishing the sharpness of resonance of the natural frequency of a system. 3. In echographic studies, any technique which uses resistive or mechanical loading to reduce the duration of an echo complex.

damping factor Ratio of any one amplitude to that next succeeding it in the same sense of direction.

daraf The unit of electrical elastance, equal to the reciprocal of the capacitance in farads.

dark resistance The resistance of a photoresistive device (such as a cadmium sulfide cell) in total darkness.

dark-trace tube Cathode-ray tube with a screen coated with a halide of sodium or potassium. The screen normally is nearly white, and whenever the electron beam strikes, it turns a magenta color which is of long persistence. The screen can be illuminated by a strong light source so that the reflected image may be made intense enough to be projected.

Darlington amplifier A transistor circuit which in its original form consists of two transistors in which the collectors are tied together and the emitter of the first transistor is directly coupled to the base of the second transistor. Therefore, the emitter current of the first transistor equals the base current of the second transistor. This connection of two transistors can be regarded as a compound transistor. This connection of two transistors can be regarded as a cascaded emitter follower, or B multiplier.

Darlington-connected phototransistor Phototransistor whose collector and emitter are connected to the collector and base, respectively, of a second transistor. The emitter current of the input transistor is amplified by the second transistor and the device has very high sensitivity to light.

Darlington connection A form of compound connection in which the collectors of two or more transistors are connected together and the emitter of one is connected to the base of the next. Two transistors connected in this way constitute a Darlington pair.

Darlington pair See *Darlington amplifier.*

D'arsonval galvanometer A device in which direct current flowing through a coil pivoted between the poles of a permanent magnet reacts with the magnetic field and produces a torque. The coil is generally connected to an indicator pointer overlying a calibrated scale. Coil springs return the coil and pointer to a zero (neutral) position when no current is applied. These springs also serve as flexible couplings between the coil ends and the meter terminals.

data A general term used to denote any or all facts, numbers, letters, and symbols, or facts that refer to or describe an object, idea, condition, situation, or other factors. It connotes basic elements of information which can be processed or produced by a computer. Sometimes data is considered to be expressible only in numerical form, but information is not so limited.

data acquisition The process by which events in the real world are translated to machine-readable signals. The term usually refers to automated systems in which sensors of one type or another are attached to machinery.

data acquisition and conversion system A method of processing analog signals and converting them into digital form for subsequent processing or analysis by computer or for data transmission.

data acquisition system 1. A computer system designed to gather, process, and record data from multiple remote locations at a central computing facility. 2. A combined multilevel system in which analog biomedical data is presented to a computer by instruments that interface to the patient. The computer processes data, as needed, and feeds results to the user through terminals or the instrument which originated the data. The user, however, can choose to use any combination of information sources he chooses, ignoring the computer completely when its functions are unnecessary.

data base The entire body of data on one or more related subjects. Typically, a collection of data files stored in a computer system in a manner designed for easy access (e.g., a company's entire collection of personnel records including payroll, job history, accrued vacation time, etc.; or a state's Department of Motor Vehicle records).

data block Typically, a total group of data for one item that is entered into a computer for processing, or the output from a computer as the result of processing. For example, an input data block could represent an individual shipping list and an output data block could represent a check to be sent.

data break A facility which permits input/output (I/O) transfers to occur on a cycle-stealing basis, without disturbing program execution in a computer.

data code A structured set of characters used to represent the data items of a data element. For example, the data codes 1, 2, . . . , 7 may be used to represent the data items Sunday, Monday . . . , Saturday.

data communications 1. The technology covering the transfer of data over relatively long distances 2. Transmission of data on both directions between a central location (host computer) and remote locations (terminals) through communication lines. To facilitate this, interfaces such as modems, multiplexers, concentrators, etc. are required at each of the lines.

data communications processor A small computer used to control the flow of data between machines and terminals over communications channels. It may perform the functions of a concentrator, handshaking, and formating, but does not include long-term memory or arithmetic functions.

data compression 1. Compacting of telemetry data prior to transmission. 2. The process of reducing the number of digital data samples, which are recorded or transmitted, by excluding redundant samples.

data converter A machine which changes information in one kind of language, acceptable to one type of data processing machine, into corresponding information in another kind of language acceptable to another type of data processing machine; e.g., a card-to-magnetic-tape converter is one which takes information expressed in punched cards and produces the same information expressed in magnetic tape.

data display Visual presentation of processed data by specially designed electronic or electromechanical devices through interconnection (either on-line or off-line) with digital computers or component equip-

ment. Although line printers and punch cards may display data, they are not usually categorized as displays, but as output equipments.

data element A specific item of information appearing in a set of data. For example, in the following set of data, each item is a data element; the quantity of a supply item issued, a unit rate, an amount, and the balance of stock items on hand.

data entry device An electromechanical device to allow manual input of data to a display system. Examples of data entry devices are: alphanumeric keys, data tablet, function keys, joystick, mouse, and trackball.

data-flow diagram An illustration that suggests a certain amount of circuit operation by its configuration.

data generators Specialized word generators in which the programming is designed to test a particular class of device, in which the pulse parameters and timing are adjustable, and in which selected words may be repeated, reinserted later in the sequence, omitted, etc.

data item The simplest type of information with which a computer system deals (e.g., a name or employee number); a collection of data items makes up a record (e.g., payroll data on one employee), and a number of related records make up a file (e.g., payroll data on all the employees of the company).

data link Equipment which permits the transmission of information in data format.

data management A general term that collectively describes those functions of the control program that provide access to data sets, enforce data storage conventions, and regulate the use of input/output devices.

data-phone 1. A trademark of American

Telephone and Telegraph Co. to identify the data sets made and supplied for use in the transmission of data over the regular telephone network. See *electrocardiogram data set*. 2. A telephone equipped with a modem and appropriate switching for both voice and data transmission.

data processing 1. A systematic sequence of operations, performed in accordance with precise rules of procedure, on data (e.g., the sorting of the names of chemical compounds in accordance with some common characteristic; or the translation of a scientific text). Data processing includes sorting, merging, computing, assembling, compiling, translating, extracting, storing, and retrieving data. 2. A generic term to encompass all business data applications. 3. Elaborate computer techniques to record and analyze data such as neuroelectric data from studies on neurophysiological systems.

data processing equipment Equipment designed especially for mass processing of data. It also includes transcription and transmission devices that are designed especially for producing media for mass data processing (such as punched cards, paper, or magnetic tapes).

data processing machine A general name for a machine which can store and process numeric and alphabetic information.

data processing system A network of components, typically a digital computer capable of accepting information, processing it according to a plan, and producing the desired results.

data processor Any device capable of performing operations on data (e.g., a desk calculator, a tape recorder, an analog computer, a digital computer).

data reduction 1. The transformation of raw data into a more useful form. 2. The drawing of a smooth curve through plotted positions of empirical data. 3. The process of transforming masses of data from experiments into smaller, simplified, summarizing sets of information. For example, neurological data is often reduced automatically by computer equipment before it is evaluated, since interpretation of the original data would be difficult and time consuming.

data reduction system Automatically-operated equipment engineered to simplify the use and interpretation of a large quantity of data gathered by instrument installation.

dataset 1. A modem. 2. A device that converts the signals of a business machine to signals that are suitable for transmission over communication lines and vice versa. It may also perform other related functions.

data sink In communications, a memory or recording device capable of accepting data signals from a transmission device and storing data for future use. It may also check these signals and originate error control signals. (Contrast with data source.)

data source In communications, a device capable of originating data signals for a transmission device. It may also accept error control signals. (Contrast with data sink.)

data tablet A data entry device consisting of a stylus and a graphic recorder with a coordinate grid similar to the number space of a CRT screen. By pressing the stylus on the tablet, an interrupt is generated and the coordinates of the stylus are stored in special x-y input registers. The registers are then read by the host computer. The tablet generally replaces the light-pen for cursor and tracking symbol movements, and is used extensively in storage tube display systems where light-pen tracking and identification are impossible.

data terminal 1. Data processing equipment that serves as a junction point for the collection and transfer of data from various sources. It may include or connect with several types of data processing equipment. 2. That part of a system that is concerned with the functions of generating data and/or recording or display of data, together with the control equipment and/ or software necessary to control these functions. 3. A device which modulates and/or demodulates data between one input/output device and a draft transmission link.

data transmission Transmitting information from one place to another or one part of a system to another.

data transmission system Means for transmitting (data-phone, radio, etc.). See also *telemetry*.

data word A word which may be primarily regarded as part of the information manipulated by a given program. A data word may be used to modify a program instruction, or to be arithmetically combined with other data words. Typical large-computer dataword formats are: (a) Alphanumeric: the 48 informations bits are arranged in 8 groups of 6 bits each. Each six-bit code group can represent one of 56 alphanumeric characters; (b) Binary: the 48 information bits represent 44 signed or 48 unsigned binary digits; (c) Decimal: the 48 information bits are arranged in 12 groups of 4 bits each. Each four-bit code group can represent a decimal digit or a hexadecimal letter (also Numeric); and (d) Mixed: the 48 information bits are arranged to represent some combination of alphanumeric characters and decimal and binary digits.

dB Abbreviation for decibel, which is one-tenth of a bel. A unit expressing the ratio of two voltages, currents, or powers. It is equal to 20 times the common logarithm of the ratio of the two voltages or two currents, and 10 times the common logarithm of the ratio of the two powers.

dB meter Meter having a scale calibrated to read directly in decibel values at a reference level that must be specified (usually 1 milliwatt across 600 ohms equals 0 decibels). Used in audio frequency amplifier circuits or broadcast stations, public address systems, and receiver output circuits to indicate volume level.

dc Abbreviation for direct current. An electric current that flows in one direction.

dc amplifier An amplifier which has a frequency response band that goes down to dc. This definition includes direct-coupled amplifiers but is not restricted to them.

dc continuity 1. An established pathway for conduction of current from a direct current source, neglecting pathway resistance. A reading on a conventional ohmmeter applied across the terminals of a circuit with dc continuity will result in a deflection of the meter.

dc coupled The connection by which a device passes the steady-state characteristics of a signal as well as the alternating component, transient, or oscillating characteristics of the signal.

dead 1. Not connected to any source of electric potential. 2. Having the same potential as the earth. 3. The absence of life, as revealed by cessation of vital signs and electroencephalographic waves relating to conscious thought.

dead band Specific range of values in which the incoming signal can be altered without also changing the outgoing response.

deadbeat Coming to rest without vibration or oscillation. Thus, the pointer which a highly damped meter or galvanometer moves to a new position without overshooting and vibrating about its final position.

dead room Room so thoroughly soundproofed that practically all sound is absorbed, and there is little reflection or echo.

dead-short Short circuit having minimum resistance.

dead time Any definite delay deliberately placed between two related actions in order to avoid overlap that might cause confusion or to permit a particular, different event such as a control decision, switching event, or similar action to take place.

death instinct (thanatos) In Freudian theory, the unconscious drive toward dissolution and death. Coexists with and is in opposition to the life instinct (eros).

debilitated Weak, lacking in strength.

decimal See *number system.*

decubitus ulcer A bedsore.

debug 1. To detect and correct malfunctions in an electronic circuit or system. 2. To locate and correct any errors in a computer program. 3. To detect and correct malfunctions in the computer itself. 4. To test-run and check out a program of machine instructions for a digital computer for the purpose of eliminating mistakes. 5. To examine or test a procedure, routine, or equipment for the purpose of detecting and correcting errors. 6. To locate and neutralize hidden listening devices.

debugger A computer program that helps in finding errors (bugs) in a program by stopping the program at predetermined points, noticing when certain instructions are executed, etc.

debugging 1. A shakedown or burn-in process undertaken to carry an equipment or a population of parts through its period of infant mortality during which catastrophic type failures occur at an abnormally high rate. 2. Correcting errors ("bugs") in hardware or software.

debugging phase The run-in time required on new equipment to eliminate the initial failures due to faulty components or workmanship.

decade counter A logic device that has 10 stable states. The device may be cycled through these 10 states with 10 clock or pulse inputs. A decade counter usually counts in a binary sequence from state 0 through 9 and then cycles back to 0. Sometimes referred to as a *divide-by-10 counter.*

decay 1. The degenerative process associated with necrosis. Also called *catabolism.* 2. The decrease in the radiation intensity of any radioactive material with respect to time. 3. In a shortage tube, a change in magnitude or configuration of stored information by any cause other than erasing or writing.

decay rate The time rate of the disintegration of radioactive material generally accompanied by the emission of particles and/or gamma radiation.

decelerated electrons Electrons which after traveling at a great rate of speed, strike a target, become quickly decelerated, and cause the target to emit X-rays.

decibel 1. One-tenth of a bel. 2. A unit of change in sound intensity. A three-decibel (two-to-one) change is approximately the smallest change that the ear can perceive. Larger decibel increments reflect the fact that sound intensity must be squared in order for the ear to perceive a doubling of intensity. An increase in intensity is expressed as a positive number of dBs, a decrease as a negative value. No change in intensity is 0 dB, and 0 is also used to indicate a starting point, from which changes are measured. 3. A unit used to measure and compare gains, losses, and signal levels on a logarithmic scale. An amplifier with a gain of 20 dB provides a 100-fold increase in signal strength. 4. Unit for expressing the ratio of two values, the number of decibels being 10 times the logarithm to the base 10 of a power ratio, or 20 times the logarithm to the base 10 of a voltage or current ratio.

decimal code   A code in which each allowable position has one of ten possible states. The conventional base ten or decimal number system is a decimal code.

decimal-coded digit   One of 10 arbitrarily selected patterns of 1's and 0's used to represent the decimal digit.

decimal numbering system   The widely used numbering system having Arabic numerals 0 through 9. It uses 10 marks (0 through 9), thus having a base or radix of 10. In the decimal system we think in tens. For example, here's how we arrive at the decimal number 2,345: $2000 + 300 + 40 + 5 = 2,345$ or: $2(10^3) + 3(10^2) + 4(10^1) + 5(10^0) = 2,345$. In the decimal system, all numbers are obtained by using the radix (total number of marks), in this case, 10, raised to various powers.

decimal to binary conversion   The process of converting a number written to the base of ten, or decimal, into the equivalent number written to the base of two, or binary.

deck   1. In computer usage, a collection of cards, commonly a complete set of cards which have been punched for a definite service or purpose. 2. The transport mechanism of a magnetic tape recording instrument.

decode   1. To translate digital data into alphanumeric characters. 2. In a translator,

to provide an output signal which is logically derived from several input signals plus information in the translator's memory. 3. To apply a set of unambiguous rules specifying the way in which data may be restored to a previous representation, i.e., to reverse some previous encoding.

decoder   1. A device that translates a combination of signals into a single signal representing that combination. Often used to extract information from a complex signal. 2. A device which accepts information coded in one system and converts it to an equivalent representation in another system (e.g., BCD-to-seven segment decoder).

decompensation   Failure of physiological compensation to some stimulus; for example, failure of hypertrophy of a chamber of the heart to overcome obstruction due to a defective valve.

decoupling network   A filter network placed in leads (such as power supply) which serve several circuits to prevent coupling or interaction between the circuits.

decremental conduction   Impulse conduction with ever-decreasing action-potential size, eventually leading either to an impulse of constant size or to extinction of the impulse.

deemphasis network   A network having a transmission characteristic which is the inverse of a preemphasis network. It restores the spectrum of the signal to its original condition.

defecation   The evacuation of feces.

defense mechanism   A specific process, operating unconsciously, that is employed to seek relief from emotional conflict and freedom from anxiety.

defibrillate   To   terminate   fibrillation

(spasmodic twitching) of muscular tissue by chemical, physical, or electrical means.

defibrillation The stoppage of fibrillation of the heart.

defibrillator Apparatus which applies large-scale electrical impulses to the heart. Designed to stop fibrillation and restore the normal cardiac cycle. The defibrillator, as devised by Lown, employs a dc power supply, storage capacitor, series-discharge inductor, and precordial or internal paddle electrodes. Applied to chest or myocardium, the paddles deliver the energy stored in the capacitor, according to a prescribed waveform determined by the capacitor, inductor, and patient impedance. The applied shock energy typically lasts for 10 milliseconds and can be as much as 400 joules. It causes simultaneous depolarization of the myocardial musculature, followed by repolarization as the defibrillating wave passes. In most instances, the next sinus impulse should regain capture of the ventricular muscles and a restoration of natural heart rhythm and pumping action. Energy dose deliverable to the patient is adjustable and most defibrillators are also equipped for use in synchronized cardioversion.

defibrillator paddle(s) One or both electrodes, by which the defibrillating energy pulse is applied to the patient. Because voltage is so high and energy appreciable, paddles are carefully insulated so as to offer minimal risk to the user. One or two pushbutton switches are provided on the paddles, so that the user can discharge the energy pulse while holding the paddles in contact with the patient.

deflection coil Inductor used to produce a magnetic field that will bend the electron beam a desired amount in a cathode-ray of an oscilloscope, patient monitor, television receiver, or television camera.

deflection plates In an electrostatic cathode-ray tube, two pairs of parallel electrodes; the pairs set one forward of the other and at right angles to each other, parallel to the axis of the electron stream. An applied potential produces an electric field between each plate pair. By varying the potential, a desired horizontal, vertical, or angular displacement of the electron stream can be achieved, so as to trace complex waves on the phosphor screen.

deflection yoke Assembly of one or more electromagnets to produce deflection of one or more electron beams. Also called *scanning yoke*.

degeneration See *negative feedback*.

degradation failure Failure of a device because of a shift in a parameter or characteristic which exceeds some previously specified limit.

dehydration Loss of water from the blood cells and tissues.

deionization time Time required for the control grid of a gaseous tube to regain control after plate current has been interrupted.

delay 1. Time required for a signal to pass through a device or a conducting medium. 2. Time which elapses between the instants a signal wave passes any two designated points of a circuit. Such delay is primarily determined by the constants of the circuit. 3. An intentional waiting period interposed between command and response. 4. The time by which the occurrence of an event is retarded.

delay distortion Distortion caused by the fact that the higher frequency components of a signal travel slower over the transmission facility than the lower frequency components, therefore arrive later and out of phase. Also called *phase distortion*.

delay equalizer A network that introduces an amount of phase shift complementary to that present in the circuit, at all frequencies within the equalized band. Generally consists of a number of delay sections connected in series. Each section is resonant at a particular frequency, and produces a certain amount of delay for energy at the resonant frequency. The particular resonant frequencies of the sections are either fixed or adjustable, depending upon the type of equalizer.

delay/frequency distortion That form of distortion which occurs when the envelope delay of a circuit or system is not constant over the frequency range required for transmissions.

delay line 1. Real or artificial transmission line or equivalent device designed to introduce delay. 2. Any device for producing a time delay of a signal. May employ active or passive components, or combinations of these. 3. A sequential logic element with one input channel in which the state of an output channel at any one instant, T is the same as the input channel state at the instant T-N, where N is a constant interval of time for a given output channel (i.e., an element in which the input sequence undergoes a delay of N time units). 4. A device capable of causing energy impulses to be retarded in time from point to point, thus providing a means of storage by circulating intelligence-bearing pulse configurations and patterns. Examples of delay lines are material media such as mercury, in which sonic patterns may be propagated in time; lumped constant electrical line; coaxial cables, transmission lines, recirculating magnetic drum loops, and shift registers.

delay-line shortage Electronic computer storage or memory device consisting of a delay line and means for regenerating and reinserting information into the delay line.

delta Brain wave signals whose frequency is approximately 0.2 to 3.5 Hz. The associated mental state is usually a deep sleep, or a trance-like state.

delta connection A method of connecting a three-phase power circuit in which each of three transformer windings is connected across one of the three phases, so that the winding connection diagram looks like the Greek letter delta, an equilateral triangle.

demand pacemaker An atrioventricular assist device used on cardiac patients suffering from ventricular slowing, which in turn has resulted in reduced cardiac output or cardiac arrest. It generates controllable, periodic, direct electric potentials (pulse voltages) through electrodes buried in a patient's chest (across the heart area). Thousands of encapsulated, implantable pacemakers are in use today as they have been widely employed in hospital intensive care units. For patient use, two basic modes are provided: Continuous operation, in which the pacemaker provides preset pulses of fixed frequency and current capabilities. Demand operation, in which the instrument supplies pulses (on demand) only during slowing of the heart (bradycardia), missed beats, or a systole.

dementia precox Obsolete term for schizophrenia.

demodulator 1. Device which operates on a carrier wave to recover the intelligence with which the carrier was originally modulated. 2. Device which detects an amplitude modulated signal and produces the modulating frequency as a direct current of varying amplitude. 3. Device which receives tones from a transmission circuit and converts them to electrical pulses, or bits, which may be accepted by a data acquisition system.

demultiplexer A device used to separate two or more signals that were previously combined by a compatible multiplexer and are transmitted over a single channel.

demultiplexing See *multiplexer*.

dendrite Any of the extending branches of a nerve cell which receive sensor or effector impulses and conduct them to the cell body; or, conduct impulses originating in the cell body to a synopse or muscle.

densitometer An instrument which measures optical density by comparing the light transmission or reflection properities of a substance against standard media. The densitometer is used in the determination of cardiac output, wherein indocynanine green dye is injected via a catheter into the left heart, and a blood sample withdrawn from a vein through the densitometer. The integral of the optical density change thereby obtained is a measure of cardiac output.

dentaphone A device by which deaf persons can hear sound via sonic conduction through the teeth and bone structures leading to the ear.

depletion layer A zone, several atoms thick, at the junction of N-type and P-type semiconductor materials in which there are no current carriers, either free electrons or holes, unless biased by a dc voltage. Free electrons in the N-type material are repelled by negative charges in the P-type material are repelled by the positive nucleous of atoms in the N-type material.

depolarization A decrease or sudden change in the membrane potential of a cell. The contraction of a muscle is accompanied by depolarization to zero, from the normal resting potential, and its relaxation is accompanied by repolarization.

depolarize To cause to become partially or wholly unpolarized.

depolarizer A chemical substance, usually manganese dioxide, added to a LeClanche dry cell to reduce the polarizing chemical products (typically, hydrogen, which insulates the carbon center post) resulting from discharge, and thus to keep the discharge rate constant.

depolarizing current An applied electrical current in the outward direction through the cell membrane causing depolarization.

depressed state A state following a subthreshold stimulus in which the threshold is greater than the resting value.

depression Psychiatrically, a morbid sadness, dejection, or melancholy; to be differentiated from grief, which is realistic and proportionate to what has been lost. A depression may be a symptom of any psychiatric disorder or may constitute its principal manifestation.

depth In echographic studies, the distance from the transducer to the interfaces from which echoes are reflecting. Measured along the horizontal axis of an A-mode oscilloscope, in which the constant sweep timebase is converted to depth, assuming the velocity of propagation of the ultrasound wave to be constant.

derate To use a device at a lower current or voltage than it is capable of handling in order to reduce the probability of failure, or to permit its use under a condition of high ambient temperature.

dermatitis Inflammation of the skin. The numerous causes may be of external or internal origin.

dermohygrometer Instrument for measuring skin resistance without inducing constant current into the skin.

-desis Suffix meaning binding or fixation.

destructive read A process of reading out stored information that results in erasure of the data in the memory.

destructive readout The reading of a memory cell, such as a toroidal ferrite core, where the contents of the memory cell are destroyed when it is interrogated by a *read pulse*. A *write pulse* is then required to reinsert the bit in memory.

destructive readout memory Abbreviation DRO. A memory type in which reading the contents of a storage location destroys the contents of that location. Data must be rewritten after reading, if the data is to be returned. An example of a DRO memory is the core memory used in digital computers.

detection (demodulation) Process of recovering from modulated radio-frequency signals the sound, biophysical, or picture signals modulating the carrier frequency. Usually carried out by a system of rectifications. The names used to denote this special application of rectification are *detection* and *demodulation*. The diode, which will only conduct current in one direction, is one of the simplest detectors. By applying the modulated signals to it with a suitable resistor-capacitor load, the voltage across the capacitor is arranged to vary at the modulation frequency rather than the carrier frequency. More sophisticated detectors are the discriminator and ratio detector (for detection of frequency-modulated and phase-modulated signals), and the phase-locked loop.

detent 1. A mechanical element used to lock and unlock a rotary mechanism. Usually a toothed dog which drops into a depression (ratchet) in the moving member.

2. A notch or latch on a control knob which holds it in a fixed position until it is moved intentionally.

detune 1. To adjust a circuit so that it does not respond to (is not resonant at) a particular frequency. 2. To change the inductance or capacitance of a tuned circuit so that it is not resonant at the frequency of the applied signal.

device 1. The physical realization of an electrical element, circuits, or functions in a physically independent body which cannot be further reduced or divided without destroying its stated function. This term is commonly applied to active and passive components. Examples are integrated circuits, transistors, PNPN structures, tunnel diodes, and magnetic cores, as well as resistors, capacitors, and inductors. 2. An assemblage of components each of which loses its identity in contributing to the functioning of a larger unit and being physically combined into it. 3. A mechanical, electrical, and/or electronic contrivance with a specific purpose.

dew point Is an absolute measure, found as the temperature to which a gas must be cooled at constant pressure to be saturated with respect to water. Dew point is normally expressed in degrees Centigrade.

D flip-flop (D stands for delay.) A flip-flop whose output is a function of the input which appeared one pulse earlier; for example, if a 1 appeared at the input, the output after the next clock pulse will be a 1.

dia Prefix meaning through.

diabetes Syndrome caused by deficient secretion of antidiuretic hormone (ADH) by the pituitary gland, and characterized by polyuria, the urine being of low specific gravity.

diabetic neuropathy A symmetrical affection of periphernal nerves which is a complication of unregulated diabetes.

DIAC A bidirectional breakdown diode which conducts only when a specified breakdown voltage is exceeded in either direction. Unlike the zener diode, which has a comparatively slow increase in current characteristic after breakdown, the DIAC has a sharp knee, thus yielding sufficient pulse current for reliable triggering of thyristors, such as the TRIAC and SCR.

diagnosis 1. The process of identifying a disease from its signs and symptoms. A differential diagnosis is made by comparing a case with other diseases producing similar signs and symptoms. 2. Process of locating and explaining malfunctions in equipment, or detecting errors in a computer routine.

diagnostic check A specific routine designed to locate a malfunction in a computer.

diagnostic routine 1. A specific procedure or program for the detection and isolation of a circuit malfunction or processing error in a computer. 2. A series of procedures and tests conducted for the purpose of identifying a disease by assembling a mosaic of symptoms. See also *differential diagnosis*.

diagnostics Information on what tests a device has failed and how they were failed, used to aid in troubleshooting.

diagnostic team A group of qualified specialist personnel formed to diagnose a failure or disease. The medical and technical personnel for such a team are determined by the analysis requirements in each such case.

diagonal pliers Pliers with cutting jaws at an angle to the handles to permit cutting off wires close to terminals.

diagram 1. A schematic representation of the sequence of subroutines, designed to solve a problem. 2. A coarser and less symbolic representation than a flow chart, frequently including descriptions in English words. 3. A schematic or logical drawing showing the electrical circuit or logical arrangements which constitute a specific functional area of equipment, or the total system.

dialysis 1. Method of separating small molecules (crystalloids) from colloids by placing the mixture in a container made of a membrane which is permeable only to small molecules (semipermeable membrane). The container is placed in water into which the small molecules diffuse, leaving the colloids in the container. 2. The process of separating solutes of low molecular weight from those of high molecular weight and colloids, through the use of a permeable membrane. This is the principle underlying the function of the artificial kidney.

dialyzer An extracorporeal *artificial kidney* which assumes the renal function, either temporarily or periodically, for individuals whose kidneys are diseased or damaged.

diamagnetic Describing a metal, such as antimony, bismuth, or zinc, which has a magnetic permeability less than one and is therefore repelled by a magnet.

diamagnetic bodies Those bodies which, when placed in an inhomogeneous magnetic field, tend to move toward its weaker regions.

diamagnetic material A material which is less magnetic than air, or in which the intensity of magnetization is negative. There is no known material in which this effect has more than a very feeble intensity.

Bismuth is the leading example of materials of this class.

diaphoresis Profuse perspiration.

diaphragm 1. The muscular septum separating the chest from the abdomen. 2. A sensing element (as in a pressure transducer or stethoscope) consisting of a membrane placed between two volumes The membrane is deformed by the pressure differential applied across it, and this deformation can be realized as a proportional electrical signal by its action upon strain-sensitive elements, a solid-state device, or a piezoelectric crystal. 3. The moving member of a loudspeaker, such as a cone, or an alternately shaping vibrating piece.

diaphragm-type artificial heart A heart in which only one side of the ventricle moves like a diaphragm upon introduction of air pressure between a rigid outer housing and the diaphragm itself.

diarrhea Frequent passage of loose, watery stools.

diastole That part of the cardiac cycle when the ventricles fill with blood.

diastolic Referring to that portion of the cardiac cycle when the ventricle muscles are at rest (as contrasted with the systolic phase, when the ventricular volume is diminishing by reason of the contractile force exerted by those same muscles).

diastolic pressure The minimum value of blood pressure, occurring during the relaxation and atrial-filling phase of the heart muscle. This pressure can be measured at various points in the systemic or pulmonic circulations. See *systolic pressure.*

diathermal apparatus Apparatus for generating heat in body tissue by use of high-frequency electromagnetic radiation. See *diathermy.*

diathermy The passage of high-frequency

electric current through a tissue. Because of the electrical resistance of the tissue, heat is generated. This is diffuse when large electrodes are used, and as such constitutes a physiotherapeutic aid. Tissues may be cauterized by using a small electrode. Passing a high-frequency electric current through body tissues generates heat in the latter. If matters are so arranged that the current is concentrated at one point, the heat is enough to kill the tissue at this point and, for example, eradicate small tumors (*surgical diathermy*). If the current is more spread out, gentle heating may be obtained, for example, of rheumatic joints (*medical diathermy*).

diathermy interference A form of radio-frequency interference caused by diathermy equipment. In television reception, it manifests as a horizontal herringbone pattern across the picture, of intensity proportional to proximity to the source.

diathermy machine A medical apparatus consisting of an RF oscillator frequently followed by RF amplifier stages, used to generate high-frequency currents that produce heat within some predetermined part of the body for therapeutic purposes.

dicrotic 1. Having two beats. Usually applies to secondary pulse wave due to the closure of the semilunar valves since, when this is marked as in conditions associated with vasodilatation, e.g., high fever, it gives the impression of a double pulse. The dicrotic notch (sometimes called aortic notch) is clearly visible in intracardiac pressure traces, as well as in arterial pulse traces recorded at the body surface. 2. The second expansion of the artery that occurs during the diastole of the heart.

die 1. The process of ceasing to live. 2. Sometimes called *chip,* a tiny piece of semiconductor material, broken from a semiconductor slice, on which one or more active electronic components are formed. Plural: dice. 3. A portion of a wafer bearing an individual circuit or device cut or broken from a wafer containing an array of such circuits or devices.

dielectric 1. A material which will not conduct electricity, but will contain an electrostatic field. 2. Insulating material used between the metal plates of a capacitor which can store electric charges in the form of dielectric stress. 3. The insulating material between the metallic elements of an electromechanical component or any of a wide range of thermoplastics or thermosetting plastics.

dielectric absorption Property of an imperfect dielectric whereby all electric charges within the body of the material caused by an electric field are not returned to the field.

dielectric breakdown voltage Voltage between two electrodes at which electric breakdown of a specific specimen occurs under prescribed conditions of test. Also called electric breakdown voltage, breakdown voltage, or hi-pot.

dielectric constant That property of a dielectric which determines how much electrical energy it can store up compared with the quantity of electrical energy that could be stored by a vacuum (or air) if it were substituted directly into the space occupied by the subject dielectric material. Hence, the ratio of the capacitance of a capacitor with a given dielectric material, to the capacitance of the same capacitor with air or vacuum as the dielectric. (Also called permittivity, specific inductive capacity, or capacitivity.) Some representative values of dielectric constant are:

Air	1.0
Paraffin	2.1
Mica	5.8
Glass	8.0

dielectric dissipation See *loss tangent*.

dielectric hysteresis A lagging of an electric field in a dielectric behind the alternating voltage which produces it. It causes a loss similar to that of magnetic hysteresis.

dielectric loss The time rate at which electric energy is transformed into heat in a dielectric when it is subjected to a changing electric field.

dielectric loss factor The product of the dielectric constant of a material and the tangent of its dielectric loss angle. Also called *dielectric loss index*.

dielectric loss index Dielectric loss factor.

dielectric phase angle The angular phase difference in degrees or radians between the voltage applied to a capacitor dielectric and the alternating current through it.

dielectric strength Maximum potential gradient that a dielectric material can withstand without failure. Value obtained for the dielectric strength will depend on the thickness of the material and on the method and conditions of test. Also called *electric strength*.

diethanolamine A radiopaque dye injected into the human body for X-ray study. Used in the early days of radiology.

differential amplifier An amplifier having two similar input circuits connected so as to respond to the difference between two voltages or currents and effectively suppress voltages or currents which are alike in the two input circuits. ECG, EEG, and EMG potentials are typically amplified by this method, in order to reduce or eliminate artifact and interference signals picked up by the patient's body or by electrode wires.

differential comparator A circuit using differential amplifier design techniques to compare an input voltage with a reference voltage. When the input voltage is below the reference, the circuit output is in one state; when the input voltage is above the reference, the output is in the opposite state. Commonly used for pulse amplitude detector circuits. A/D conversion, and for data transmission in electrically noisy environments.

differential diagnosis Discrimination between diseases with similar symptoms.

differential input An input circuit that rejects voltages which are the same at both input terminals and amplifies the voltage difference between the input terminals. May be either balanced or floating and may also be guarded.

differential–input capacitance The capacitance between the inverting and noninverting input terminals of a differential amplifier.

differential–input impedance The impedance between the inverting and noninverting input terminals of a differential amplifier.

differential–input measurement One in which the two inputs to a differential amplifier are connected to two points in a circuit under test and the amplifier displays the difference voltage between the points. In this type of measurement, each input of the amplifier acts as a reference for the other and ground connections are only used for safety reasons. Synonymous with *floating input.*

differential–input rating The maximum differential input which may be applied between the two terminals of an operational amplifier.

differential–input resistance The resistance between the inverting and noninverting input terminals of a differential amplifier.

differential–input voltage The maximum voltage that can be applied across the input terminals of a differential amplifier without causing damage to the amplifier.

differential–input voltage range The range of voltages that may be applied between input terminals without forcing the circuit to operate outside its specifications.

differential–input voltage rating The maximum allowable signal that may be applied between the inverting and noninverting inputs of a differential amplifier without causing damage to the amplifier.

differential instrument Galvanometer or other measuring instrument having two circuits, or coils, usually identical, through which currents flow in opposite directions. The difference, or differential, effect of these currents actuates the indicating pointer.

differential–mode gain The ratio of the output voltage of a differential amplifier to the differential–mode input voltage.

differential–mode input The voltage difference between the two inputs of a differential amplifier.

differential–mode signal A signal that is applied between the two input terminals of a balanced three-terminal system (the third terminal being the common reference for both inputs).

differential modulation A type of modulation in which the choice of the significant condition for any signal element is dependent on the choice for the previous signal element.

differential output voltage The difference between the values of two ac voltages, 180° out of phase, present at the output terminals of an amplifier when a differential input voltage is applied to the input terminals of the amplifier.

differential pressure transducer Pressure transducer that simultaneously accepts two independent pressure sources, the output of which is proportional to the pressure difference between the sources. This transducer type is employed in pneumotachography and body plethysmography.

differential stage A symmetrical amplifier stage with two inputs balanced against each other so that with no input signal or equal input signals, no output signal exists. A signal to either input, or an input signal unbalance, produces an output signal proportional to the difference.

differential transducer A device which is capable of following, simultaneously, the voltages across or from two separate signal sources and providing a final output proportional to the difference between the two signals.

differential transformer 1. A transformer which connects two or more signal sources to a single circuit, keeping them isolated from each other, and with an output proportional to the difference between the signals. 2. An electromechanical device which converts a physical change or movement to a linear electrical voltage.

differential voltage gain Ratio of the change in output signal voltage at either terminal, or in a differential device, to the change in signal voltage applied to either input terminal, all voltages being measured to common reference.

differentiating circuit A circuit whose output voltage is proportional to the time rate of change of the input voltage. The output waveform is then the time derivative of the input waveform, and the phase of the output waveform leads that of the

input by 90°. An RC circuit gives this differentiating action. Also called *differentiating network* and *differentiator*.

differentiator 1. Analog computer device which produces an output proportional to the derivative of one variable with respect to another, usually time. 2. Transducer or circuit whose output waveform is the time derivative of its input waveform. Such a transducer or circuit preceding a frequency modulator makes the combination a phase modulator; or following a phase detector makes the combination a frequency detector. Its ratio of output amplitude to input amplitude is proportional to frequency and its output phase leads its input phase by 90°. 3. Circuit which produces an output voltage substantially in proportion to the rate of change of the input voltage or current. Differentiating circuits employ time constants that are short compared to the duration of the pulse applied, thus differentiating the input pulse.

diffraction grating An optical device consisting of an assembly of narrow slits or grooves which produce a large number of beams that can interfere to produce spectra.

diffused device One where a base, usually of silicon, has successive layers of P and N characteristics diffused upon and into the base by means of a series of masks, and around which P and N materials, usually phosphorus and boron, adhere to the base by gaseous diffusion in a high temperature furnace. It is possible to build areas of resistance, capacitance and active diodes and transistors into the base, creating entire circuits. Performance is poor in presence of radiation.

diffused junction A junction that has been formed by solid or vapor diffusing an impurity (in) a small controlled amount within a semiconductor crystal wafer without heating.

diffused-junction rectifier A semiconductor diode in which the PN junction is produced by diffusion.

diffused-junction transistor A transistor in which the emitter and collector electrodes have been formed by diffusion of an impurity into the semiconductor wafer without heating.

diffused-layer resistor A resistor formed by including an appropriate pattern in the photomask to define diffusion areas.

diffused mesa transistor A three terminal semiconductor created by subjecting a silicon wafer to gaseous diffusions of both N and P type impurities to form both emitter-base and collector-base junctions. For germanium diffused mesa transistors, the collector-base junction is formed by gaseous diffusion, but the emitter-base junction is formed by an evaporated metallic strip. The collector-base junction is then defined by etching away undesired portions of emitter and base regions, exposing a mesa.

diffused planar transistor A three terminal semiconductor made by two gaseous diffusions (as is the mesa) but the collector-base junction is defined by oxide masking. Junctions are formed beneath this protective oxide layer resulting in a device with lower reverse currents and good dc gain at low currents.

diffused transistor A transistor in which the emitter and collector junctions are both formed by diffusion.

diffusion 1. The gradual assumption of an even distribution of molecules of gases or fluids within a given volume brought about by their random movement. 2. In a transistor, the movement of carriers into a region of fewer carriers. 3. The movement of donors and acceptors at extremely high temperatures, or a technique for producing diode or transistor junctions by the introduction of donors and acceptors into a semiconductor at elevated temperatures. 4. A process used in the production of semiconductors which introduces minute amounts of impurities into a substrate material such as silicon or germanium and permits the impurity to spread into the substrate. The process is very dependent on temperature and time.

diffusion process The process of doping semiconductor materials by injecting an impurity into the crystal lattice at an elevated temperature. This process is usually performed by exposing the semiconductor crystal to a controlled surface concentration of dopants.

diffusion transistor A transistor in which current flow is a result of diffusion of carriers, donors, or acceptors, as in a junction transistor.

digestion Process by which the food constituents, carbohydrates, fats, and proteins, are broken down in the body to simple chemical substances which are soluble and can be absorbed and used for the production of energy and other metabolic purposes (see *metabolism*). Digestion takes place in the alimentary tract which is comprised of the mouth, pharynx, esophagus, stomach, and intestines, and is brought about predominately by the action of enzymes aided by the movements of the parts through which the food passes.

In the mouth the food is moistened by saliva, which contains an enzyme, ptyalin. Ptyalin acts on starch, a carbohydrate present in food, and initiates its breakdown into simpler substances. Fats and proteins are not affected by saliva. The food then passes through the pharynx and down the esophagus into the stomach. The most important enzyme in the stomach is pepsin. This enzyme, together with the hydrochloric acid in the stomach, acts on proteins and breaks them down into less complex substances, proteoses and peptones. The rhythmical movements of the stomach wall cause intimate mixing of the food when the pepsin and hydrochloric acid, thus ensuring greater digestion of the proteins present. When the food has reached a fluid consistency it passes out of the stomach into the first part of the small intestine. Here the last stages of digestion of proteins and starch, and the greater part of the digestion of fats and carbohydrates takes place, together with the absorption of the products of digestion. The numerous enzymes present are derived from glands in the wall of the intestine, and from the pancreas through a small duct joining the intestine. Bile from the liver flows into this part of the intestine, and although not containing enzymes, helps in both the digestion and absorption of food. The proteoses and peptones formed in the stomach are here broken down into amino acids, the end products of protein digestion, which are absorbed into the bloodstream through the intestinal wall, and pass to other parts of the body. The simple sugars which are the end result of carbohydrate breakdown are absorbed in a like manner. Fats are broken down into fatty acids, which combine with substances present in the bile before absorption. The absorption of digested food is confined to the small intestine, and takes place through the villi. The muscular contractions of the intestinal wall greatly aid both the digestion and the absorption of food.

digit   One of the symbols 0, 1, 2, 3, 4, 5, 6, 7, 8, and 9 used in numbering in the scale of 10. One of these symbols, when used in a scale of numbering to the base (or radix) n, expresses integral values ranging from 0 to n-1, inclusive. 2. Any one of the fingers of the human hand. 3. A character used to represent one of the nonnegative integers smaller than the radix (e.g., in binary notation, either 0 or 1). 4. In telephonics, one of the successive series of incoming pulses which operate a switching train.

digital   1. Pertaining to discrete variables (as opposed to analog). Digital computers produce very close to exact solutions of complicated problems, since they retain numbers to several decimal places. 2. Using numbers expressed in digits and in a certain scale of notation to represent all the variables that occur in a problem. 3. Of, or pertaining to, the general class of devices or circuits whose output varies in discrete steps (i.e., pulses or on-off characteristics). 4. Elements or circuits whose output is utilized as a discontinuous function of its input. 5. Circuitry in which data-carrying signals are restricted to either of two voltage levels, corresponding to logic 1 or 0.

digital circuit   A circuit which operates like a switch (it is either on or off), and can make logical decisions. It is used in digital computers and similar equipment. The more common families or digital integrated circuits (called *logic forms*) are TTL, CMOS, PMOS, DTL, SOS, HTL, ECL, I²L, and RTL.

digital communication   The transmission of intelligence by the use of a signal bearing encoded binary numbers. The transmission means is a nominally discontinuous signal that changes in polarity, frequency or amplitude.

digital computer   A device which processes information represented by combinations of discrete or discontinuous data as compared with an analog computer for continuous data. More specifically, it performs sequences of arithmetic and logical operations, not only on data but on its own program. Still more specifically, it is a stored program digital computer capable of performing sequences of internally stored instructions, as opposed to calculators, on which the sequence is impressed manually.

digital data   Data represented in discrete, discontinuous form, as contrasted with analog data represented in continuous form. Digital data are usually represented by means of binary coded characters, (e.g., numbers, signs, symbols, etc.).

digital data handling system   The electronic equipment which receives digital data, operates on them in a suitable manner, records them on a suitable medium and presents them directly to a computer or a display.

digital device   Typically, an integrated circuit (IC) that switches between two exclusive states or levels, usually represented by logical 1 or 0.

digital differential analyzer   Special-purpose digital computer that performs integration, by means of a suitable integration code, on incremental quantities and that can be programmed for the solution of differential equations in a manner similar to an analog computer.

digital format   Use of discrete integral numbers in a given base to represent all the quantities that occur in a problem or calculation. All information stored, transferred, or processed by a dual-state (binary) conduction may be expressed in digital form.

digital IC   1. A switching type integrated circuit. 2. An IC that processes electrical signals that have only two states, such as

on or off, high or low voltages, or positive or negative voltages. In electronics, digital normally means binary or two-state.

digital information display   Presentation of digital information in visual form on the face of a cathode-ray tube, or on a matrix of light-emitting diodes, arranged to form characters.

digital integrator   Device for summing or totalizing areas under curves, that gives numerical readout. See also *integrator*.

digitalis   A powerful cardiac stimulant and diuretic, derived from the dried leaf of the purple foxglove plant (*digitalis purpurea*). The extract contains digitoxin and gitoxin. Used in the treatment of heart failure to increase the cardiac output and relieve edema.

digital logic modules   Circuits which perform basic logic decisions AND/OR/NOT; used widely for arithmetic and computing functions, flip-flops, half-adders, multivibrators, etc. See also *logic system*.

digital magnetic tape recording   A method of recording binary coded data, using two separate and distinct magnetic-flux levels.

digital output   Transducer output that represents the magnitude of the measured in the form of a series of discrete quantities coded in a system of notation. Distinguished from analog output.

digital phase shifter   Device which provides a signal phase shift by the application of a control pulse. A reversal of phase shift requires a control pulse of opposite polarity.

digital readout indicator   An indicator that reads directly in numerical form, as opposed to an analog indicator pointer and scale, which provide an interpretative or extrapolated reading.

digital recording   A method of recording

in which the information is first coded in a digital form. Most commonly, a binary code is used and recording takes place in terms of two discrete values of residual flux. See *digital magnetic recording*.

digital resolution The ability of a digital computer to approach a truly correct answer, generally established by the number of places expressed, and the value of the least significant digit in a digitally coded representation.

digital signal Typically, an electrical signal having two states—on or off, high or low, positive or negative—such as could be obtained from a telegraph key or two-position toggle switch. Digital normally means binary or two-state. However, digital signals can be made up of discontinuous pulses whose information is contained in their durations, periods, and/or amplitudes.

digital telemetry A method of transmitting signals (such as ECG) by means of radio waves or line carrier, in which the analog values of the phenomena are converted to equivalent digital signals and impressed upon a carrier. Such systems are characterized by higher noise-immunity and greater dependability than those in which the analog wave is transmitted.

digital thermometer Electronic temperature measuring device that reads and/or prints out numerically, as contrasted to an analog type (such as the traditional glass clinical thermometer) in which the length of a mercury column must be read against a scale of corresponding temperature values.

digital-to-analog conversion The generation of representative analog (usually variable-voltage) signals in response to a digital code.

digital-to-analog converter Abbreviated DAC or D/A converter; 1. A unit or device that converts a digital signal (byte or word) into a voltage or current whose magnitude is proportional to the numeric value of the digital signal. For example:

Digital Input	Analog Output
00001 (binary 5)	2 volts
01010 (binary 10)	4 volts
10100 (binary 20)	8 volts

2. A computing device which changes digital quantities into physical motion, such as turns of a potentiometer. See also *converter*.

digital transmission A mode of transmission in which all information to be transmitted over the facility is first converted into digital form and then sent down the line as a stream of pulses. (Such transmission may imply a serial bit stream, but parallel forms are also possible.) When noise and distortion threaten to destroy the integrity of the pulse stream, the pulses are detected and regenerated, thus eliminating the degrading effects.

digital voltmeter An electronic instrument which converts an applied analog voltage of unknown magnitude into a direct-reading display of known numeric value.

digitize 1. To convert an analog potential into a representative digital signal. 2. To change an analog measurement into a number expressed in digits.

digitizer A device which converts analog data into corresponding digital value.

digitizing The process of converting an analog signal to a digital signal.

dilatation Increase in size; enlargement. The operation of stretching.

diode A device which permits the flow of current in one direction only. Diodes are used in both vacuum tube and semiconductor form to control the passage or non-passage of signals to another element in an electrical system. This characteristic makes the diode capable of rectifying alternating current to direct current, detecting modulated radio waves, and as switching elements in logic circuits. See also *junction diode*.

diode bridge A series-parallel configuration of four diodes, whose output polarity remains unchanged whatever the input polarity. Provides full-wave dc rectification of alternating current, and also is used for amplitude and phase detection, power and logic switching, and in balanced modulators.

diode limiter Peak-limiting circuit employing a diode that becomes conductive when signal peaks exceed a predetermined value, thereby shunting the excess signal away from the protected input. Typically used in the isolated patient input circuits of physiological monitoring amplifiers to guard circuits against excessive voltages from electrosurgical apparatus and defibrillator potentials.

diode logic An electronic circuit using current-steering diodes, such that the relations between input and output voltages correspond to AND or OR logic functions.

diode matrix A two-dimensional array of diodes used for a variety of purposes such as decoding and read-only memory.

diode rectifier A half-wave rectifier of two elements, between which current flows in only one direction. It may be (a) an electron-tube diode in which current flows only from cathode to anode, (b) a point contact semiconductor diode, or (c) a semiconductor junction diode.

diode-switch Diode which is made to act as a switch by the successive application of positive and negative biasing voltages to the anode (relative to the cathode), thereby allowing or preventing, respectively, the passage of other applied waveforms within certain limits of voltage.

diode transistor logic (DTL) A logic circuit family that uses diodes at the input to perform the electronic logic function that activate the circuit transistor output. In monolithic circuits, the DTL diodes are a positive level logic AND function or a negative level OR function. The output transistor acts as an inverter, with the result that the circuit becomes a positive NAND or a negative NOR function.

DIP Abbreviation for dual in-line package.

direct access device See *random access device*.

direct-coupled transistor logic (DCTL) One of the earliest forms of monolithic logic circuits. This family consists solely of transistors that are directly coupled and have no buffering components or resistors between circuit stages. Relies heavily on the relative V_{BE} and V_{CE} of the transistors for its operation. More reliable circuit operation evolved DCTL into the RTL (resistor transistor logic) family.

direct coupling Connecting two circuits, or stages in an amplifier, with a wire or resistor so that the coupling will pass all frequencies down to zero (dc) with no discrimination. These circuits are especially suited to use in the display and recording of low-frequency biophysical phenomena, where absence of phase shifting and differentiation provide a reliable replication of the applied signal.

direct current (dc) An electrical current of essentially constant average value which flows in one direction only. An intermittent or varying current in one direction is called *pulsating dc*.

direct-current amplifier   Amplifier that is capable of amplifying direct current voltages and slowly varying voltages.

direct current restorer   A device by which a direct current component or reference is added to an alternating current signal after its processing. See also *clamping*.

directly grounded   See *solidly grounded*.

direct material   A semiconductor material in which electrons drop from the conduction band directly to the valence band to recombine with holes. The recombination process conserves energy and momentum.

direct memory access   Direct communication between memory and peripherals. In computers where this is not inherent, a device that bypasses the central processor unit.

direct writing galvanometer recorder   Recorder using a pen attached directly to a galvanometer movement, which writes traces on paper or signal at frequencies up to about 300 Hz.

disc   1. Large, rigid discs which are coated with magnetic material. Can be magnetically read or written on by a computer. Typically, a rigid disc stores one million or more words and can be accessed in 20 to 30 milliseconds. 2. Flexible (or "floppy") discs are similar in application to rigid discs, but are made of magnetic-oxide-coated vinyl material, with consequent cost saving. The flexible disc is jacketed within an enclosing envelope to give sufficient rigidity for handling and repeated use.

discharge breakdown   Breakdown resulting from degradation of material by gas discharges present in voids or on the surface. See also *breakdown*.

discrete   Made of distinct parts; also, numerals or other symbols which stand for quantities or single values. An abacus is a simple computer which handles discrete information, while a slide rule is a simple computer which handles analog information and from which discrete answers may be estimated. The word disrete is always associated with a finite number of items in a set; cursive or analog is always associated with an infinite number of points in a line or curve. 2. As opposed to integrated or combined components, discrete devices are separate, individual components which, when combined, make up a circuit or unit. The term is also used in the sense that pulses may be discrete, being separated, and each giving its own piece of information.

discrete circuit   Circuit built of separate, individually manufactured, tested and assembled electronic components (transistors, diodes, resistors, etc.).

discrete component   An individual non-integrated circuit component, either active or passive, complete in itself; such as a resistor, diode, capacitor, or transistor used as an individual and separable circuit element.

discrete device   A traditional electrical component such as a resistor capacitor, or transistor. Contrasted with an integrated circuit which contains or is functionally equivalent to the combination of many discrete components.

discrete element   See *discrete component*.

discrete part   See *discrete component*.

discrete thin-film component   An individually packaged electronic component having one or more thin films serving as resistive, conductive, and/or insulating elements. Resistors and potentiometers having thin-film metallic resistance elements are typical examples.

discriminator   1. Device in which amplitude variations are derived in response to frequency or phase variations. 2. Part of a receiver circuit which removes the desired signal from an incoming frequency or phase modulated carrier wave by changing modulation in terms of frequency variations into amplitude variations. 3. Circuit, the output voltage of which varies in amplitude and polarity according to the frequency of the applied signal. It is used as a detector in a frequency–modulated receiver.

disc storage The storage of data on the surface of magnetic discs, thus forming a random-access auxiliary computer memory which holds vastly greater amounts of data than core. Discs are removable and storable with data intact.

disinfection The destruction or removal of pathogenic organisms, especially by means of chemical substances.

diskette See *flexible disk.*

display A visual presentation of information, as on a cathode-ray tube, digital register, liquid crystal display unit, light-emitting diode matrix, or any of several forms of alphanumeric devices.

dissecting aneurysm Dilation of the wall of an artery occurring when blood is forced into the potential space between the layers of the arterial wall.

dissipation factor Of a dielectric material, the ratio of the energy lost in heat to the energy stored in the dielectric. Some typical values of dissipation factor at one kilohertz are:

Polyethylene	0.0002
Paraffin	0.0006
Mica	0.0006
Polyvinyl chloride	0.0185

distal 1. Situated away from the center. 2. Peripheral; at the greatest distance from some reference point; away from the center of the body.

distortion An unwanted change in waveform. Principal forms of distortion are inherent nonlinearity of the device, nonuniform response at different frequencies, and lack of constant proportionality between phase-shift and frequency. (A desired or intentional change might be identical, but it would be called modulation.) Harmonic distortion disturbs the original relationship between a tone and other tones naturally related to it. Intermodulation distortion (IM) introduces new tones caused by mixing of two or more original tones. Phase distortion, or nonlinear phase shift, disturbs the natural timing sequence between a tone and its related overtones. Transient distortion disturbs the precise attack and decay of a musical sound. Harmonic and IM distortion are expressed in percentages; phase distortion in degrees; transient distortion is usually judged from oscilloscope patterns.

distributed capacitance Capacitance that exists between the turns in a coil or choke, or between adjacent conductors or circuits, as distinguished from the capacitance which is concentrated in a capacitor.

distributed constants Constants such as resistance, inductance, or capacitance that exist along the entire length or area of a circuit, as distinguished from constants concentrated in circuit components.

distributed inductance The inductance which is spread uniformly along the entire length of a conductor, as distinguished from lumped inductance which is concentrated at a point.

distribution amplifier 1. An RF power amplifier used to feed a speech or music distribution system and having sufficiently low output impedance so changes in load do not appreciably affect output voltage. 2. An RF power amplifier used to feed television, radio, or telemetry signals to a number of receivers in a hospital.

distributor See *memory register.*

dither An oscillation which is introduced to overcome or smooth the effects of friction, hysteresis, or cogging.

diuresis An increased secretion of urine.

diuretic A substance which increases the secretion of urine in the kidneys.

diverticulm A pouch-like process from a hollow organ, e.g., esophagus, intestine, urinary bladder.

document reader A general term referring to OCR (optical character reading) equipment which reads a limited amount of information (one to five lines). Generally operates from a predetermined format and is therefore more restricted in the location of information to be read. The forms involved are generally tab card size or slightly smaller or larger.

donor 1. The contributor of a tissue, fluid, or organ used for transplantation or transfusion into the body of a recipient. 2. An impurity, usually antimony, arsenic, or phosphorous, which is added to a germanium or silicon semiconductor material to increase the number of free electrons. A material thus doped is called an N-type semiconductor.

donut See *land.*

dope 1. To add impurities (dopants) to another substance, usually solid, in a controlled manner to produce certain desired properties. For example, silicon is doped with small amounts of other semimetallic elements to increase the number of electrical carriers. Ruby is aluminum oxide doped with chromium oxide. 2. A street expression for heroin, morphine, and other hard drugs of an addictive nature.

doping Diffusion of impurities into semiconductor materials such as silicon or germanium to change the material from an insulator into a conductor of electricity.

doping agent An impurity element added to semiconductor materials used in crystal diodes and transistors. Common doping agents for germanium and silicon include aluminum, antimony, arsenic, gallium, and indium.

Doppler effect A change in an observed frequency of a wave caused by some change in the effective length of the path of travel between the source and the observer. An example of this effect is an ambulance moving along a street with its siren sounding. As the ambulance approaches the observer, the pitch of the siren seems quite high but, on passing, the pitch decreases rapidly to a lower value. In medical use, ultrasonic impulses are used to noninvasively measure the flow of blood within the vessels by measuring the Doppler shift in echoes reflecting from the pulsating vessel.

Doppler shift Magnitude, in cycles per second, of the change in the observed frequency of a wave caused by the Doppler effect. See *Doppler effect.*

dorsal A position more toward the back of some point of reference (opposite of ventral).

dose 1. According to current usage, the radiation delivered to a specified area or volume or to the whole body. Units for dose are roentgens for X or Gamma ray, reps or equivalent roentgens for Beta rays. No statement of dose is complete without specifying location. It is usually specified as the amount of energy absorbed by tissue at the site of interest per unit mass. 2. The specified quantity of medication to be administered, compensated for patient age, sex, condition, and coexisting circumstances, adjudged appropriate to achieve a therapeutic effect.

dose equivalent A radiological term used to express the amount of effective radiation when modifying factors have been considered. The product of absorbed dose multiplied by a quality factor multiplied by a distribution factor. It is expressed numerically in rems.

dosimeter Instrument for measuring total (integrated) amount of nuclear radiation incident on an object. See also *film badge*.

dot When drawn on the input of a logic symbol, indicates that the active signal input is a negative input. The lack of a dot indicates a positive active signal. Also called *bubble*.

dot AND See *wired AND*.

dot matrix display A display format consisting of small light emitting elements arranged as a matrix. Various elements are energized to depict a character. A typical matrix is 5×7.

dot OR See *wired OR*.

double-diffused transistor A transistor in which two PN junctions are formed in the semiconductor wafer by gaseous diffusion of both P-type and N-type impurities. An intrinsic region can also be formed.

double pole A term applied to a contact arrangement to denote that it includes two separate contact forms; that is, two single-pole contact assemblies which can be opened or closed as one, but which serve separate circuits.

double-pole switch Switch that operates simultaneously in two separate electric circuits or in both lines of a single circuit.

double-precision arithmetic Arithmetic which is necessary to obtain more accuracy than a single word of computer storage will provide. This is accomplished by using two computer words to represent one number.

double throw A term applied to a contact arrangement to denote that each contact form included is a breakmate.

double-throw switch Switch, by means of which a change in circuit connections can be obtained by closing the switch blade into either of two sets of contacts.

downtime 1. The period during which an equipment is malfunctioning or not operating due to mechanical, electrical, or electronic failure, but not for lack of work or absence of an operator. 2. The total time during which a system is not in condition to perform its intended function. Downtime can, in turn, be subdivided into a number of categories such as active maintenance downtime, supply time and waiting time. 3. The period during which a computer is malfunctioning or not operating correctly; contrasted with available machine time, idle time, or standby time.

dps Abbreviation for disintegrations per second.

drain 1. The working-current terminal at one end of the channel in a FET that is the drain for hole-electrons or free electrons from the channel. Corresponds to the collector terminal of a bipolar transistor. 2. A conductive material or surface, usually connected to earth ground, which drains away static charges from persons and objects. The conductive floor of an operating room is a drain, designed to prevent spark discharges from static buildup in the presence of explosive anesthetics.

drain wire Metallic conductor frequently used in contact with foil-type signal-cable shielding to provide a low-resistance ground return at any point along the shield.

drift 1. The movement of electrons or holes in a semiconductor. 2. A slow change in frequency baseline, or other parameter, usually attributable to gain changes or component value shift in the presence of heat.

drift transistor 1. Transistor having two plane parallel junctions, with a resistivity gradient in the base region between the junctions to improve the high-frequency response. Also called *diffused-alloy transistor*.

driver 1. Electronic circuit which supplies input to another electric circuit. 2. Stage of amplification which precedes the power output stage. 3. Unit (or stage) driving another unit (or stage). 4. A device in a logic family controlled with normal logic levels whose output has the capability of sinking or sourcing high current. The output may control a lamp, relay or a very large fanout of other logic devices. 5. A device driving a higher output device or transistor by supplying power, voltage, or current to it.

DRO Abbreviation for destructive readout memory.

dropping resistor A resistor placed in series between a voltage source and a load for the purpose of reducing the voltage supplied to the load.

dropsy Edema; accumulation of fluid in a body cavity, especially the abdomen.

drug pacemaker The use of certain pharmaceutical agents that increase the ventricular rate in a diseased heart.

drum storage A type of addressable data storage unit used in some digital computers, consisting of a cylindrical drum coated with a magnetic material and continuously rotated. Data are recorded on and read from parallel tracks on the drum surface by a number of recording-playback heads.

dry 1. Said of circuits or contacts which neither break nor make a direct current flow. 2. Free from water or humidity.

dry cell Cell in which the electrolyte exists in the form of a jelly, or is absorbed in a porous medium, or is otherwise restrained from flowing from its intended position. Energy is derived from the reaction of an acid or alkaline paste on dissimilar metals, or on a metal and a carbon electrode. The normal open-circuit voltage is 1.5 volts, the paste is sealed in normal use. and the cell cannot be recharged.

dry circuit A circuit wherein the open circuit voltage is 0.03V or less and the current 200 ma or less. The voltage is most important, because at this level the voltage is not great enough to break through most oxides, sulfides, or other films which can build up on contacting surfaces.

dry contacts Contacts which neither break nor make current.

dry disc rectifier A rectifier using metal discs as rectifying elements. See *rectifier, cuprous oxide* and *rectifier, selenium*.

dry electrolytic capacitor An electrolytic capacitor in which the electrolyte is a paste rather than a liquid. The dielectric is a thin oxide film formed on one of the plates by chemical action.

dry-reed relay A relay that consists of one or more capsules containing contact mechanisms that are generally surrounded by an electromagnetic coil for actuation. The capsule consists of a glass tube with a flattened ferromagnetic reed sealed in each end. These reeds, which are separated by an air gap, extend into the tube so as to overlap. When placed in a magnetic field they are brought together and close a circuit.

D-type flip-flop A flip-flop that will propagate whatever information is at its D (data) conditioning input prior to the clock pulse, to the 1 output on the leading edge of a clock pulse.

dual in-line package (DIP) 1. The most widely used type of IC package for industrial-grade ICs. The standard DIP is a molded plastic package about ¾ in. long and ⅓ in. wide, with two rows of pins spaced on 0.1 in. centers. This package is more popular than the flat pack or TO can for industrial use because it is relatively inexpensive and is easily dip-soldered into pc boards. 2. A package for electronic components that is suited for automated assembly into printed circuit boards. The DIP is characterized by two rows of precisely spaced external connecting terminals, or pins, which are inserted into the holes of the printed circuit board.

dual-trace A mode of operation in which a single beam in a cathode-ray tube is time shared by two signal channels. See *alternate mode* and *chopped mode*.

ductus A duct; a little canal of the body.

ductus arteriosus An arterial canal or duct that connects the pulmonary artery and the aorta in fetal life. Occasionally this remains patent.

dump 1. To remove all power, intentionally, accidentally, or conditionally, from a system or component. 2. To intentionally transfer all or part of one section of a computer memory into another section. Also called *core dump, memory dump, memory printout,* and *storage dump.*

dump check Computer check that usually consists of adding all the digits during dumping, and verifying the sum when retransferring.

Dunmore cell See *lithium chloride sensor.*

duodecimal See *number system.*

duodenostomy Surgical establishment of a communication between the duodenum and another structure.

duodenum The first 30 cm (12 in.) of the small intestine, beginning at the pyloric orifice of the stomach.

duplex 1. Two conductors twisted together, usually with no outer covering. This term has a double meaning and it is possible to have parallel wires and jacketed parallel wires, and still refer to them as duplex. 2. An electrical modifier for terms such as *switch, outlet,* or *receptacle,* specifying that the device is a gang of two identical units.

duplexed system A system with two distinct and separate sets of facilities, each of which is capable of assuming the system function while the other assumes a standby status. Usually both sets are identical in nature. Also called *redundant system.*

duty cycle 1. The ratio or percentage of *on* time (period during which current flows) to *off* time (period during which current does not flow). In pulsed circuits, especially with semiconductors, duty cycle determines the device dissipation rating required for reliable operation. 2. A measure of the effect of a pulsed input to a lump. Expressed as a percentage of on time as compared to total time.

duty factor In pulsed circuits, the product of the pulse duration and the pulse repetition frequency of a wave composed of pulses that recur at regular intervals.

df/dt The first derivative of a function; its rate of change with respect to time.

dp/dt The rate of change of blood pressure with respect to time; the first derivative of the pressure function. This function provides information regarding cardiac performance and is often used to determine effects of drugs on myocardial contractility. It is obtained by differentiating the signal representing the pressure wave, using a passive (resistor/capacitor network) or active, differentiator circuit.

dv/dt The rate of change of voltage with respect to time. Proportional to current flow in a capacitor.

dynamic behavior A description of how a system or an individual unit functions with respect to time.

dynamic braking A means of stopping a motor by reconnecting it as a generator to dissipate its stored mechanical energy in a resistor as electrical energy. At zero speed no holding force exists.

dynamic characteristic See *load characteristic*.

dynamic check Check used to ascertain the correct performance of some or all components of equipment or a system under dynamic or operating conditions.

dynamic electricity See *electricity*.

dynamic memory The storage of data on a device or in a manner that permits the data to move or vary with time, and thus the data is not always available instantly for recovery; e.g., acoustic delay line, magnetic drum, or circulating or recirculating of information in a medium.

dynamic MOS array A circuit made up of MOS devices which requires a clock signal. The circuit must be tested at its rated (operating) speed. Known as *clock-rate testing*.

dynamic problem check Of an analog computer, any dynamic check used to ascertain that the computer solution satisfies the given system of equations.

dynamic programming In operations research, a procedure for optimization of a multistage problem solution, wherein a number of decisions are available at each stage of the process.

dynamic range Of a transmission system, the difference in decibels between the noise level of the system and its overload level.

dynamic register A memory in which the storage takes the form of capacitively charged circuit elements and therefore must be continually refreshed or recharged at regular intervals.

dynamic relocation The ability to move computer programs or data from auxiliary memory into main memory at any convenient location. Normally, the addresses of programs and data are fixed when the program is compiled.

dynamic scattering LCD See *LCD*.

dynamic shift register A shift register that stores information using temporary charge storage techniques. The major drawback of this technique is that the information is lost if the clock repetition rate is reduced below a minimum value.

dynamic storage Information storage using temporary charge storage techniques. It requires a clock repetition rate high enough to prevent loss of information.

dynamic storage allocation A storage allocation technique in which the location of programs and data is determined by criteria applied at the moment of need.

dynamic subroutine A digital computer programming subroutine which involves parameters (such as decimal point position) from which a properly coded subroutine is derived. The computer itself generates the subroutine according to the parametric values chosen.

dynamo A machine having an armature and a field, one rotating and one stationary. May be either a generator or motor, but usually the term is intended to mean a direct current generator.

dynamometer Instrument for measuring force, such as force of muscular contraction. Also applied to instruments for measuring power (horsepower).

dynamotor A rotary electrical machine used to convert from direct current to alternating current. The machine has a single field structure and a single rotating armature having two windings, one equipped with a dc commutator and the other with ac slip rings.

dyne The standard centimeter-gram-second unit of force; the force that produces an acceleration of one centimeter per second on a mass of one gram.

dys A prefix meaning bad, difficult, painful, or abnormal.

dysentery Inflammation of the large intestine. There are two kinds of dysentery, bacillary and amoebic; the former due to a bacillus, the latter to the entamoeba hisolytica.

dysfunction Abnormal operation of an organ or portion of a bodily system.

dysphagia Difficulty in swallowing.

dysrhythmia An abnormality of rhythm. Typically used in describing a disordered pattern in the brain waves or the ECG.

dysuria Difficulty in voiding or pain on voiding.

EA Abbreviation for electrical anesthesia.

E and M leads The output and input leads, respectively, of a signaling system. The E lead provides an open or ground. The M lead accepts open or ground, or battery or ground, as the circuit may require.

eardrum See *tympanic membrane*.

earth currents 1. Currents flowing through the ground due to natural causes, affecting the magnetic field of the earth and sometimes causing magnetic storms. Also called *telluric current*. 2. Return, fault, leakage, or stray current passing through the earth from electrical equipment. Also called *ground current*.

earth ground A connection from an electrical circuit or equipment to the earth through a water pipe or a metal rod driven into the earth. This connection reduces shock hazards from faulty equipment. Water pipes may no longer be reliable grounds because of the use of transite pipe, neoprene gaskets, and other nonconducting links. Ground rods driven under the interior of a large building may gradually become ineffective because the building may drive the local water table down so far that the rod is essentially surrounded by dry soil. Ground requirements are defined in the U.S.A. by the National Electrical Code.

EC Abbreviation for extension cylinder.

ecchondroma A tumor composed of cartilage at the junction of cartilage and bone.

ECG Abbreviation for electrocardiograph, electrocardiography, and electrocardiogram.

ECG analyzer Instrument for scanning tape-recorder ECG for significant diagnostic information.

ECG (clinical) The strip chart record of the patient's electrocardiogram, taken with five ECG electrodes (four applied to the limbs by straps, the fifth, exploring electrode, applied to the precordium at each of six specified positions). Twelve ECG leads are typically recorded: the bipolar extremity leads (1, 2, and 3), the unipolar extremity leads (AvR, AvL, and AvF), and the six unipolar precordial leads (V_1

through V_6). The recordings are made, using equipment having diagnostic-quality ECG frequency response. See *ECG, diagnostic*.

ECG (diagnostic) The multiple-lead electrocardiographic waves of a patient, as conditioned for recording of the clinical scalar electrocardiogram. The diagnostic-quality ECG reproduces the full frequency spectrum of the cardiac waves, faithfully reporting even the most subtle anomalies. Frequencies in the range of 0.05 to 100 hertz are present, without distortion, as required by the Committee on Electrocardiography of the American Heart Association. See also *ECG (clinical)*.

ECG electrode In unipolar ECG, the active (exploring) electrode that is on the chest near the heart, and is used with a remote (indifferent) electrode on a limb. When three or four limb electrodes are used, they are joined to form a central electrode which becomes the *indifferent electrode*. Also called *Wilson center terminal*.

ECG (monitoring) The single-lead electrocardiographic waves of a patient, as conditioned for continuous display on a cardioscope. Typically, a monitoring ECG does not reproduce the full frequency spectrum of the cardiac waves, in order to obtain a stable, or *quiet trace,* free of annoying artifact from patient movement. This is achieved by filtering out low and high frequency components. Thus, a typical monitoring ECG presents frequencies in the range of 1 to 25 hertz, with some distortion, particularly in the S-T segment of the QRS complex. The display quality is generally adequate for recognition of major arrhythmias.

echo An acoustic signal that has been reflected or otherwise returned with sufficient magnitude and time delay to be detected as a signal distinct from that which was transmitted. In ultrasonics, the echo cannot be heard, but is detected by means of electronic devices. The interval of time, between transmission of the acoustic signal and the reception of the echo, is proportional to the distance between the transmitted and the reflecting surface.

echocardiography A sonar-like noninvasive method of diagnosing cardiac malfunctions. A pulsed ultrasound beam is directed through the chest wall and echoes reflected from differing tissue interfaces (e.g., soft tissue and blood, tissue and bone) are detected by the excited barium titanate transducer which originated the pulse. Echoes are recorded against a linear time base to provide a hard-copy record of cardiac structure movement in a given time span, generally defined by the ECG.

echo-complex A group of echo signals reflected by closely spaced tissue.

echoencephologram The record produced by electroencepholography, which see.

echoencephalograph Device using ultrasonic energy and echo-ranging technique to determine brain midline, hematoma, tumor.

echoencephalography Visualization of the fluid filled cavity of the brain by means of ultrasonic diagnostic devices. Since tumors are usually unilateral, they produce an asymmetry of the cerebral hemispheres and a movement of the midline toward one direction. Echoes reflected from the midline are, therefore, not equal bilaterally and a diagnosis is possible.

echoencephaloscope An ultrasonic diagnostic instrument designed specifically for brain studies. A transducer placed against the patient's head, generates a series of ultrasonic pulses and also detects the returning echoes. Each pulse and its associated echoes are displayed on a cathode-ray tube.

echogram The hard-copy record of an echographic study. Generally recorded by either Polaroid photography or a cathode-ray tube-type strip chart recorder. An echogram of the heart reveals relative movement of cardiac structures (i.e., changes in ventricular dimensions, motions of mitral valve leaflets) in a pictorial manner. Tracings are created by ultrasound echoes returned from these structures within a specific time frame.

echosonogram Graphic display obtained with ultrasound pulse reflection techniques. (An echocardiogram and echoencephalogram are examples of echosonograms.)

ECL Emitter-coupled logic (also known as *CML, current-mode logic*). Digital logic is performed by emitter-coupled transistors operating in unsaturated mode, amplification by transistors. Favored for high speed, adaptability to large-scale arrays, wide range of available logic functions, excellent speed/power product. Disadvantages include high cost, poor interface with saturated-logic circuits, moderate noise immunity, fairly high power dissipation.

ecology Study of the linkage of all living things to others and to their environment.

ecmnesia A lapse in memory, the memory before and after the lapse being normal.

-ectasis Suffix meaning stretching or dilatation.

-ectomy Suffix meaning cut out, remove or excision.

ectopic That which is situated at a place away from the normal location (e.g., an ectopic pregnancy, in which the fetus de-

velops in a fallopian tube rather than in the uterus).

ectopic beat Contraction of the heart (heartbeat) which occurs outside the normal rhythm of the heart.

ectopic focus An irritable spot of tissue generating impulses which may cause extra beats (premature contractions) of the heart.

ectopic kidney Kidney which is abnormally situated in the abdominal cavity.

ectopic viscera Organs which are situated in an abnormal place as a result of a developmental anomaly.

eczema Inflammatory skin disease that produces a great variety of lesions, such as vesicles, thickening of skin, watery discharge, and scales and crusts, with itching and burning sensations. Eczema is caused by allergy, infections, and nutritional, physical, and sometimes unknown factors.

eddy current A circulating current which is induced within transformer cores or any conductor which is exposed to a varying magnetic field. It is a loss, and causes heating of the conductor.

edema Excessive accumulation of water and salt in the tissue spaces, caused by kidney or heart disease (*generalized edema*) or by local circulatory impairment stemming from inflammation, trauma, or neoplasm (*localized edema*).

edge effect Nonuniformity of electric fields between two parallel plates caused by an outward bulging of electric flux lines at the edges of the plates.

Edison effect Phenomenon wherein electrons, emitted from a heated element within a vacuum tube, will flow to a second element that is positive with respect to the heated element. See also *Richardson effect* and *thermionic emission*.

edit To arrange or rearrange digital computer output information before printing it out. May involve deleting unwanted data, selecting pertinent data, inserting invariant symbols such as page numbers and typewriter characters, and applying standard processes such as zero-suppression.

editor A computer program for text editing. An editor makes it easy to use a video terminal to write programs.

EDP Electronic data processing.

EEG Abbreviation for electroencephalography, electroencephalograph, or electroencephalogram.

EEG electrode Electrode that attaches to the scalp for detecting EEG (brain waves).

effective address In computers, the memory address upon which the instruction operates.

effective capacitance Total capacitance existing between any two given points of an electric circuit.

effective current Value of alternating current which will give the same heating effect as the corresponding value of direct current. Effective value is 0.7071 times the peak value in the case of sine-wave alternating currents.

effectively grounded Grounded through a ground connection of sufficiently low impedances (inherent and/or intentionally added) so that fault grounds which may occur cannot build up voltages dangerous to connected personnel or other equipment.

effective resistance The increased resistance shown by a conductor to the flow of an alternating current, as compared to direct current. It is due to the fact that, as the frequency increases, the current is not distributed uniformly throughout the conductor but tends to flow in the surface layer. See also *skin effect.*

effective value 1. Of an alternating current or voltage, the square-root-of the mean-square value. 2. Of an alternating current, that value which will cause the same heat in a resistor as the corresponding value of direct current. If the alternating current or voltage is a sine wave, the effective value is the peak value multiplied by 0.7071. See *rms amplitude.*

effective voltage Of an alternating voltage, the square-root-of the mean-square voltage. If the alternating voltage is a sine wave, the effective value is the peak value multiplied by 0.7071.

effector impulse See *nerve impulse.*

effects of electric shock The probable effects of electric shock (depending on points of contact) to the human body are:

Current in mA	Effects
0-1	Perception
1-4	Surprise
4-8	Mild discomfort
8-20	Serious discomfort
21-40	Muscular inhibition
40-up	Respiratory block and/or ventricular fibrillation, leading to death.

efferent 1. Conveying from the center; e.g., the motor nerves which convey impulses from the brain and spinal cord to muscles and glands. Also *afferent.* 2. In neurology, referring to the direction of an impulse away from or out of a specified center, such as the cell body of a neuron.

efficiency The ratio, expressed in percentage, of signal output to input. (Often used to estimate the power needed to drive a loudspeaker.)

ego In psychoanalytic theory, one of the

three major divisions of human personality, the others being the id and superego. The ego, commonly identified with consciousness of self, is the mental agent mediating among three contending forces: the external demands of social pressure or reality; the primitive instinctual demands arising from the id imbedded as it is in the deepest level of the unconscious; and the claims of the superego, born of parental and social prohibitions and functioning as an internal censor or conscience.

eight-level code   A computer code that utilizes eight impulses for describing a character. Start and stop elements may be added for asychronous transmission. Often used to refer to the *USASCII code.*

EKG   Germanic abbreviation of *electrokardiogram,* generally less favored in written use today than in Einthoven's time. Anglicized form is *ECG.*

electret   A permanently polarized piece of dielectric material, produced by heating the material and placing it in a strong electric field during cooling. Some barium titanate ceramics can be polarized in this way, as also can carnauba waxes. The electric field of an electret corresponds somewhat to the magnetic field of a permanent magnet.

electric   Containing, producing, arising from, actuated by, or carrying electricity, or designed to carry electricity and capable of so doing. Examples: electric eel, energy, motor, vehicle, wave.

electrical   Relative to, pertaining to, or associated with electricity but not having its properties or characteristics. Examples: electrical engineer, handbook, insulator, rating, unit.

electrical anesthesia (EA)   The use of electrical current, passed through the head, to reduce responsiveness of the patient.

Although still in the experimental stages of development, electrical anesthesia has been used during major surgery with promising success. Frequencies in the range of 700 to 4000 Hz have been employed, and current levels up to approximately 130 milliamperes are typical.

electrical charge   Symbol Q, q. 1. Quantity of electricity on (or in) a body. 2. The excess of one kind of electricity over the other kind. A plus sign indicates that the positive electricity is in excess, a minus sign indicates that the negative electricity is in excess. (Compare with *electric charge.*)

electrical degree   One-360th part of a cycle of alternating current.

electrical element   The concept in uncombined form of any of the individual building blocks from which electric circuits are synthesized.

electrical filter   Device for rejecting or passing a specific band of signal frequencies.

electrical-impedance cephalography   A method of evaluating blood circulation in the brain by measuring the impedance changes resulting from volume changes of blood in the brain. The electrical impedance of the head is measured through a pair of surface electrodes, and this impedance decreases with an increase of blood volume in the brain. The technique is also known as *rheoencephalography.*

electrically connected   Connected by means of a conducting path, or through a capacitor as distinguished from connection merely through electromagnetic induction.

electrical shock   Combined reaction of bodily tissues to the flow of electrical current. See *effects of electric shock.*

electric charge   1. A property of electrons and protons. Similarly charged particles repel one another. Particles having opposite charges attract one another. 2. Electric energy stored as stress on the surface of a dielectric.

electric circuit A path consisting of conductors and/or interconnected circuit elements through which an electric current can flow.

electric contact A separable junction between two conductors which is designed to make, carry, or break (in any sequence or singly) an electric circuit.

electric current Electricity in motion. In the atoms of metallic substances there are a number of *free electrons* or negatively charged particles which wander in the spaces between the atoms of the metal. The electron movement is normally without any definite direction and cannot be detected. The connection of an electric battery produces an electric field in the metal and causes the free electrons to move or drift in one direction, and it is this electron drift which constitutes an electric current. Electrons being of negative polarity are attracted to the positive terminal of the battery and so the actual direction of flow of electricity is from negative to positive, that is, opposite to the conventional direction usually adopted.

electric field Field of force which exists in the space around electrically charged particles. Lines of force are imagined to originate at the protons or positively charged particles and to terminate on electrons or negatively charged particles.

electric force Electric field intensity measured in dynes.

electricity Property of certain particles to possess a force field which is neither gravitational nor nuclear. The type of field force associated with electrons is defined as negative and that associated with protons and positrons as positive. The fundamental unit is the charge of an electron: $1,60203 \times 10^{-19}$ coulomb. Electricity can be further classified as static electricity or dynamic electricity. Static electricity in its strictest sense refers to charges at rest as opposed to dynamic electricity, or charges in motion. Static electricity is sometimes used as a synonym for triboelectricity or frictional electricity.

electric pile Invented by Volta in 1800, it was the first primary battery known to the modern world. The pile consisted of an arrangement of pairs of discs of copper and zinc, each pair separated by a disc of moistened pasteboard. Later Volta arranged a series of cups filled with brine, each containing a zinc and copper plate, and by connecting these together obtained an electric current.

electric shield Housing of metal, usually aluminum or copper, placed around a circuit. The housing prevents interaction between circuits by providing a low resistance and reflecting path to ground for high-frequency radiations. See also *magnetic shield.*

electroacoustic transducer Transducer for receiving waves from an electric system and delivering waves to an acoustic system, or vice versa.

electroanesthesia See *electrical anesthesia.*

electrobiology The biology of electrically active tissues. See *bioelectricity.*

electrocardiogram (ECG) 1. A hard-copy record of heart action potentials obtained by measuring instantaneous potential differences at the surface of the body. In general, the recording describes the depolarization of myocardial muscle cell masses, providing a graphic, but indirect view of the heart's competence. The most common electrocardiogram, is the so-called lead II, in which the potentials appearing between the right arm and left leg are recorded. The principal components of the lead II ECG are the P wave, the QRS complex and the T wave. The P wave signals the origination of a pacing impulse in the specialized conductive system. While this impulse is en route to the heart's musculature, the ECG is silent. Upon its arrival and with resulting depolarization of muscle tissue, the QRS complex appears. The T wave signifies repolarization of the cardiac muscles in preparation for another stimulus. A host of other electrocardiographic leads are often used to gain a more complete view of cardiac electrophysiologic action. In each lead, potentials at different prescribed points on the body surface are recorded. Twelve such recordings normally constitute the *clinical scalar electrocardiogram;* these are: leads I, II, and III, AvF, AvR and AvL and V1 through V6. 2. The tracing made by an electrocardiograph.

electrocardiogram data set   A special data set that enables the acoustical coupling of any telephone handset to an electrocardiogram recorder for the transmission of the cardiogram over the telephone line. The data set provides a 2025 Hz signal for echo suppressor disabling, and modulates the 0 to 100 Hz ECG signal to the 1726 to 2250 Hz band for transmission over the telephone line.

electrocardiograph   An instrument for studying heart action and detecting human heart irregularities by measuring and recording voltage changes accompanying each heartbeat. The potentials which occur represent depolarization of the major muscle masses of the heart and these are recorded by sensing voltage differences on the body surface, using electrodes of pre-

scribed composition in specified positions. A differential amplifier lead network, filter, and chart recorder capable of 0 to 100 Hz frequency response with 25 mm/sec paper speed are essential parts of the modern electrocardiograph.

electrocardiographic simulator   Device that produces simulations of ECG waveforms for research and education purposes. One technique in use is storage of data bits in an integrated circuit read-only memory (ROM). When read and combined, these bits create a representation of an ECG wave. Altering bits changes the wave, making many precise simulations of dysrhythmic events possible.

electrocardiography   The process of recording and interpreting the electrical activity of the heart muscle for diagnosis and treatment purposes. The heart-generated voltage wave is picked up by means of surface electrodes applied to the limbs and chest at specified locations. After suitable amplification, the ECG voltage is applied to a strip-chart recorder. The measurement and interpretation of potentials associated with the action of the heart was originated early in the 20th century by Dr. Wilhelm Einthoven, who devised methods for measuring cardiac action potentials at the body surface, by means of electrodes in contact with the limbs, and their appropriate connection to a sensitive galvanometer.

electrocardiophonograph   An instrument that records graphic traces of heart sounds for careful study. The recordings fix precisely the times at which valve action occurs and reveal valvular defects which affect blood flow. Use of selective filters to reject unwanted frequencies allows the trace to record suspicious sounds more clearly. Visual analysis thus augments the auscultation of the heart, giving visible meaning to such acoustical phenomena as murmurs, splits, runs, and gallops.

electrocardioscope See *patient monitor.*

electrochemical transducer A device which uses a chemical change to measure the input parameter and the output of which is a varying electrical signal proportional to the measurand.

electrochemical valve Electric valve consisting of a metal in contact with a solution or compound, across the boundary of which current flows more readily in one direction than in the other direction, and in which the valve is accompanied by chemical changes.

electrochemistry Department of chemistry which studies the production of electricity by chemical means, and the effects of electricity on chemical reactions. See *accumulator* and *electrolysis.*

electroconvulsive therapy Therapy used especially in depressive mental illness, consisting of passing a low amperage electric current between electrodes placed on the side of the head. Ordinarily this would cause a convulsive motor discharge, but the reaction is modified, i.e., abolished by general anesthesia and muscle relaxant drugs administered before the shock is applied.

electrode 1. Conductor that provides a current path between the desired signal voltage and impedance changes that occur within the body and a signal conditioner (amplifier, modulator, etc.). Electrodes should be made of conducting material that permits good current transfer between the subject and the signal conditioner. The material must be inert to the chemicals on or in the body. Active metals such as zinc and nickel replace hydrogen when in contact with the skin. This causes ions to gather on the electrodes, producing polarization that blocks the signal. A less active

metal, such as silver, produces fewer ions and less polarization. Thus, silver-plated electrodes with a silver-chloride coating (silver/silver-chloride) provides good current transfer with minimum polarization in use on human skin. 2. Terminal at which electricity passes from one medium into another. 3. In a vacuum tube, the conducting element that performs one or more of the functions of emitting, collecting, or controlling electrons. Electrodes include cathodes, grids, and plates. 4. Of a semiconductor device, an element that performs one or more of the functions of emitting or collecting electrons or holes, or of controlling their movements by an electric field. 5. A conductor, by means of which a current passes into or out of a fluid or an organic material such as human skin, often one terminal of a lead. 6. A metallic conductor such as in an electrolytic cell, where conduction by electrons is changed to conduction by ions or other charged particles. 7. A conductor used to establish electrical connection with the patient.

electrode potential The potential in volts which an electrode has when immersed in an electrolyte, compared to the zero potential of a hydrogen electrode. It depends upon the material of which the electrode is made. See also *electrode-potential series.*

electrode-potential series A series of chemical elements arranged in the order of their electrode potentials. Any metal will replace any other metal below it in the series in a chemical or electrolytic action.

Electrode	Electrode potential (volt)
Magnesium	+2.400
Zinc	+0.762
Chromium	+0.557
Ferrous Iron	+0.441
Tin	+0.136

Lead(ous)	+0.122
Ferric Iron	+0.045
Hydrogen	0.000
Oxygen	−0.397
Copper	−0.470
Silver	−0.798
Mercury	−0.799
Lead(ic)	−0.800
Gold	−1.500

electrodermal See *electrodermography*.

electrodermography The recording of the electrical resistance of the skin, which varies with the amount of sweating, and constitutes a sensitive index to the activity of the autonomic nervous system.

electrodynamometer Instrument in which the mechanical reactions between two parts of the same circuit are used for detecting or measuring an electric current.

electroencephalogram (EEG) 1. Recording of electrical events occurring in the brain obtained from signals received by electrodes placed at various points on the head. The signals are amplified and recorded as with the ECG. The waveforms comprise the average effect of the discharges of thousands of nerves in the cerebral cortex and the brain stem and can therefore be interpreted only at a gross level. An electroencelphalogram is not necessarily a periodic function, although it can be—particularly if the patient is unconscious. These voltages are of extremely low level and require recording apparatus which displays excellent noise rejection. 2. The tracing of brain waves made by an electroencephalograph.

electroencephalograph An instrument that records the minute electrical potentials produced by the brain. Mental states and brain condition are associated with certain alterations in frequency of the EEG waves, leading to the classifications such as alpha (8.5–13 Hz), beta (13–22 Hz), delta (0.2–3.5 Hz), and theta waves (3.5–7.5 Hz). Analysis of the EEG provides important contributing information about brain function and is a major determinant in ascertaining clinical death. See *electroencephalography*.

electroencephalography 1. The process of recording and interpreting the electrical activity of the brain. Voltage picked up by electrodes placed on the scalp is typically 50 microvolts in amplitude. After sufficient amplification, the EEG signal is applied to a strip-chart recorder. 2. Graphic recording of electric currents developed in the brain, by electrodes applied to the scalp, to the surface of the brain, or placed within the brain.

electroencephaloscope An instrument for detecting brain potentials at many different sections of the brain and displaying them on the cathode-ray tube.

electrogastrogram The graphic record obtained by the synchronous recording of the electrical and mechanical activity of the stomach.

electrogastrograph Instrument for recording gastric motility, uses external electrodes and/or ingested pressure-sensitive transducer capsule.

electrogram Any plot, graph, or tracing which results from the action of an electrically driven stylus, pen, or light spot on prepared paper. A scalar electrogram is the plot of a function relative to time. A vector electrogram is the plot of x and y inputs with respect to one another.

electrograph Any instrument used to make a plot, graph, or tracing on prepared sensitized paper (or other chart material), by means of an electrically controlled stylus, pen, or light spot.

electroluminescence 1. The glow of a

phosphor caused by an electrical current or field. 2. Direct conversion of electrical energy into light in a liquid or solid substance. One example involves the photon emission resulting from electron hole recombination in a PN junction. This is the mechanism involved in the light emitting diode.

electrolysis 1. Process of splitting up a compound in a state of solution (or when molten) by the passage of an electric current through it. Pure water is a good electrical insulator, but when certain chemicals (common salt, sulfuric acid, sodium hydroxide, etc.) are dissolved in it, the solution will conduct an electric current. The elements of the compound are normally held together by chemical affinity which is really an electric bond, but when dissolved in water it is split up into charged particles or ions. This splitting up into ions is known as dissociation, and the theory assumes that normally as many molecules of the compound are being reformed as are being split up. When an EMF is applied to two electrodes immersed in the solution (the electrolyte), the positive ions move to the negative electrode (the cathode) and the negative ions move to the positive electrode (the anode). The charges are neutralized at the electrodes and the particles are released to be deposited on the electrode or react chemically with whatever is present. The process of electrolysis has many industrial applications. Aluminum is obtained cheaply from its ores; copper is refined; electroplating is carried out; chlorine is obtained from seawater; and heavy water is prepared in bulk. 2. The chemical destruction of metals resulting from a galvanic reaction. 3. Separation of ions by placing them in an electric field. 4. Destruction of hair follicles by the passage of an electric current.

electrolyte 1. A material which, in solution form, will conduct electricity. 2. The liquid or solid conductor between the electrodes of a battery or other electrons. 3. A nonmetallic electric conductor in which the current is carried by the movement of ions. Examples are (a) the sulfuric acid solution in storage batteries, (b) the sodium hydroxide solution in counter–EMF cells, and (c) water with dissolved salts.

electrolytes Substances which ionize in solution. Potassium and sodium are chief electrolytes in the body.

electrolytic capacitor A fixed capacitor, having a relatively high capacitance due to a very thin, electrically-formed nonconducting chemical dielectric film. To maintain this dielectric film, the electrolytic capacitor must always be connected with correct polarity to the supply voltage.

electrolytic conduction Current flow due to movement of ions in an electrolyte when a voltage is applied between electrodes immersed in an electrolyte.

electrolytic potential Difference in potential between an electrode and the immediately adjacent electrolyte, expressed in terms of some standard electrode difference.

electrolytic rectifier A rectifier consisting of metal electrodes in an electrolyte, in which rectification of alternating current is accompanied by electrolytic action. A polarization film formed on one of the electrodes permits current flow in one direction but not in the other.

electromagnet Consists of a number of turns of wire wound on a soft-iron core of high permeability. When an electric current is passed through the coil, a magnet is formed, the flux strength of which is proportional to the product of the current flow and the number of turns of wire (ampere-turns). The disconnection of the current reduces the magnetic effect. Examples of the use of the electromagnet are the lifting magnets fitted to cranes in steel foundries, magnetic clutches, and telephone and telegraph relays.

electromagnetic Pertaining to the mutually perpendicular electric and magnetic fields associated with the movement of electrons through conductors, as in an electromagnet.

electromagnetic compatibility (EMC) Refers to electrical equipment able to tolerate the EMR (electromagnetic radiation) produced by other equipment and which does not itself produce EMR causing an interference (EMI) to other equipment.

electromagnetic coupling Coupling that exists between circuits when they are mutually affected by the same electromagnetic field.

electromagnetic energy Forms of radiant energy, such as radio waves, heat waves, light waves, X-rays, gamma rays, and cosmic rays.

electromagnetic field 1. Field of influence which an electric current produces around the conductor through which it flows. 2. Rapidly moving electric field and its associated magnetic field located at right angles to both electric lines of force and to their direction of motion. 3. Magnetic field resulting from the flow of electricity.

electromagnetic flowmeter A device that measures blood flow by a technique that depends upon the energizing of an electromagnetic coil, placed so that its magnetic field is at right angle to the blood flow. Blood motion induces a voltage in a transducer placed at right angles to both the blood flow and the magnetic field.

electromagnetic focusing Use of magnetic coils about a cathode-ray tube to direct a magnetic or magnetostatic focusing field on the electron beam within the tube.

electromagnetic interference (EMI) Electromagnetic phenomena which, either directly or indirectly, contribute to a degradation in performance of an electronic receiver or system. (The phrases *radio interference*, *radio-frequency interference*, and *noise* have also been employed at various times in the same context.)

electromagnetic radiation (EMR) Radiation made up of oscillating electric and magnetic fields and propagated with the speed of light. Includes gamma radiation, X-rays, ultraviolet, visible and infrared radiation, and radar and radio waves.

electromagnetic spectrum 1. The frequencies (or wavelengths) present in a given electromagnetic radiation. A particular spectrum could include a single frequency or a wide range of frequencies. 2. The continuous range of frequencies, from 0.1 to 10^{22} Hz, of which a radiated signal is composed. Spectral dimensions are more conveniently described in terms of wavelength (Angstroms) where one Angstrom is equivalent to 10^{-7} mm. The electromagnetic spectrum includes radio-frequency waves, light waves, microwaves, infrared, X-rays (Roentgen rays) and gamma rays.

electromagnetic waves Circuits carrying high-frequency oscillating currents radiate energy in the form of two interdependent fields, an electric one and a magnetic one, each field tending to maintain the other: the changing magnetic field giving rise to the electric field, and the changing electric field creating the magnetic field. The radiation is said to consist of electromagnetic waves which travel through space at the speed of light, i.e., 186,284 miles per second or 299,796 million meters per second. James Clerk Maxwell, by means of applied mathematical methods, and on the basis of the experimental work carried out by Faraday (who earlier had demonstrated the relation between magnetic and electric phenomena), postulated the existence of electromagnetic waves and that they were of the same nature as light, conversely that light waves were a form of electromagnetic waves. This theory was experimentally demonstrated by Heinrich Hertz seven years after Maxwell's death; these experiments also formed the foundation of radio and television.

electromechanical The interaction between an electrical current and a mechanical force or motion. Usually involves the movement of a ferromagnetic object (lever, etc.) caused by an electromagnetic field. It may also involve the generation of an electrical current by the motion of a permanent magnet relative to an electrical conductor.

electromechanical transducer A transducer that receives mechanical excitations and supplies electrical signals, or to which an electrical signal is applied with the result that a proportional mechanical output is obtained.

electrometer Electrostatic instrument for measuring charge, or high-impedance-input device for measuring potential. Originally used only for static charge-measuring instruments; now used to refer to special high-impedance amplifiers, such as used with pH electrodes. Types include vacuum tube, vibrating capacitor, chopper, and field-effect transistor.

electrometer amplifier An amplifier circuit having sufficiently low-current drift and other noise components, sufficiently low amplifier input-current offsets, and

adequate power and current sensitivities to be usable for measuring current variations of considerably less than 10^{-12} amperes.

electromotive force (EMF) 1. The force that produces a movement of electric charges resulting in a current flow. Electromotive force is commonly called voltage. 2. The ability of diffusing ions to charge the cell membrane with an electrical potential difference. It is believed that sodium, potassium, and chloride ions are the major ions participating in this process.

electromyogram 1. Classically, a waveform of the contraction of a muscle as a result of electrical stimulation. Usually, stimulation comes from the nervous system (normal muscular activity). The record of potential difference between two points on the surface of the skin resulting from the activity or action–potential of a muscle. 2. The tracing of muscular action–potentials by an electromyograph.

electromyograph (EMG) An instrument for measuring and recording potentials generated by muscles.

electromyography (EMG) The technique of monitoring or recording electrical activity in muscle tissue. A single electrical spike potential is generated when a muscle fiber contracts. This electrical activity can be detected by placing electrodes either within the muscle, or on the skin surface directly over the muscle being monitored. Magnitude of the spike potentials is roughly proportional to the amount of muscular tension. The EMG, obtained by surface electrodes, is a record of the electrical spikes from many muscle fibers, averaged together.

electron The smallest known negatively charged stable particle, discovered by Sir J. J. Thomson in 1897. It has a charge of 1.6022×10^{-19} coulombs, and all electric charges are presumed to be integral multiples of this number. Electrons constitute the extranuclear structure of atoms, and hence are present in all matter. High-speed electrons emitted during radioactive decay are called beta rays. Electrons released from a negatively charged electrode by the action of heat, light, ions or intense electrical fields constitute cathode rays.

electronarcosis Anesthesia induced by the passage of electrical current through the brain. Although used only infrequently in most medical centers at the present time, electronarcosis shows great promise as a simple, extremely safe, and easily controlled method of introducing anesthesia. In the past, this term and electroshock have been used interchangeably.

electron beam machining The process of using a controlled stream of electrons to weld or shape a piece of material.

electron charge Charge of a single electron. Its value is 1.6022×10^{-19} coulomb. The fundamental unit of electrical charge. Also called *elementary charge*.

electronegative gas A gas in which electron attachment takes place, a process by which free electron becomes attached to a neutral molecule to form a negative ion. These ions are immobile and massive and thus are incapable of collisional ionization themselves.

electron gun 1. Electrode structure which produces and may control focus, deflect, and converge one or more electron beams. 2. Portion of a cathode-ray tube or camera tube which emits a beam of controlled electrons.

electronic Pertaining to the application of that branch of science which deals with the motion, emission, and behavior of currents of free electrons, especially in vacuum, gas, or phototubes and special conductors or semiconductors. Contrasted with electric, which pertains to the flow of large currents in wires or conventional conductors.

electronic cardiotachometer An instrument that converts the time interval between successive heartbeats into an equivalent voltage. This time interval is often determined by allowing a capacitor to charge at a known rate during this interval. At the end of this interval, the resultant capacitor voltage is measured and is the voltage analog of period. This voltage is linear with respect to period, but nonlinear with respect to frequency or heart rate, since frequency is equal to the reciprocal of period ($f = 1/t$). When this voltage is applied to a galvanometer, the scale is not linear and is compressed at the higher rates. The nonlinear scale is sometimes difficult to read, especially at the higher rates. This compression or nonlinearity is often compensated for by using a linearizing amplifier between the capacitor and the meter, or by sampling and holding a value, as in a digital display cardiotachometer.

electronic data processing (EDP) The compilation, storage, manipulation, and interpretation of information by the use of electronic equipment, including computers.

electronic data processing center A center that maintains automatically operated equipment, including computers, designed to simplify the use and interpretation of the mass of data-gathered by modern instrumentation installation or information collection agencies. Abbreviation: EDP center.

electronic data processing system The general term used to define a system for data processing by means of machines using electronic circuitry at electronic speed, as opposed to electromechanical equipment. See also *automatic data processing system*.

electronic depilation Removal of hair by use of electronic instrumentation.

electronic digital computer A machine which uses electronic circuitry in the main computing element to automatically perform arithmetic and logical operations on data, by means of an internally stored program of machine instruction. Such devices are distinguished from calculators on which the sequence of instructions is externally stored and is impressed manually (desk calculators) or from tape or cards (card programmed calculators).

electronic gate A device whose input-output relations correspond to a Boolean algebra function (AND, OR etc.) and which accomplishes this by using diodes and/or transistors, rather than relays or mechanical switching means.

electronic industries Industrial organizations engaged in the manufacture, design, and development and/or substantial assembly of electronic equipment, systems, assemblies, or the components thereof.

electronic interference Any electrical or electromagnetic disturbance that causes undesirable response in electronic equipment. Electrical interference refers specifically to interference caused by the operation of electrical apparatus that is not designed to radiate electromagnetic energy.

electronic larynx An instrument which assumes the functions of the complex of cartilages and related structures at the opening of the trachea in the pharynx, so that intelligible speech can be produced by one in which the natural structures have lost their function or, by necessity, have been surgically removed.

electronic pacemaker An electrical device, usually with electrodes planted in the myocardium, that performs the pacing function in a diseased heart no longer capable of pacing itself. Electronic pacemakers can receive power from implanted batteries, radio frequency signals, biological energy sources, etc. See *Stokes-Adams syndrome*.

electronic packaging Coating or surrounding an assembly of electronic components with a dielectric compound.

electronic parts Basic circuit elements which cannot be disassembled and still perform their intended function, such as capacitors, connectors, filters, resistors, switches, relays, transformers, crystals, integrated circuits, electron tubes, and semiconductor devices.

electronic raster scanning Scanning by electronic, rather than mechanical, means in a predetermined pattern of scanning lines in such a way as to provide a substantially uniform coverage of an area. The scanning means may be the deflection coils surrounding the neck of a cathode-ray tube, which move the electron beam across the screen both horizontally and vertically. The beam can be turned on and off, independent of the scanning action, thus making possible a very great variety in the types of photo, character, or graphic data appearing on the screen.

electronics 1. That field of science concerned with electron flow in devices, systems, or circuits. 2. The application of devices which function through electron flow.

electronic sphygmomanometer Device that measures and/or records blood pressure electronically.

electronic stethoscope An electronic amplifier for auscultating sounds within a body. Its selective controls permit tuning for low heart sounds or high pulmonary tones. It has an auxiliary output for recording or viewing audio patterns.

electronic stimulator A device for applying electronic pulses or signals to activate muscles, or to identify nerves, or for muscular therapy.

electronic switch 1. An electronic circuit used to perform the function of a high-speed switch. Applications include switching a cathode-ray oscilloscope back and forth between two inputs at such high speed that both input waveforms appear simultaneously on the screen. 2. A diode, thyristor, transistor, tube, or integrated circuit which can simulate the on and off states of a mechanical switch.

electronic timer A device that uses an electronic circuit to operate a device at a predetermined interval of time after the circuit is energized, as in timing exposures for a photographic printer or in controlling an electronic generator.

electronic voltmeter Voltmeter which uses the rectifying and amplifying properties of electron devices and their associated circuits to secure desired characteristics, such as high-input impedance, wide-frequency range, or crest indications. When the electron devices are electron (vacuum) tubes, it is called a vacuum tube voltmeter.

electron image tube Cathode-ray tube having a photoemissive mosaic upon which an optical image is projected and an electron gun to scan the mosiac and convert the optical image into corresponding electrical current.

electron microscope A microscope in which the specimen is illuminated by electrons which are focused by means of specially shaped magnetic fields. The electrons serve the same purpose as light in the ordinary optical microscope, but are employed because, when considered as waves,

they have a very much smaller wavelength than light and can, consequently, be used to observe specimens very much smaller than could be observed with light, for example, viruses and molecules. The electron image cannot be directly seen by the eye, but is made visible by means of a fluorescent screen similar to that used in a cathode-ray tube. Photographs can be taken by allowing the electrons to fall onto a photographic plate. Resolving power is typically 30 Å (about 150 times better than optical microscopes). Magnification beyond 20,000 times is possible.

electron tubes Devices used to control the flow of electrons. They may be either gas filled, or partially or fully evacuated (vacuum). Common tubes include vacuum tubes, cathode-ray tubes, photo tubes, mercury vapor tubes, thyratrons, and microwave tubes.

electron unit Unit of charge (negative or positive) equal to the charge on an electron (1.6022×10^{-19} coulomb).

electron-volt (eV) Amount of energy gained by an electron in passing from one point to another that is one volt higher in potential (one electron-volt is equal to 1.6022×10^{-19} volt). For particles, the electron-volt is a measure of kinetic energy, while for electromagnetic radiation, it reflects the frequency of the quantum, since energy is proportional to the frequency. Typically, energies are expressed in millions of electron-volts, or MeV.

electronystagmograph (ENG) Instrument for recording eye movements induced by electrical stimulation.

electro-oculogram The record of the changes in potentials generated by the movement of the eyeball. This record can be obtained without direct contact with the eye or interference with vision. It can be made in absolute darkness and with eyes open or closed. The signal is detected by electrodes attached to the skin surface near the eye. Two electrode pairs are required to detect horizontal and vertical movement. Horizontal movement can be detected by placing electrodes on the outer corners of the eye. Vertical movement can be detected using electrodes applied above and below the eye.

electro-oculography The process of recording and interpreting the muscle voltages accompanying eye movements. Electrodes placed on the skin near the eye pick up the eye-position voltage; this voltage is amplified and applied to a strip-chart recorder.

electrophonic effect The sensation of hearing produced when an alternating current of suitable frequency and magnitude from an external source is passed through a subject.

electrophoresis 1. The migration of colloidal particles under the influence of an applied electrical field as reported by Lodge in 1886. A colloidal particle, such as a protein molecule, has large numbers of positive and negative radicals that act as if they were on the surface. Thus, since protein molecules carry electric charges, they will migrate when subjected to an electric field. The fractional nature of the net charge makes possible a wide variety of electrophoretic patterns at a given pH. Also known as *cataphoresis*.

electrophoresis apparatus An apparatus for causing migration of charged particle (ions) in solution in an electric field. Types include paper, cascading electrodes, high voltage, gel, and thin layer.

electrophoresis scanner An instrument for reading bands on paper strips or gel,

measuring particle movement due to electrophoresis.

electrophoretic pattern The graphic result of the electrophoresis of one substance. Blood-plasma proteins, for example, have a certain typical electrophoretic pattern.

electrophotometer An instrument using a photoelectric sensor for colorimetric determinations.

electrophysiological event The origination of an electrical impulse as a result of an action taking place in living tissue.

electrophysiology The science of physiology in its relations to electricity; the study of the electric reactions of the body in health.

electroplate The application of a metallic coating on a conductive surface by means of electrolytic action.

electroplating Process of depositing by electrolysis one metal upon another. An electric current is arranged to flow from immersed plates (the anodes) to the object to be plated through a metallic salt solution (the electrolyte). The anodes are of the same metal as that in the electrolyte and are slowly dissolved into it. The metal ions are attracted to the objects being plated and there give up their electric charges and deposit themselves on the surface. Silver, nickel, copper, and zinc are the metals most commonly used in this process.

electropositive 1. Referring to an electrode having a more positive potential. 2. Referring to an element which is more positive than hydrogen in the electrochemical series.

electroretinogram Recording of electrical response of the retina to light.

electroretinograph (ERG) An instrument for measuring electrical response of human retina to light stimulation.

electroretinography The process of recording and interpreting the voltage generated by the retina of the eye. A tiny electrode fitted to a plastic contact lens picks up voltage from the surface of the eyeball.

electroscope Electrostatic instrument for measuring a potential difference or an electric charge by means of the mechanical force exerted between electrically charged surfaces. In its simplest form, two gold leaves are suspended in a hollow glass chamber (often partially evacuated) and connected to an external electrode. When a charge is applied, the two leaves spring apart, because they have assumed equal charges.

electrosection A surgical cutting technique employing a radio-frequency arc.

electroshock The use of electric currents for purposes of inducing convulsions in psychotic patients. While the mechanism of electroshock is not fully understood, it is believed that relatively large currents (approximately 100 milliamperes for 0.1 seconds) passed through certain areas of the brain, are able to erase past psychic trauma and thereby act to relieve the basic condition (melancholia, mania, catatonia) underlying the psychotic state. The convulsion is not believed to be curative in itself but only to be a by-product of the therapeutic process. See also *electronarcosis*.

electroshock therapy Treatment of certain mental disorders by passing an electric current through the brain.

electrosleep See *electronarcosis*.

electrospinograph A device for detecting and recording electric signals of the spinal cord.

electrostatic Pertaining to electricity which is not flowing, and which exists only as a potential (voltage) and an electric field (i.e., static electricity).

electrostatic capacitor Two conducting electrodes separated by an insulating material such as air, ceramic, mica, gas, paper, plastic film, or glass. These are generally high impedance devices. Characteristics: low dielectric constant, high voltage, excellent high frequency response, low dissipation factor.

electrostatic charge 1. The quantity of electrical energy residing on the surface of an insulated body as a result of an excess or deficiency of electrons. 2. Electrical energy stored in a capacitor.

electrostatic coupling A form of coupling by means of capacitance so that charges on one circuit influence another circuit owing to the capacitance between the two.

electrostatic deflection In a cathode-ray tube, the deflection of the electron beam by means of pairs of charged electrodes on opposite sides of the beam. The electron beam is bent toward a positive electrode and bent away from a negative electrode.

electrostatic field A volume within which there is electric stress, produced by stationary electric charges.

electrostatic generator Apparatus used in atomic research and capable of reaching potentials of several million volts; sometimes referred to as a Van de Graaff machine after its inventor, R. J. Van de Graaff.

electrostatic induction The process of inducing stationary electric charges on an object by bringing it near another object which has an excess of electric charges. A negative charge will induce a positive charge, and vice versa.

electrostatics Study of electric charges or electricity at rest as opposed to the study of electricity in motion or electric currents.

electrostatic series See *triboelectric series*.

electrostatic shield A metallic enclosure or screen placed around a device to ensure that it will not be affected by external electrical fields.

electrostatic storage A device possessing the capability of storing changeable information in the form of charged or uncharged areas on the screen of a cathode-ray tube.

electrostimulation See *stimulation*.

electrostriction A mechanical deformation caused by the application of an electric field to any dielectric material. The deformation being proportional to the square of the applied field. This phenomenon results from the induced dipole movement caused by the applied field, resulting in the mechanical distortion.

electrosurgery The surgical use of electricity in such applications as resection, coagulation, laser heating, laser welding, diathermy, desiccation of tumors, and hemostasis.

electrosurgical unit An RF generator whose output is applied to a blade or wire loop used instead of a conventional scalpel. The RF current from the blade or loop is used for surgical incision or excision, and minimizes blood loss and infection risk by cauterizing the cut tissues.

electrotherapeutics The branch of medical science dealing with the use of electricity in the treatment of disease. Electroshock therapy is one example.

electroversion See *cardioversion*.

element 1. A substance made up of atoms with the same atomic number. 2. A circuit or device performing some specific, identifiable function. 3. A substructure, such as a component, part, or subassembly. Also, a portion of a part which cannot be renewed without destroying the part.

elementary particles The particles of which all matter and radiations are composed. All are short-lived, do not exist independently under normal conditions (except electrons, protons, and neutrinos in the form of cosmic rays), and are of less atomic size. Originally, this term was applied to any particles which could not be further subdivided; now it is applied to nucleons (protons and neutrons), electrons, mesons, antiparticles, and strange particles, but not to alpha particles or deutrons. Also called fundamental particles.

emaciation A state of wasting away.

embedding Complete enclosure of an electronic device. This is usually a free-flowing process in which compound surrounds and fills all voids throughout the assembly. Molds or forms are usually necessary during processing but are removed after solidification of the compound. Since there is no other protective surface, an embedding compound requires ample physical strength.

embolism Sudden blocking of an artery by a dislodged blood clot (after surgery), a fat globule (after a fracture), gas bubbles (after sudden decompression), bacterial clumps (bacterial endocarditis), or other foreign matter. The arteries most usually affected are those of the brain, heart, lungs, and extremities.

embolus An abnormal particle (as an air bubble or a blood clot) circulating in the blood.

embryo A human offspring prior to emergence from the womb; hence, a beginning or undeveloped stage.

emergency power supply Auxiliary source of power for emergency use.

emesis Vomiting.

EMF Abbreviation for electromotive force.

EMG Abbreviation for electromyography.

EMG electrode Electrode for use in muscle studies.

EMI Electromagnetic interference; designation for all kinds of interference occurring at any point in the electromagnetic spectrum.

-emia Suffix meaning blood.

emission beam angle between half-power points The angle centered on the optical axis of a light-emitting diode within which the relative radiant power output or photon intensity is not less than half of the maximum output or intensity.

emissivity The ratio of the radiant energy emitted by a source of radiation to the radiant energy of a perfect (blackbody) radiator of the same area, at the same temperature.

emittance Power radiated per unit area from a surface.

emitter 1. Electrode within a transistor from which carriers are usually minority carriers. When they are majority carriers, the emitter is referred to as a majority emitter. 2. Of a transistor, a region from which charge carriers that are minority carriers in the base are injected into the base. 3. A device used on a punch-card machine to give pulses at regular, timed intervals during the machine cycle.

emitter-coupled logic See *ECL.*

emitter junction Of a transistor, a junction normally biased in the low-resistance direction to inject minority carriers into a base.

emitter semiconductor Junction normally biased in the low-resistance direction to inject minority carriers into an interelectrode region.

empathy An objective awareness of the feelings, emotions, and behavior of another person. To be distinguished from sympathy, which is usually nonobjective and noncritical.

empirical Pertaining to a statement or formula based on experience or experimental evidence, rather than on mathematical conclusions.

emphysema (pulmonary) Lung disease characterized by over-distention of the air sacs of the lungs (alveoli), their loss of elasticity, and destruction of the walls separating them. It results in a reduction of the respiratory surface, chronic shortness of breath, wheezing, and cough.

EMR Abbreviation for electromagnetic radiation.

emulsion A liquid which is distributed throughout another liquid in small globules.

enabling pulse A pulse which opens a normally closed electrical gate, or that prepares a circuit for some subsequent action.

encapsulating Enclosing an article (usually an electronic component or the like) in a closed envelope of plastic, by immersing the object in a casting resin and allowing the resin to polymerize, or if hot, to cool and solidify.

encephalitis Inflammation of the brain, usually caused by a viral infection. In its epidemic form, encephalitis lethargica, the victim suffers lethargy, opthalmoplegia, hyperkinesia, and sometimes, residual neurologic disability, particularly of the Parkinsonian form with oculogyric crisis. Profound disturbances of sleep rhythm is a symptom which gives this disease its common name: sleeping sickness.

encephalogram The X-ray of the brain, after air, which is radiolucent, has been introduced into the cerebral ventricles. Contrasted with the electroencephalogram, which is the recording of brain electrical activity by means of cranial or intracranial electrodes.

encephalomyelitis Acute inflammation of the brain and spinal cord.

encephalopathy Disease affecting the brain.

encode 1. To apply a code, frequently one consisting of binary numbers, to represent individual characters or groups of characters in a message. 2. To substitute letters, numbers, or characters for other numbers, letters, or characters, usually to intentionally hide the meaning of the message except to certain persons who know the encoding scheme.

encoder 1. In an electronic computer, a network or a system in which only one input is excited at a time and each input produces a combination of outputs. See also *matrix*. 2. An electromechanical device that can be attached to a shaft to produce a series of pulses to indicate shaft position; when the output is differentiated, it is an accurate tachometer. (It is fundamentally oriented to digital rather than analog techniques.) An encoder contains a disc with a printed pattern that makes and breaks a circuit as it rotates. The more make and break cycles per revolution, the better the resolution. Also called a *digital-to-analog computer*.

encoding Translation of information to a coded form from an analog or other easily recognized form without significant loss of information.

end-around carry In a computer, the operation which adds the carried information from the left-most bit to the results of the right-most addition. End-around carry is used for 1's complement and 9's complement arithmetic.

endarteritis Inflammation of the intima or lining endothelium of an artery.

endemic 1. Occurring frequently in a particular locality. 2. Present at all times among a particular population.

endo Prefix meaning within.

endocardial electrode An electrode arranged to be applied directly to the endocardium of the heart, for purposes of conduction study, or for application of pacing pulses. See *implantable pacemaker*.

endocardium The endothelial tissue of the heart that lines the cavities of the atria and ventricles. Conduction of depolarizing waves in the ventricles is believed to start at the endocardium and progress outward through the myocardium.

endocrine The term used in describing the ductless glands giving rise to an internal secretion. Some of the organs of internal secretion have both an internal and an external section, and so may have ducts. The endocrines are: suprarenals, thymus, thyroid, parathyroid, pituitary, pancreas, ovaries, and testicles.

endogenous Developing from within.

endoradiograph Equipment for X-ray examination of internal organs and cavities by means of radiopaque materials.

endoradiosonde A physiologic transducer which can be swallowed or implanted within a subject to obtain data from within the body. There are two types of endoradiosondes—*passive* and *active*—either of which can be made sensitive to temperature, pressure, pH, or other variables. One type of passive unit consists of a resonant circuit and a transducer which varies the reactance of the circuit with changes of temperature or pressure. Data are obtained from this circuit by measuring changes in resonant frequency. An external RF source provides the variable frequency stimulus, and "ringing" of the passive circuit at a given frequency relates physiological information. The active endoradiosonde (radio "pill") contains an oscillator whose frequency varies in proportion to physiological changes affecting the transducer. Measurement of oscillator frequency by a precision frequency meter outside the body provides a quantitative measure of the physiologic conditions within.

end-organ Collection of cells connected to the peripheral nervous system which act as a transducer, transforming a stimulus into a nerve discharge (receptor) or a nerve discharge into a stimulus, e.g., end-plate.

endoscope An instrument for the inspection of the interior of a hollow organ.

endothermic reactions Reactions in which heat is absorbed from surrounding bodies or medium.

end-plate Accumulation of muscle cytoplasm and nuclei in association with the terminal branches of a motor nerve through which the nerve discharge stimulates contraction of muscle.

enema Introduction of fluid into the rectum to relieve constipation.

energized Being electrically connected to a voltage source. Also called *alive, hot,* and *live.*

energizing pulse (main bang) In echographic studies, a step electrical pulse applied to the crystal transducer which causes it to "ring" at an ultrasonic rate, thus generating the ultrasound pulse.

energy The capacity to do work or the stored work, for example, in an electric battery, a column of fluid, or a moving fluid. Energy for artificial hearts must be supplied, at present, from extracorporeal sources, such as through electric cables or fluid lines.

energy conversion devices Devices including primary and secondary cells, fuel cells, photovoltaic systems, electrochemical energy converters, radiation conversion devices, thermionic converters, converters using solar, ionic, or nuclear energy sources, devices for creating a plasma in an interaction space between an emitter and a collector, electrostatic generators for creating an electrical output, organic and inorganic ion exchange and membrane devices, electron-volt energy devices, devices for direct conversion of fuel to electricity, and electrical energy storage unit devices capable of delivering a power output.

energy level A particular value of energy of a physical system, such as a nucleus, which the system can maintain for a reasonably long time. Systems on an atomic scale have only certain discrete energy levels and cannot occupy values in between these levels.

enhanced state A state following a subthreshold stimulus in which the threshold is lower than the resting value.

enhancement MOS A type of MOS transistor in which no current flows in the absence of an input control signal on the control terminal (called the gate) of the transistor. This reduces power dissipation (power dissipation occurs only when an input signal is present) and results in excellent logic state recognition (full off being one state and on the other).

enteric Pertaining to the intestines.

enthalpy Total energy that a system possesses by virtue of its temperature. Thus, where U is the internal energy, then the enthalpy = U + PV, where PV represents the external work.

enzyme Protein which acts as a catalyst. There are many different kinds of enzymes which increase the reactivity of certain substances, with varying degrees of specificity. Enzymes are essential in order for metabolism to occur at body temperature, since most of the chemical reactions taking place in the body would occur at an imperceptibly slow rate in the absence of enzymes to activate the substrate. The precise mechanism of enzyme action is not yet known.

EOG See *electro-oculography*.

eosinophil Having an affinity for eosin (a red bacteriologic stain and diagnostic reagent) or any acid stain.

eosinophilic leukemia A form of leukemia characterized by the exhibition of eosinophils.

epi Prefix meaning upon.

epidermis Outer layer of skin. The skin of animals is composed of two layers made up of cells, the epidermis (cuticle) and the true skin (corium). The epidermis consists of several layers of fattened cells which contain no blood vessels or nerves. The innermost cells of the epidermis are nourished by lymph from the true skin while the outer cells consist of horny flakes. The term is also used in regard to the skin of plants.

epilepsy Any of various disorders marked by disturbed electrical rhythms of the central nervous system and manifested by convulsive attacks usually with clouding of consciousness. Major epilepsy (grand mal) is characterized by gross convulsive seizures, with loss of consciousness. Minor epilepsy (petit mal) is minor nonconvulsive epileptic seizures; may be limited to only momentary lapses of consciousness.

epitaxial New layers of atoms deposited on a host material in such a manner that the new (epitaxial) layer perpetuates the crystalline structure of the host substrate.

epitaxial layer A thin, precisely doped monocrystalline layer of silicon grown on a heavily doped thick wafer, into which semiconductor junctions are diffused. In conventional integrated circuit processing, the thick wafer is P-doped, the epitaxial layer N-doped.

epitaxial mesa transistor A three terminal semiconductor where a film of single crystal semiconductor material is deposited on a substrate of the same type material. Then, using this deposited region as the collector, base and emitter regions are formed by diffusion and the collector base junction is defined by etching a mesa. Epitaxial transistors have the advantages of lower saturation resistance and short storage times.

epitaxial planar transistor A three–terminal semiconductor where a thin collector region is epitaxially deposited on a low-resistivity substrate; base and emitter regions are then produced by gaseous diffusion, with the edges of the junction under the protective oxide mask.

epitaxy The oriented intergrowth between two crystals, in which the orientation of one is determined by the orientation of the other.

epithelium A sheet-like tissue that surrounds a structure, and has one surface exposed, the other attached to the inner structure. It can absorb and secrete substances and it can protect (to a degree) the body which it surrounds; it can repair itself. It is also responsive to a stimulus. Examples of epithelium are the lining of the tongue, the skin, the outer lining of the eye, the inner lining of the stomach, and the outer lining of glands.

equalization The process of reducing attenuation distortion and/or phase distortion of a circuit by introduction of networks which add compensating attenuation and/or time delay at various frequencies in the transmission band.

equalizer 1. Device designed to compensate for an undesired amplitude-frequency or phase frequency characteristic or both, of a system or transducer. 2. Network having an attenuation complementary to that of a telephone line, and inserted for the purpose of correcting frequency distortion caused by the line. 3. Network which improves the frequency response of a radio system. Usually a combination of resistors, inductors, and capacitors. 4. Series of connections made in paralleled, cumulatively-compound, direct-current generators to give the system stability.

equipment ground A connection from earth ground to a noncurrent-carrying metal part of a wiring installation of electric equipment. It reduces shock hazard and provides electrostatic shielding.

equipotential Surface in an electric field having every point at the same electric potential is said to be an equipotential surface. Imaginary lines of equipotential (equipotentials) can be drawn for electric fields in a similar manner to the imaginary lines of force. The lines of equipotential are assumed to be at right angles to the electric lines of force. Electrons tend to move along the lines of force and at right angles to the equipotentials.

equivalent circuit Arrangement of common circuit elements whose characteristics, over a range of interest, are electrically equivalent to those of a more complicated circuit or device.

equivalent differential input capacitance The equivalent capacitance looking into the inverting or noninverting inputs of a differential amplifier with the opposite input grounded. See also *equivalent differential input impedance.*

equivalent differential input impedance The equivalent impedance looking into the inverting or noninverting input, with the opposite input grounded and the operational amplifier operated in the linear amplification region.

equivalent differential input resistance The equivalent resistance looking into the inverting or noninverting input of a differential amplifier with the opposite input grounded. See also *equivalent differential input impedance.*

equivalent input offset current The difference between the two currents flowing into the inverting and noninverting inputs of a differential amplifier when the output voltage is zero.

equivalent input offset voltage In a differential amplifier, the amount of voltage required at the input to bring the output to zero. Usually this voltage is adjustable to zero by using either a built-in or external variable resistor (balance control).

equivalent input wideband noise voltage The output noise voltage of a differential amplifier with the input shorted, divided by the dc voltage gain of the amplifier. This voltage is measured with a true RMS voltmeter and is limited to the combined bandwidth of the amplifier and meter.

equivalent networks Two networks are equivalent if one can replace the other in a circuit without any effect upon the external circuit.

erase 1. To replace all the binary digits in a storage device by binary zeroes. 2. In charge-storage tubes, to charge or discharge storage elements to eliminate previously stored information. 3. To remove recorded information from magnetic tape by passing the tape through a strong, constant magnetic field (dc erase) or through a high-frequency alternating field (ac erase). Alternating current fields can also be externally applied to tape, resulting in accidental erasure.

erase head In magnetic recording, a device for obliterating any previous recordings on the magnetic media. May be used for preconditioning the magnetic media for recording purposes.

erasure The elimination of recorded signal from a tape, generally to allow for re-use of the tape for a new recording. Erasure is generally accomplished by exposing the magnetic coating to a very strong magnetic field of alternating polarity and then gradually reducing the strength of the field to zero, to leave the tape in a neutral (no magnetization) state. All recorders use a special erase head to obliterate the previously-recorded signal before the record heads lay down the new signal.

erectile Capable of erection, as in the erectile tissues of the penis, clitoris, and mammary glands.

ERG See *electroretinography*.

ergodynamograph See *ergometer*.

ergograph See *ergometer*.

ergometer Instrument for recording force and work associated with muscular contraction, or amplitude of contraction. In one form employing a treadmill or bicycle, the exerted force is applied to a generator which has an adjustable output load. The energy dissipated in the load thus gives an indication of the input energy, and the degree of required input effort can be changed, as necessary, by increasing or decreasing generator load resistance.

ergosterol A sterol, found in fats and present in the skin, which is converted into vitamin D by irradiation with ultraviolet light.

ergotism Poisoning by alkaloids present in the fungus Claviceps purpurea which is found in cereals. One of the alkaloids, ergometrine, is used in obstetrics to control postpartum hemorrahage; another, ergotamine, is used in the treatment of migraine.

error The difference between a true value and an observed or calculated one.

error correcting code 1. A code used for digital data transmission having redundant characteristics such that, by the application of specific logical procedures, a limited number of bits lost in transmission may be replaced at the receiving end. 2. Code that improves the accuracy of an incoming message by repeating coded symbols as a check. Capacity of the message channel is reduced, however. 3. Code in which the forbidden pulse combination produced by gain or loss of a bit indicates which bit is wrong.

erythema Red skin, due to vasodilatation in the dermis.

erythema multiforme An acute, inflammatory skin disease consisting of raised red lesions of varying size and shape which may blister. The cause is unknown but it may represent an abnormal immune response.

erythema nodosum Red tender skin nodules on the legs which may occur in conjunction with certain conditions such as tuberculosis and sarcoidosis.

erythroblast Nucleated red blood cell normally found in the bone marrow and which gives rise to a red blood corpuscle.

erythrocytes See *blood corpuscles*.

erythrodermia Red skin. The whole body

surface is involved in an inflammatory vasodilatation.

erythromelangia Severe pain associated with extreme dilatation of arteries, especially those supplying the limbs.

-esis Suffix meaning condition or state of.

esophageal balloon A latex balloon attached to a tube and introduced into the esophagus by way of a nostril used in measuring lung compliance and pulmonary resistance.

estrogen-progestogen contraceptives Oral contraceptive agents.

esu Abbreviation for electrostatic unit.

etiology The study of the causes of disease.

eu Prefix meaning well or good.

euphoria An exaggerated feeling of physical and emotional well-being inconsonant with reality.

eupnea Normal, regular, or effortless breathing.

eustachian tube The canal from the throat to the ear.

euthanasia A painless death procured by the use of drugs.

eV Abbreviation for electron–volt.

even harmonic Any harmonic whose frequency is the fundamental frequency multiplied by an even number. The even harmonics of 60 Hz are 120 Hz, 240 Hz, 360 Hz, 480 Hz, etc.

event recorder Recorder used to record on time of on or off type signal.

evisceration Removal of the abdominal contents.

evoked potentials Potentials recorded by placing electrodes on exposed brain tissue when stimuli are delivered to a nerve in the periphery of the organism. The delay between the onset of the stimulus and evoked response and the character of the potential with time are some of the significant variables in this type of experiment. In the absence of an applied stimulus, responses can still be observed in the nervous system which arise from sponteneous activity intrinsic in the neurons themselves or excited by internal receptors of various kinds.

ex Prefix meaning away from.

except-gate Gate in which the specified combination of pulses producing an output pulse is the presence of a pulse on one or more input lines and the absence of a pulse on one or more other input lines.

exchange register See *memory register.*

excision A cutting out.

exclusive OR A logic element with two inputs and one output. The output is true if either of the two inputs is true, but not if both are true.

exclusive OR circuit Circuit that produces an output signal when any one, but not more than one, input is in its prescribed state.

exclusive OR gate A type of gate that produces an output when the inputs are the same, but not when they are different, i.e., AB or $\overline{AB}$.

exclusive OR operator A logical operator which has the property that if P and Q are two statements, then the statement P*Q, where the * is the exclusive OR operator; the statement is true if either P or Q (but not both) is true, and false if P and Q are both false or both true, according to the following table, wherein the figure 1 signifies a binary digit or truth.

P	Q	P*Q	
0	0	0	(even)
0	1	1	(odd)
1	0	1	(odd)
1	1	0	(even)

Note that the exclusive OR is the same as the inclusive OR, except that the case with both inputs true yields no output; i.e., P*Q is true if P or Q are true, but not both. Primarily used in *compare* operations.

excoriation 1. Any superficial loss of tissue, such as that produced when scratching the skin. 2. A scratch mark.

excreta The natural discharges from the body; urine, feces, or sweat.

execute To interpret a machine instruction and perform the indicated operation(s) on the operand(s) specified.

execution time In computers, the time required to execute an instruction, usually several machine cycles.

executive routine Routine that controls loading and relocation of routines, and in some cases, makes use of instructions that are unknown to the general programmer. Effectively, an executive routine is part of the machine itself. Also called *monitor routine, master routine, supervisory program,* and *supervisory routine.*

extenteration Removal of all contents.

extenteration of orbit Removal of all the contents of the bony orbit.

extenteration of pelvis Removal of pelvic contents and transplantation of ureters onto the sigmoid colon.

exo Prefix meaning outside.

exogenous Developing from outside.

exophthalmos Protrusion of the eyeball. May accompany enlargement of the thyroid gland.

exostosis A tumor growing from bone.

exotherm Rise in temperature resulting from the liberation of heat by polymerization.

exothermic reactions Reactions in which heat is given out to surrounding bodies or media.

expectorate To bring mucus up from the lungs or trachea.

expiration 1. The outflow of breath from the lungs. (Expiration followed by inspiration of air constitutes one respiratory cycle.) 2. The passage of a time period. 3. The act of dying.

expiratory reserve volume The gas volume which can be expired at the end-expiration level in normal breathing.

expirograph An instrument which measures expired volume of gases from the lungs and supplies the means for analyzing gas samples.

exsangumate To make bloodless.

extension The act of straightening a limb.

external device An input or output device under command of the computer. Examples: paper tape reader, line printer, magnetic tape recorder.

external memory An auxiliary storage unit, such as punched or magnetic tape, which is external to a computer's internal memory.

external pneumocardiograph A pneumocardiograph that measures changes in the dimensions of the thorax.

external storage See *internal storage.*

extracellular Situated or occurring outside a cell or the cells of the body.

extracorporeal Outside of the body.

extrapyramidal Motor nerve tracts and

associated centers which do not directly communicate with the main pathway (pyramidal tract). It is a very complex system with many functions, one of which is to regulate muscle tone.

extrasystole An extra contraction of the heart, in addition to those constituting the normal rhythm. Extrasystoles may be caused by ectopic foci in either the atrial or ventricular regions.

extravasation The escape of blood or lymph into the tissues.

extremely low frequency (elf) A frequency below 300 hertz.

extrinsic semiconductor A semiconductor whose electrical properties are dependent upon impurities.

extroversion A state in which attention and energies are largely directed outward from the self, as opposed to interest primarily directed toward the self, as in introversion.

eye Receptive organ of sight. The eyeball, embedded in a padded socket of the skull, is directly connected with the brain by the optic nerve. Further protection is given by the eyebrows, the eyelashes, and eyelids. Lachrymal fluid, produced by the lachrymal gland, keeps the eye moist. The fluid is led to the lachrymal sac, in the inner corner of the eye, whence the tear duct leads to the nose.

The eyeball is formed of several layers, with different functions. A layer of the conjunctive membrane, which also lines the eyelids, covers the tough outer layer or sclerotic coat. In front of the eye, this coat forms the transparent cornea and the white of the eye is part of the same coat.

Underneath lies the choroid coat, equipped with many nerves and blood vessels, and a lining which absorbs confusing rays of light. In the very front of the eye, underneath the cornea, the choroid coat forms the iris. At the side of the iris, the choroid forms the ciliary body with the ciliary muscle controlling contractions of the eyeball.

The iris itself is colored and has the pupil in the middle. This small hole becomes smaller or bigger according to the intensity of light. Directly behind the iris lies the transparent crystalline lens. The lens is convexly curved in front and behind and its convexity can be increased or diminished by muscles. Between cornea and lens is the anterior chamber, filled with fluid and embedding the iris. Behind the lens lies the larger posterior chamber, filled with a semiliquid substance.

The innermost lining of the eye forms the most important part of the eye, the retina. This nervous cell net is highly sensitive to light and transmits impressions of light through the optic nerve to the brain where they are interpreted.

eyeball potential The tiny potentials measurable at the eyeball surfaces by means of special electrodes, resulting from depolarization of muscles controlling eye position.

eye movement monitor Device that measures and records horizontal and vertical eye movement by noncontacting photoelectric technique and/or by electrodes on the eyelid.

F Abbreviation for fahrenheit.

f Abbreviation for femto (10^{-15}).

fa Abbreviation for femtoampere.

facial nerve The seventh cranial nerve, arising in the pons and innervating the musculature of the chin, nose, and lips.

-facient Suffix meaning make.

-fact Suffix meaning make.

facultative Able to live under varying conditions.

Fahrenheit Pertaining to a temperature scale based on the definition of the boiling point of water as 212° and the freezing point as 32° under normal atmospheric pressure. It is symbolized by F.

fail-safe Describing a circuit or device which fails in such a way as to maintain circuit continuity or prevent damage.

failure mechanism The physical cause of a circuit failure.

failure rate The probability of failure per unit of time of items in operation; sometimes called hazard. A good estimate of failure rate from a test is:

$$\lambda = \frac{r}{t}$$

where λ is the failure rate, r is the number of failures, and t is accumulated operating time for the items. Failure rate is usually expressed as the percentage of failure during 1000-hour intervals. For an exponential distribution of failure, or constant failure rate, it is the reciprocal of mean-time-between-failures.

falciform Sickleshaped. Applied to certain ligaments and other structures.

fall time Measure of time required for a circuit to change its output from a high level (1) to a low level (0). Typically applied to describing the electrical waveshape of the output, in which case it is the time interval during which the trailing edge of the wave decreases from 90% to 10% of its maximum amplitude.

false Statement for a zero in the Boolean algebra.

fan-in 1. The number of inputs that can be connected to a logic circuit. 2. The number of operating controls in a single device which individually and in combination will produce the same output.

fan-out The number of parallel loads within a given logic family that can be driven from one output mode of a logic circuit.

farad The basic unit of capacitance. That capacitance which permits the storing of one coulomb of electricity for each volt of applied potential difference. An impractically large unit. The unit commonly used is the microfarad, equal to one-millionth of a farad.

faraday effect The effect of magnetism on light, occurring when a plane-polarized beam of light passes through certain transparent substances in a direction parallel to the lines of a strong magnetic field. The plane of polarization is rotated a certain amount.

faraday shield Network of parallel wires connected to a common conductor at one end to provide electrostatic shielding without affecting electromagnetic waves. The common conductor is usually grounded. Also called *Faraday screen* or *Faraday cage*.

Faraday's Law In electromagnetic induction, the electromotive force induced in a circuit is proportional to the rate at which the flux linkages of the circuit are changing. Also called *law of electromagnetic induction*.

faradic current Intermittent, asymmetic alternating electric current obtained from the secondary winding of an induction coil.

faradism A rapidly interrupted current, the impulses of short duration (0.1 to 3.0 ms).

far field In echographic studies, the region of the ultrasonic beam beyond the distance of a^2/λ where a is the radius of the transducer and λ is the wavelength of the ultrasound frequency.

fascia Sheet of connective tissue, e.g. Superficial fascia which is connective tissue separating the dermis from underlying structures. Deep fascia: condensations of connective tissue investing muscles and forming compartments for certain structures.

fascicle A small bundle of fibers.

fast-access storage See *rapid storage*.

fastigium The high point of a fever.

fast neutron A neutron with energy greater than approximately 100,000 electron volts. (Compare *intermediate neutron, prompt neutron, thermal neutron*.)

fast reactor A reactor in which the fission chain reaction is sustained primarily by fast neutrons rather than by thermal or intermediate neutrons.

fault current A current which flows from a conductor to ground or to another conductor during an accidental short, cross or ground, including that caused by an arc. Fault currents are of particular concern to the biomedical engineer working in the hospital environment, because these currents pose unusual risks to the safety of electrically susceptible patients.

fault electrode current Peak current that flows through an electrode under fault conditions, such as arc-backs and load short circuits.

fc Abbreviation for footcandle.

febrile Pertaining to fever.

feces Excreta discharged from the intestines.

feedback 1. The returning of a fraction of the result of a process to affect that mechanism which controls the process. In physiologic systems, feedback plays a dominant role in maintaining the homeostasis of the organism. For example, a small amount of the hormone secreted by each endocrine gland is carried by the circulation back to that gland where the level of the hormone is detected and a correction in the gland output made. Experiments have also shown that instrumentation can be used to close this loop of an individual, to help the individual learn how to control bodily functions once though to be beyond conscious control (see biofeedback). 2. The act of returning a portion of the output voltage of a circuit which includes amplification to the input of that circuit. 3. In a control system, the signal (or signals) fed back from a controlled process to denote its response to the command signal. Feedback is derived by comparing actual response to desired response, and any variation is fed as an "error signal" into the original control signal to help enforce proper system operation. Systems, employing feedback are termed *closed loop* systems, with feedback closing the loop. 4. So-called negative feedback requires that the output signal returned to the input be 180° out of phase with respect to the input signal. The effect is to diminish distortion and noise accompanying the input signal, although overall gain is also reduced. 5. So-called positive feedback is used to regeneratively increase amplification, or to cause oscillation. Here, the output signal is fed back in phase with the input signal, so that loop gain exceeds unity. A narrowing of amplifier response usually results as feedback increases.

feedback amplifier An amplifier that uses a passive network to return a portion of the output signal to modify the performance of the amplifier.

feedback attenuation An attenuation factor in the feedback loop of an operational amplifier by which the output voltage is attenuated to produce the input error voltage.

feedback control A type of system control obtained when a portion of the output signal is operated upon and fed back to the input in order to obtain a desired effect.

feedback control signal That portion of the output signal which is returned to the input in order to achieve a desired effect, such as fast response.

feedback loop The components and processes involved in correcting or controlling a system by using part of the output as an input.

feedback oscillator Amplifier in which the output is coupled back to the input, the oscillation being maintained at a frequency determined by the parameters of the amplifier and the feedback circuits.

feedthrough capacitor A feedthrough insulator that provides a desired value of capacitance between the feedthrough conductor and the metal chassis or panel through which the conductor is passing. Used chiefly for bypass and decoupling purposes in VHF and UHF circuitry.

female contact A contact located in an insert or body in such a manner that the mating contact is inserted into the unit. This is similar in function to a socket contact.

femoral artery The artery of the thigh, from the groin to the knee.

femoral canal The small canal internal to the femoral vein. The site of a femoral hernia.

femto (f) Prefix meaning 10^{-15}.

femtoampere Abbreviated fa. 10^{-15} ampere.

femtovolt (fv) 10^{-15} volt.

femur The thigh bone. Longest bone in the human body.

fenestration Making an artificial window; an operation performed in certain types of deafness.

-ferent Suffix meaning bear or carry.

ferrite Magnetic core materials consisting of powdered, compressed, and sintered ferric oxide combined with other oxides, such as of nickel, nickel-cobalt, and yttrium-iron. The three most common ferrites are called spinels, garnets, and hexagonal ferrites. Has a high resistivity, so eddy-current losses at high frequencies are very low. Typically used in inductors and more memories.

ferrite bead Magnetic information storage device consisting of ferrite powder mixtures in the form of a bead fired on the current carrying wires of a memory matrix.

ferrite-core memory Magnetic memory consisting of readin and readout wires threaded through a matrix of tiny toroidal cores molded from a ferrite.

ferroelectric materials Those materials in which the electric polarization is produced by cooperation action between groups or domains of collectively oriented molecules. See *ferromagnetic*.

ferromagnetic Pertaining to a phenomenon exhibited by certain materials in which the material is magnetically polarized in one direction or the other, or reversed in direction by the application of a positive or negative magnetic field of magnitude greater than a certain amount. The material retains the magnetic polarization unless it is disturbed. The polarization can be sensed, by reason that a change in the material's field induces an electromotive force, which can cause a current flow in associated windings.

ferromagnetic amplifier A parametric amplifier based on the nonlinear behavior of ferromagnetic resonance at high RF power levels. In one version, microwave pumping power is supplied to a garnet or other ferromagnetic crystal mounted in a cavity containing a strip line. A permanent magnet provides sufficient field strength to produce gyromagnetic resonance in the garnet at the pumping frequency. The input signal is applied to the crystal through the strip line, and the amplified output signal is extracted from the other end of the strip line. Sometimes incorrectly called a garnet maser, but the operating principal differs from that of the maser.

ferromagnetic resonance Resonance at which the apparent permeability of a magnetic material at microwave frequencies reaches a sharp maximum.

ferromagnetics Science that deals with storage of information and control of pulse sequences by the magnetic polarization properties of materials.

FET Abbreviation for field-effect transistor. 1. A transistor controlled by voltage rather than current. The flow of working current through a semiconductor channel is switched and regulated by the effect of an electric field exerted by electric charge in a region close to the channel, called the gate. Also called a unipolar transistor, a FET has either P-channel or N-channel construction. 2. A transistor whose internal operation is unipolar in nature. The metal oxide semiconductor FET (MOSFET) is widely used in integrated circuits because the devices are very small, have low power needs, and can be manufactured with few steps. 3. A solid-state device in which current is controlled between source terminal and drain terminal by voltage applied to a nonconducting gate terminal.

fetal cephalometry　The measurement of the fetal skull by ultrasonic echo detection. Since both sides of the skull produce an echo, the biparietal diameter of the fetal head can be measured *in utero* by timing the two echoes. Thus, it is possible to determine the optimum time for intervention in patients scheduled for elective termination of their pregnancy.

fetal monitor　An instrument for displaying or recording the fetal electrocardiogram or other indicator of heart action. Some instruments simultaneously record the maternal electrocardiogram. Depending upon the primary purpose of the instrument, it may be referred to more definitively as a fetal electrocardiograph, fetal cardiotachometer or fetal phonocardiograph.

fetch　That portion of a computer cycle during which the location of the next instruction is determined.

fetus　The name given to a developing human organism after the second month of pregnancy.

fiber　An elongated, threadlike structure, such as a muscle fiber or nerve fiber.

fiber optics　An assemblage of transparent glass fibers all bundled together parallel to one another. The length of each fiber is much greater than its diameter. This bundle of fibers has the ability to transmit a light image from one of its surfaces to the other around curves and otherwise inaccessible places with an extremely low loss of definition and light by a process of total reflection. Sometimes also called *optical fibers* or *optical fiber bundles*.

fibrillation　Uncoordinated contraction of heart muscle. May affect the atria only (atrial fibrillation) or the ventricles (ventricular fibrillation). Because fibrillation prevents efficient contraction of the heart muscle it is rapidly fatal when it affects the ventricles. When only the atria are involved the ventricles beat separately at their own rate. In the presence of ventricular fibrillation, emergency measures are needed to sustain life and to terminate the lethal arrhythmia. Cardiopulmonary resuscitation (CPR) is generally applied without delay and a defibrillator is used to countershock the muscle fibers, in hope that spontaneous return to normal sinus rhythm can be secured.

fibrin　1. Long chain protein which forms a fibrous gel matrix as a basis for a blood clot. It is formed from the soluble plasma protein, fibrinogen, by the action of thrombin. 2. A protein substance which forms an essential part of a blood clot.

fibrin foam　A preparation of human fibrin used as hemostatic pack.

fibroadenoma　A tumor composed of mixed fibrous and glandular elements.

fibrocystic disease　Mucoviscidosis. A disease due to a defect in salt retention in mucous and sweat glands. As a result the mucopolysaccharide secretions, e.g., in the respiratory tract and from the pancreas, become excessively viscous resulting in the formation of retention cysts which become fibrosed, hence the term, fibrocystic disease.

fibroelastosis　A rare disorder affecting principally the heart in which an excess of collagen and elastin is formed under the endocardium. This impairs the efficiency of the heart with resultant cardiac enlargement and cardiac failure. Occasionally the heart valves are also involved.

fibromyoma　A tumor composed of mixed muscular and fibrous tissue. Especially common in the uterus, and commonly spoken of as *fibroids*.

fibrosarcoma　Malignant tumor of fibroblasts.

fibrosis Decomposition of fibrous connective tissue and usually occurring in regions which have suffered some trauma.

fibrositis Inflammation of connective tissue, although the term is not only used in this strict sense.

fibula The small bone on the outer side of the leg.

fibular Pertaining to the fibula.

fidelity Degree to which a system accurately reproduces at its output the essential characteristics of the signal impressed on its input.

field 1. Region containing electric and/or magnetic lines of force. 2. One of the two equal parts into which a frame is divided in interlaced scanning for television. 3. Portion of a data word which is set aside for specific information content in computer operation. 4. The area in which surgery is taking place, bounded on all sides by sterilized tissue or drapes. Typically called a *sterile field*.

field density In a particular cross-sectional area of a magnetic or electric field, the number of magnetic or electric lines of force or the magnetic or electric flux passing through it.

field distortion Change from the normal direction of the lines of force in a magnetic or electric field due to an external influence.

field-effect transistor A transistor in which current carriers (holes or electrons) are injected at one terminal (the source) and pass to another (the drain) through a channel of semiconductor material whose resistivity depends mainly on the extent to which it is penetrated by a depletion region. The depletion region is produced by surrounding the channel with semiconductor material of the opposite conductivity and reverse-biasing the resulting PN junction from a control terminal (the gate). The depth of the depletion region depends on the magnitude of the reverse bias. As the reverse-biased junction draws negligible current, the characteristics of the device are similar to those of a vacuum tube.

field maintenance That maintenance, exclusive of rebuilding, authorized and performed by designated maintenance units as a support on activity. This category of maintenance normally is limited to maintenance consisting of replacement of unserviceable parts, assemblies or subassemblies.

filament 1. In an incandescent lamp, a loop of resistance wire which reaches temperatures up to 3200°K, and glows as a result, releasing light energy. 2. In an electron tube, a loop of resistance wire which emits electrons when heated by a current. 3. A thin, flexible, hair-like structure, composed of organic or inorganic material.

file A collection of related records, e.g., in inventory control, one line of an invoice containing data on the material, the quantity and the price forms an item, a complete invoice forms a record, and the complete set of such records forms a file.

filiform Threadlike.

filiform bougie A slender bougie.

filipuncture A method of treating aneurysm by inserting a fine wire thread. This acts as a foreign body, and the blood inside the sac clots.

film badge A device worn by persons who may be daily exposed to the radiation of X-ray and nuclear machines. Radiation, impinging upon the sensitive film, fogs the film in proportion to the degree of radiation received. This makes it possible to keep track of incidental radiation exposure for each individual and to prevent cumulative exposures which may pose a hazard to the individual's health.

film integrated circuit A circuit consisting of elements that are films, all formed *in situ* upon an insulating substrate. Also called *film microcircuit.*

film microcircuit See *film integrated circuit.*

film resistor A component in which the resistance element is a thin layer of conductive material on an insulated form. The conductive material does not contain either binders or insulating material.

filter 1. Arrangement of resistors, inductors, capacitors, and/or active circuit elements, which offers comparatively little opposition to certain frequencies while blocking the passage of other frequencies. 2. Combination of a high impedance with a short-circuiting capacitor for suppressing ripples in a battery charging circuit. 3. Tuned circuit designed to pass alternating current of a specified frequency. 4. Arrangement of optical elements which permit the passage of certain frequencies of the light spectrum, while suppressing transmission of other frequencies. 5. A porous membrane which allows separation of minute bodies from suspension in liquids. 6. A material having large surface area (such as activated charcoal), over which a liquid containing dissolved gases is passed for purposes of separation. Gas molecules are absorbed by the charcoal, effectively cleansing the liquid.

filter attenuation Loss of power through a filter due to absorption of power in resistive materials, to reflection, or to radiation, usually expressed in decibels.

filter choke Normally, an iron-core coil which allows direct current to pass while opposing the passage of pulsating or alternating current.

filum A structure resembling a thread.

filum terminale The tapering end of the enlargement of the lumbar spinal cord.

finger plethysmograph An instrument for detecting and displaying volume changes of blood in the finger during the cardiac cycle. Some types employ a light source and photocell on opposite sides of the finger; the amount of light reaching the photocell varies with the volume of blood in the finger. In another type, an electric current is passed through the finger between a pair of electrodes. During each contraction of the heart, blood is forced through the circulatory system and the volume of blood in the finger increases. The increased blood reduces the impedance of the finger, and this change is recorded, or displayed on a cathode-ray tube.

firing potential 1. Controlled potential at which conduction through a gas-filled tube begins. 2. The critical voltage impressed between anode and cathode of a silicon controlled rectifier, wherein internal leakage causes the device to pass from non-conduction to conduction.

firmware Software that is stored in a fixed (wired-in) or firm way, usually in a read-only memory. Changes can often only be made by exchanging the memory for an alternative unit.

first intention A surgical term for aseptic healing of a wound by bringing the edges directly together.

fission A nuclear reaction in which an atom is split into two or more lighter parts. Great quantities of energy are emitted during the fission process, since the sum of the masses of the two new atoms is less than the mass of the original heavy atom.

fissionable material Any material fissionable by slow neutron. The three basic ones are uranium 235, plutonium 239, and uranium 233.

fissure A relatively deep cleft or groove, such as the fissures of the brain. Some are abnormal, such as fissures of the skin.

fistula Pathological communication between two epithelial surfaces or cavities, e.g., rectovaginal fistula.

five-level code A code that utilizes five impulses for describing a character. Start and stop elements may be added for asynchronous transmission. A common five-level code is Baudot.

fixed capacitor Capacitor having a value that cannot be adjusted.

fixed logic Circuit logic computers, or peripheral devices that cannot be changed by external controls. Connections must be physically broken to rearrange the logic.

fixed memory A nondestructive readout memory into which information normally can be written only once. The ROM is a fixed program memory. Programs are usually stored in fixed memories. Fixed memories, typically, can be altered only by mechanical change.

fixed point arithmetic A system in which the location or symbol that separates the integral and fractional parts of a numerical expression remains fixed with respect to one end of the numerical expression.

fixed rate (asynchronous) pacemaker A pacemaker delivering impulses to the heart continuously at a present rate.

fixed storage A storage device that stores data not alterable by computer instructions, e.g., magnetic core storage with a lockout feature or punched paper tape. Synonymous with *nonerasable storage, permanent storage, read-only storage*.

flaccid Limp, flabby, or soft.

flagellum Fine, thread-like structure projecting from surface of certain cells, e.g., spermatozoa, which by a lashing movement propels the cell. See *cilia*.

flame photometer Apparatus used to measure very small quantities of metals by the brightness of their characteristic flame.

flash lamp A device which converts a large amount of stored electrical energy into light by means of sudden electrical discharge. The excitation is started by a high-voltage discharge pulse. An avalanche of excited atoms produces a light-emitting plasma; essentially, a long arc device, operated on a pulsed basis. Maximum plasma emission is obtained by operating the lamp at full current for a short period of time. The flash is accomplished by storing electrical energy in a capacitor and allowing the capacitor to discharge through the lamp.

flashover A disruptive discharge around or over the surface of a solid or liquid insulator.

flat pack 1. Semiconductor network encapsulated in a thin-rectangular package, with the necessary connecting leads projecting from the edges of the unit. 2. Any small, flat, square, or rectangular integrated or hybrid circuit package, with leads coming from the sides of the package, in the same plane as the package.

flatulence Distention of the stomach or the intestinal tract with gas.

flatus Gas in the intestinal tract.

flexible disc Small, flat, flexible disc approximately seven inches in diameter and removable, which can store about one hundred thousand words and be accessed in about 100 miliseconds. Also called *floppy disc* or *diskette*.

flexion   The act of bending.

flip-flop   1. Device having two stable states and two input terminals (or types of input signals) each of which corresponds with one of the two states. The circuit remains in either state until caused to change to the other state by application of the corresponding signal. 2. A similar bistable device with an input which allows it to act as a single-stage binary counter. Flip-flops find principal uses in digital computers and in physiological instrumentation. They are also extensively used in patient monitoring. Of interest, the term is often used to describe neurons, which exhibit a flip-flop response.

flip-flop multivibrator   Biased rectangular wave generator which operates for one cycle when a synchronizing trigger signal is applied. Also called *start-stop multivibrator.*

float   To move or shift one or several characters into positions to the right or left as determined by data structure or programming desires; e.g., to float asterisks to the right or left of numerical fields; dollar signs to the rightmost nonspace positions.

floating charge   Continuous charging of a storage battery at a low current value to keep the battery fully charged while it is standing idle or on light duty.

floating ground   A reference ground that is not earthed.

floating input   1. An isolated input circuit not connected to ground at any point (the maximum permissible voltage to ground is limited by electrical design parameters of the circuit involved). It is understood that in a floating input circuit, both conductors are equally free from any reference potential, a qualification which limits the types of signal-sources which can be oper-

ated floating. See *differential input measurement.*

floating-point arithmetic   A system in which the location or symbol that separates the integral and fractional parts of a numerical expression is regularly recalculated.

floating-point calculation   Calculation in a computer which takes into account the varying location of the decimal point (if base 10) or binary point (if base 2). The sign and coefficient of each number are specified separately.

floating potential   The dc voltage between an open circuit terminal and a reference point when a dc voltage is applied to the third terminal and the reference terminal.

floppy disc   See *flexible disc.*

flow chart   1. A graphic representation of the major steps of work in process. The illustrative symbols may represent documents, machines, or actions taken during the process. The area of concentration is on what is done or where it is done rather than how it is to be done. Synonymous with process chart and flow diagram. 2. A graphical representation of a computer programming operation sequence, in which symbols are used to show such operations as read, write, compare, etc.

fluctuation noise   See *random noise.*

fluid amplifier   A mechanical amplifying device with a supply at one end and a choice of flow paths at the other end. The flow enters as a jet and is deflected into one output or another by (1) a jet of fluid which interacts with and deflects the main stream, or (2) the creation of a low-pressure area at either side of the stream to induce it to deflect, or (3) the introduction of any disturbance that can cause the main stream to deflect.

fluid computer Digital computer constructed entirely from fluid logic elements. All logic functions are carried out by interaction between jets of air or liquid.

fluidics The technology wherein sensing, control, information processing, and/or actuation functions are performed solely through use of fluid dynamic phenomena.

fluorescein A coal-tar derivative which stains the cornea a vivid green if there is any loss of surface epitelium, e.g., in an abrasion or ulcer.

fluorescence The property of some chemical elements, particularly gases, to emit light of a longer visible wavelength than the light with which they are illuminated.

fluorocarbons 1. A family of gases (also known commercially as Freons) characterized by combination of halogen and carbon molecules. 2. The family of plastics including polytetrafluoroethylene (PTFE); polychlorotrifluoroethylene (PCTFE); polyvinylidene and fluorinated ethylene propylene (FEP). They are characterized by properties including good thermal and chemical resistance and nonadhesiveness, and possess a low dissipation factor and low dielectric constant.

fluoroscope An instrument with a fluorescent screen suitably mounted with respect to an X-ray tube, used for immediate indirect viewing of internal organs of the body, internal structures in apparatus or masses of metals, by means of X-rays. A fluorescent image (really a kind of X-ray shadow picture) is produced.

fluoroscopy Process of using a radiologic instrument to observe the internal structure of an opaque object (as the living body) by means of X-rays.

flutter Rapid regular contraction of the atrial muscle of the heart. The rate can reach about 300 beats per minute. Because of the recovery time required by the ventricular myocardium between successive beats, it only responds to every second or third atrial contraction, i.e. 2:1 or 3:1. See *heart block.*

flux For electromagnetic radiation, the quantity of radiant energy flowing per unit time. For particles and photons, the number of particles or photons flowing per unit time. Neutron flux is a measure of the intensity of neutron radiation. It is the number of neutrons passing through 1 square centimeter of a given target in 1 second. Expressed as nv, where n = the number of neutrons per cubic centimeter, and v = their velocity in centimeters per second.

flux density A measure of the strength of a wave; flux per unit area normal to the direction of the flux; number of photons passing through a surface per unit time per unit area. Expressed in watts/cm^2 or lumens/ft.^{2}

fluxmeter Instrument that measures intensity of a magnetic field.

flyback 1. The shorter of the two time intervals comprising a sawtooth wave. 2. As applied to a cathode ray tube, the return of the spot to its starting point after having reached the end of its trace. This portion of the wave is usually not seen because of blanking circuits or shortage of time. Also called *retrace.*

focusing anode One of the electrodes in a cathode-ray tube, the potential of which may be varied to focus the electron beam.

foil A very thin sheet of metal, such as tin or aluminum. Used in the construction of fixed capacitors. Also used in describing the conductor pattern on a printed circuit board.

follicle 1. Hair follicle. A pit-like structure in the epidermis in which the hair grows and which receives the duct of sebaceous glands. 2. Ovarian follicle. Also called Graafian follicle. A fluid-filled vesicle in the ovary containing a maturing ovum which at ovulation is discharged by rupture of the follicle. The ovaries contain many ripening follicles which are under the control of the pituitary glands, but normally only one ruptures at each ovulation. The epithelium of the follicle produces oestrogens. After ovulation the follicle becomes a corpus luteum.

fontanelle A soft space in the skull of an infant before the skull has completely ossified. The anterior fontanelle, or bregma, is where the coronal, frontal and sagittal sutures meet. The posterior fontanelle is where the lambdoid and sigittal sutures meet. The anterior fontanelle should be closed in 2 years of age. Delay is a sign of rickets.

foot-candle A unit of luminance on a surface that is everywhere one foot from a uniform point source of light of one candle and equal to one lumen per square foot. Numerically, daylight is 1000 foot–candles. Normal room light is typically between 50 to 100 foot–candles.

foot-lambert Unit of luminance or brightness. Defined as the uniform luminance of a surface emitting or reflecting light at the rate of one lumen per square foot.

foramen An opening or hole in a body part, especially a bone.

foramen magnum Opening in the back of the skull through which the spinal cord passes.

foramen ovale Opening between the right and left atria in the fetus which allows oxygenated venous blood from the placenta to pass into the left side of the heart and thus bypass the pulmonary circulation. Normally it closes at birth when the pressure in the left atrium rises.

forbidden gap In the band theory of solids, the range of energies between the conduction band and the valence band; electrons cannot exist at energies in this range.

force-balance transducer A transducer in which the output from the sensing member is amplified and fed back to an element which causes a force-summing member to return to its rest position. The magnitude of the signal fed back determines the output of the device like the error signal in a servo system.

forced expiratory flow Average flow rate for a specified portion of the forced expiratory volume, usually between 200 and 1200 milliliters.

forced expiratory volume Exhaled gas volume over a specified time interval, during the performance of a forced vital capacity measurement.

forced midexpiratory flow Average flow rate determined at the midpoint of the forced expiratory volume.

forced vital capacity Maximum gas volume which can be expired, as quickly and forcibly as possible, after a maximum inspiration.

forceps Instrument used in assisting passage of the infant through the birth canal. Attributed to Italian physicians of the twelfth century.

force-velocity relation A relationship established by A. V. Hill which states that the force exerted by a contracting muscle is inversely proportional to the velocity of shortening of the contractile element of the muscle tissue. This relationship has been extended to cardiac muscle by the work of Dr. Edmund H. Sonnenblick and others, who view measurements of the force-velocity relation as a key to devising a performance index for the heart's musculature.

forebrain See *brain*.

forensic medicine Medicine insofar as it has to do with the law.

-form Suffix meaning shape or form.

format In a computer, a specified grouping of data to facilitate its storage and flow through various devices in a system. A specific format may include control codes, record marks, block marks, and tape marks located in a prearranged sequence. The format tells the operator or the system how to control the transfer, processing, and printing of data. Format also describes the layout of characters on printed copy (which would be directly related to the data format).

forming Application of voltage to an electrolytic capacitor, electrolytic rectifier, or semi-conductor device to produce a desired permanent change in electrical characteristics as a part of the manufacturing process.

Forssman antibody Antibody specific for antigen unrelated to that providing immunizing stimulus.

FORTRAN Acronym for *for*mula *trans* lation; A procedure-oriented computer language designed for problems that can be expressed in algebraic notation. Exists in several forms of FORTRAN II, FORTRAN IV, etc.

forward bias Connecting a voltage to a semiconductor diode with a polarity such that the voltage aids the movement of carriers across the potential barrier. With forward bias the P-type material is positive and N-type material is negative.

forward recovery time Of a semiconductor diode, the time required for the forward current or voltage to reach a specified value after instantaneous application of a forward bias in a given circuit.

Fourier analysis The process of analyzing a complex wave by separating it into a plurality of component waves, each of a particular frequency amplitude, and phase displacement.

four-layer diode A semiconductor diode that has three junctions with connections made only to the two outer layers that form the junctions. An example is a Shockley diode.

four-layer transistor A junction transistor that has four conductivity regions but only three terminals. An example is a thyristor.

four-level system A laser involving four electronic energy levels. The ground state (level 1) is pumped to level 4, from which the excited electrons made a downward transition to the upper laser (level 3, or Metastable level 3). Then, stimulated transition to the lower laser level 2 occurs, followed by rapid decay to the ground state. The four-level system has the advantage that the pump level and ground state are isolated from the laser action.

fovea Shallow depression in the retina in which there is an absence of rods and a concentration of cones, and where there are no intervening nerves and blood vessels. It is the site of maximum visual stimulation and the region on which the image is focused when the eyes are fixed on an object. See *eye*.

fractional distillation Type of distillation in which substances are separated by their differing boiling points.

fraenum A small membranous fold attached to certain organs, and acting as a check.

fraenum linguae A small membrous fold under the tongue.

frame frequency 1. In television, the number of times per second that the frame is scanned. In the United States, the frame frequency is 30. 2. In computers, the number of frames per unit time. 3. In telemetry, the number of times per second that a frame of pulses is transmitted or received.

FRC 1. Abbreviation for Federal Radiation Council. 2. Abbreviation for functional residual capacity.

free association In psychoanalytic therapy, spontaneous, uncensored verbalization by the patient of whatever comes to mind.

free electrons Electrons which are not bound to a particular atom, and are free to move under the influence of an electric field.

free impedance Impedance at the input of the transducer when the impedance of its load is made zero. Also called *normal impedance.*

free motional impedance Of a transducer, the complex remainder after the blocked impedance has been subtracted from the free impedance.

free oscillation Oscillation of some physical quantity connected with the system under the influence either of internal forces only, or of a constant force having its origin outside the system, or both.

free running See *multivibrator.*

free-running frequency Frequency at which a normally driven oscillator operates in the absence of a driving signal.

free-running multivibrator A multivibrator which oscillates without triggering pulses. See *multivibrator.*

free-wheeling diode A special application of a rectifier diode. It is normally connected across an inductive load to carry a current proportional to the stored energy in the inductance. It carries this current when no power is being supplied to the load by the power source, and until all of the energy in the inductive load has dissipated to zero, or until the next voltage pulse is applied.

fremitus A vibration perceived by palpation, always applied to a vibration in the chest.

frenulum A small fold of mucous membrane or skin that acts to limit the movement of an organ, as in the frenulum of the tongue. Also called *fraenum.*

freq Abbreviation for frequency.

frequency 1. The number of events that occur during a given time period. For example, in power lines, the voltage on one wire rises and falls 60 times each second. We therefore speak of the power frequency as being 60 cycles per second. The frequency of blood pressure alterations in a certain artery might be 75 pulses per minute, which is also known as the pulse rate. 2. Repetition rate of any periodic phenomenon, as an alternating electrical signal or sound vibration, expressed in cycles per second (Hz). Bass frequencies in music extend from about 20 to 200 Hz, while treble range extends from 2 to 3 kHz to the limitations of audibility (18,000 to 20,000 Hz). Middle or midrange frequencies occupy the 200 Hz to 3,000 Hz range.

frequency compensation 1. Technique of modifying an electronic circuit or device to improve or to broaden the linearity of its response characteristic with respect to frequency. 2. The compensation required in feedback amplifiers to ensure stability and prevent unwanted oscillations.

frequency converter 1. A circuit or device which changes a signal from one frequency to another. May either translate the frequency by heterodyne means, or multiply it as in a doubler. 2. In a superheterodyne receiver, the section consisting of the oscillator, mixer, and first detector stages.

frequency counter An instrument which measures frequency by counting the number of cycles (pulses) occurring during a certain time interval.

frequency deviation A measure of the output frequency excursion around the carrier, caused by modulating the oscillator's tuning input which produces a frequency modulated output signal.

frequency distortion 1. Distortion in which there is change in the relative magnitudes of the different frequency components of a complex wave, providing that the change is not caused by nonlinear distortion. 2. Impairment of fidelity introduced by a transducer as a result of the unequal transfer of frequencies, e.g., unequal amplification of frequencies with the passband of an amplifier.

frequency divider Device for delivering an output wave whose frequency is a proper fraction, usually a submultiple, of the input frequency.

frequency-division multiplexing Taking the frequency spectrum of available channel and subdividing it into a series of lower frequency bands, each of which will transmit the data of an associated low-speed device.

frequency meter Instrument for measuring the repetition rate of a recurring phenomenon, as the cycles per second of a sinusoidal waveform.

frequency-modulated signal A signal where the intelligence is contained in the deviation from a center frequency. In a recording instrument, this deviation is proportional to the applied stimulus.

frequency modulation A process whereby the frequency of a previously single-frequency carrier wave is varied in step with the amplitude of a complex modulating wave. Higher order sidebands result from this type of modulation.

frequency multiplier 1. Device for delivering an output wave whose frequency is a multiple of the input frequency. 2. Amplifier circuit which amplifies a harmonic. Its output frequency is some multiple of the original frequency.

frequency response 1. The transmission gain or loss of a system, measured over the useful bandwidth, compared to the gain or loss at some reference frequency. 2. The portion of the frequency spectrum which can be passed by a device within specified limits of amplitude error. 3. A graphical characteristic showing relative signal levels at different frequencies with respect to a given reference level. A flat frequency response is one that has a uniform level at all frequencies within a given bandwidth. 4. A measure of the ability of a device to take into account, follow, or act upon a rapidly varying condition (e.g., as applied to amplifiers). 5. The measure of any component's ability to pass signals of different frequency without affecting their relative strengths. This is shown as a graph or curve which assumes input signals equally strong at all frequencies and plots their output intensities against a decibel scale. The ideal curve is a straight line. Frequency response may also be stated as a frequency range, but with specified decibel limits indicating the maximum deviations from flat response. For instance, 30 to 20,000 Hz $\pm$ 2 dB means the component will not change the relative intensities of any frequencies within that range by more than 2 dB above or 2 dB below the ideal zero dB (volume unchanged) point. 6. The range of frequencies over which an amplifier or recording system responds within defined limits of amplification, or signal output.

frequency response analysis The use of alternating or pulsating signals to excite a control system that is being studied. The system's response to different frequencies permit analysis of its operating characteristics.

frequency stability The ability of a device, such as an oscillator, transmitter, or receiver, to maintain its designed frequency for a long period of time.

frequency swing The instantaneous departure of the frequency of the emitted wave from the center frequency resulting from modulation.

frequency synthesizer A highly precise crystal oscillator with frequency divider used to provide the exact radio frequency for radio transmission. A typical synthesizer can be set to any 100 Hz step within the 2 to 30 MHz band, and will produce an output accurate in frequency to one part in 100 million.

frequent ectopics Three or more ectopic beats occurring within a minute. This minute is a *sliding minute* which is measured backward from the most recent QRS complex for that patient.

frictional electricity See *electricity*.

frit Metal powders fused in glass binder.

frontal Relating to the forehead.

frontal plane Section drawing, etc., parallel to the main axis of the body, and at right angles to the sagittal plane.

front end Refers to the first stage of an amplifier or tuner.

front-to-back ratio 1. The ratio of forward resistance to back resistance of a diode. 2. The ratio of field strength in the front lobe of a transmitting antenna to that in the back lobe. 3. The ratio of the forward sensitivity of a receiving antenna to its backward sensitivity.

frost point An absolute measure, found as the temperature to which a gas must be cooled at constant pressure to be saturated with respect to ice. Frost point is normally expressed in Fahrenheit or Centigrade degrees.

-fug(e) Suffix meaning avoid.

fugue 1. A fleeing from reality as in hysteria. The patient has no recollection of his actions during this time. 2. A major state of personality dissociation characterized by amnesia and actual physical flight from the immediate environment.

full adder A solid-state logic device which adds binary digits and can accept a *carry* from the previous stage. It consists of two half-adders and an inverter. It has two binary inputs and a *carry input,* and provides a *sum output* and a *carry output.* Used as an adder in all stages except the units digit.

full-duplex operation Communications between two points in both directions simultaneously. Also called *duplex operation.*

full-scale value The largest value of applied electrical energy which can be indicated on a meter scale. When zero is between the ends of the scale, the full-scale value is the arithmetic sum of the values of the applied electrical quantity corresponding to the two ends of the scale. On a suppressed meter the full-scale value is the largest value of applied electrical input less the smallest value of applied electrical energy input.

full-wave rectification The process of inverting the negative half cycle of current of an alternating current input, so that it flows in the same direction as the positive half-cycle. A way to accomplish this is to use four diodes placed in a bridge configuration.

full-wave rectifier A device which rectifies both half cycles of an alternating current to produce direct current.

fulminating Severe and rapid in its course.

function 1. A means of referring to a type or sequence of calculations within an arithmetic statement. 2. A relationship is a function if, and only if, for every combination of input conditions there is one unique output (e.g., the AND function will be valid or present if all the input variables examined are valid).

functional design The specification of the working relations between the parts of a system in terms of their characteristic action.

functional device See *integrated circuit*.

functional diagram A diagram that represents the functional relationships among the parts of a system, ignoring the practical means used to achieve the functions (i.e., a block diagram or flow chart).

functional residual capacity 1. (FRC) A measure of lung function. 2. That volume of gas which remains within the lungs at the resting expiratory level. The end of expiration is chosen as the baseline valve because it is more consistent than the end-inspiratory level.

functional testing Testing objective is to determine whether the device under test reacts correctly to inputs, qualitatively.

function switch In computer systems, a network having a number of inputs and outputs, and so connected that signals representing information expressed in a certain code, when applied to the inputs, cause output signals to appear which are a representation of the input information in a different code.

function table 1. In computers, a routine by which a machine can determine the value of a dependent quantity from the values and independent variables. 2. Subroutine that can either decode multiple inputs into a single output or encode a single input into multiple outputs.

fundamental frequency Of a periodic quantity, the frequency of a sinusoidal quantity which has the same period as the periodic quantity.

fundamental particles See *elementary particles*.

fundus The enlarged part of a hollow organ farthest removed from the orifice; thus the fundus oculi is the interior of the eye behind the lens and pupil, visible with an ophthalmoscope; the fundus uteri is the top of the uterus.

funnel chest Also called pectus excavatum. A developmental deformity in which the sternum is depressed and the ribs and costal cartilages curve inwards.

fuse A device used for protection against excessive currents. Consists of a short length of fusible metal wire or strip which melts when the current through it exceeds the rated amount for a definite time. Placed in series with the circuit it is to protect.

fusible link readout memory A large

array of prediffused cells, interconnected by a fixed metallization pattern which the user can tailor to his own requirements simply by selectively burning out, interconnections.

fusiform Spindleshaped.

fusion A nuclear reaction in which smaller atomic nuclei or particles are combined into larger ones, with the fusion process resulting in a release of energy.

G

G Abbreviation for giga or 10^9.

gage (gauge) Literally, an instrument or means for measuring or testing. By extension, the term is often used synonymously with transducer. More specifically, this term refers to a pressure transducer in which the output represents the value of the applied pressure with reference to the ambient pressure.

gain An increase in signal power between input and output of an electronic circuit or device. It is usually expressed in decibels.

gain margin The amount of gain change of an operational amplifier at 180 degree phase shift angle frequency that would produce instability.

galvanic 1. Relating to production of a direct current of electricity. 2. A term for chemically generated current, such as that from a battery. Galvanic action is the effect which occurs when two dissimilar conductors are immersed in an electrolyte and connected by a closed electrical circuit. A current will flow in the circuit and the more easily oxidized conductor will be oxidized.

galvanic anode A source of emf for cathodic protection provided by a metal less noble than the one to be protected (i.e., magnesium, zinc, or aluminum) as used for cathodic protection of steel.

galvanic cell An electrolytic cell that is capable of producing electric energy by electrochemical action.

galvanic corrosion Accelerated electrochemical corrosion produced when one metal is in electrical contact with another more noble metal, both being in the same corroding medium or electrolyte, with a current flowing between them. Corrosion of this type usually results in a higher rate of solution of the less noble metal and protection of the more noble metals.

galvanism The use of direct electric current, i.e., continuous, or interrupted in therapy. Interrupted direct current consists of current pulses of duration 30 to 100 msec. It is useful for the stimulation of a muscle which has been deprived of normal nerve control.

galvanometer Instrument for indicating or measuring the flow of a small electric current, or a function of the current, by means of a mechanical motion derived from electromagnetic or electrodynamic forces which are set up as a result of the current.

galvanometer constant Number by which a certain-function of the reading of a galvanometer must be multiplied to obtain the current value in ordinary units.

galvanometer light beam recorder Recorder using mirror on galvanometer to deflect light beam to photosensitive paper. Although some delicate laboratory instruments can write waveform tracings at frequencies up to 5,000 hertz, the more usual upper limit of a galvanometer recording instrument is several hundred cycles.

galvanometer shunt Resistor connected in parallel with a galvanometer to increase its range under certain conditions. It allows only a known fraction of the current to pass through the galvanometer.

galvanometric controller A temperature indicator which has been converted to a temperature controller. The indicator operates directly off the senor's input signal. The controller portion detects the mechanical position of the pointer, and varies the output to keep the pointer at the desired temperature.

gain function A transfer function that relates either a pair of voltages or a pair of currents.

gamma 1. Unit of magnetic intensity. 2. Definite numerical indication of the degree of contrast in a photograph, facsimile reproduction, or video picture.

gamma (γ) radiation Electromagnetic disturbance (photons) emanating from an atom nucleus. This type of radiation travels in waveform much like X-rays or light, but has a shorter wavelength (approximately 1Å or 10^{-7} mm). It is very penetrating.

gamma ray A quantum of short wavelength electromagnetic radiation emitted by a nucleus in its transition from a higher to a lower energy state. Gamma rays have zero rest mass and zero charge but energies in the range of approximately 1 meV. The intensity of 1 meV of gamma is halved in 4 inches of water. Gamma radiation is used for dry, cold sterilization of any articles which would be destroyed by moisture or heat. Also used in radiotherapy.

ganglion 1. A group of nerve cells forming a very nearly independent neurologic center in the sympathetic nervous system, or in other parts of the nervous system. 2. A chronic synovial cyst connected with a tendon sheath, most commonly found on the back of the hand, near the wrist.

gangrene Tissue death (*necrosis*) with putrefaction.

gap 1. Space where radiation fails to meet minimum coverage requirements; this might be space not covered or space where the minimum specified overlap was not obtained. 2. Portion of a magnetic circuit that does not contain ferromagnetic material, such as an air gap. 3. Interval of space or time used as an automatic sentinel to indicate the end of a word, record, or file of data on a tape.

gas diode A tube having a hot cathode and an anode in an envelope containing a small amount of an inert gas or vapor. When the anode is made sufficiently positive, the electrons flowing to it collide with gas atoms and ionize them. As a result, anode current is much greater than that for a comparable vacuum diode.

gas discharge device A device utilizing the conduction of electricity in a gas, due to movements of electrons and ions produced by collision.

gas discharge display   A device containing an inert gas that gives off light when a high voltage is applied to break down (ionize) the gas. Gas discharge displays have a seven-segment format.

gas discharge laser   See *gas laser.*

gaseous discharge   The state of a gas or gas mixture in which a conduction current can be maintained in the gas by ionization. This ionization is caused by collisions between electrons and atoms or molecules of the gas, the energy being supplied by an external source such as an electric field.

gaseous electronics   A field of study which involves the conduction of electricity through gases such as the Townsend, glow, and arc discharges, and a study of all the collision phenomena on an atomic scale.

gaseous tube   An electronic tube into which a small amount of gas or vapor is admitted after the tube has been evacuated. Ionization of the gas during operation of the tube gives greatly increased current flow.

gas laser   An optical oscillator or amplifier employing as the active medium a gaseous discharge in a suitable gas or vapor. The discharge, usually contained in a glass or quartz tube with a Brewster angle window at each end, can be excited by a high-frequency oscillator (radio-frequency discharge) or by direct current between electrodes placed inside the tube. The function of the discharge is to pump the medium in order to obtain a population inversion. Also called *gas discharge laser.* See also *laser.*

gas maser   A maser in which the microwave electromagnetic radiation interacts with the molecules of a gas such as

ammonia. Use is limited chiefly to highly stable oscillator applications, as in atomic clocks.

gastric   Relating to the stomach.

gastric aspiration   A form of aspiration performed postoperatively after operations on the alimentary tract to prevent dilation of the stomach. A Ryle's tube is passed and the contents aspirated at frequent intervals or continuously. Also called *gastric suction.*

gastric juice   The digestive fluid of the stomach.

gastric lavage   Washing out the stomach; a procedure used in the treatment of poisoning.

gastric suction   See *gastric aspiration.*

gastritis   Acute or chronic inflammation of the epithelium lining of the stomach. It may be caused by the ingestion of alcohol, spices, medicines, chemicals, foods, as well as by infections or allergy.

gastroduodenostomy   The operation of making an artificial passage direct from the stomach to the duodenum or the jejunum.

gastroenteritis   Inflammation of the stomach and intestines.

gastrointestinal tract   The continuous pathway for ingestion, digestion and excretion, including the mouth, esophagus, stomach, small and large intestines, rectum, anus, and associated structures.

gastroscope   An instrument for inspecting the cavity of the stomach. It has a light at the end and is passed through the esophagus.

gastroscopy   Direct visualization of the stomach interior by means of an optical instrument called a gastroscope.

gate   1. Circuit with an output and a multiplicity of inputs so that the output is energized only when certain input conditions are met. These include both AND gate, OR gates, NANDs, and NORs, which see. 2. Circuit in which a signal (generally square wave) switches another signal on or off. 3. One of the electrodes in a field-effect transistor. 4. An output element of a cryotron. 5. To control the passage of a pulse or signal.

gate controlled switch A three junction, three-terminal solid-state device, constructed very much like a silicon controlled rectifier except that it has a turn-off ability, which is controlled by a negative-current pulse applied to the gate. Also called *gate turn-off switch*.

gate electrode The control electrode of a FET, silicon controlled rectifier, or triac.

gate trigger current In a controlled rectifier, the minimum gate current required to cause switching from the off-state to the on-state for a stated anode-to-cathode voltage.

gate trigger voltage In a controlled rectifier, the voltage between the gate and the electrode of the adjacent region required to produce the gate trigger current.

gate turn-off current In a controlled rectifier, the minimum gate current required to cause switching from the on-state to the off-state for a stated collector current in the on-state.

gate turn-off switch See *gate controlled switch*.

gate turn-off voltage The voltage between the gate and the electrode of the adjacent region in a controlled rectifier, required to produce the gate turn-off current.

gating 1. Process of selecting those portions of a wave which exist during one or more selected time intervals or which have magnitudes between selected limits. 2. Application of a specific waveform to perform electronic switching. 3. The process of selecting for use only those portions of a wave between selected time intervals or between selected amplitude limits.

gating pulse A pulse which causes operation (opening) of a gate for the period of the pulse.

gauss The electromagnetic unit of magnetic induction, which is one maxwell per square centimeter.

gaussian distribution A distribution of random variables characterized by a symmetrical and continuous distribution decreasing gradually to zero on either side of the most probable value.

gaussian noise Noise that has a normal distribution, i.e., follows the Gaussian (Laplacian) curve.

GB Abbreviation for gallbladder.

Gc Abbreviation for gigacycles (10^9 cps) (an obsolete term replaced by GHz).

Geiger counter Gas chamber-type radiation counter in which the chamber operates in avalanche region for high amplification and sensitivity.

Geiger-Muller counter Device for detecting and counting certain fundamental particles such as beta particles and alpha particles. It consists essentially of a cylinder containing a wire electrode along its axis. A high voltage is applied across the gap from cylinder wall to wire. Some gas at low pressure is let into the cylinder before it is sealed. If a beta particle enters the cylinder it collides with gas molecules and creates many electrons, all of which move toward the wire because of the electric field created by the applied voltage. This movement constitutes an electric current. The resultant effect of one beta particle, therefore, is a surge of current. This can be made to operate a mechanical counter or a light lamp. The Geiger-Muller or, simply, the Geiger counter is probably the most widely used piece of apparatus in the physics of radiation and atomic particles.

generalized edema See *edema*.

general-purpose digital computer A digital computer designed to solve a large variety of problems; that is, a computer that can be adapted to a large class of applications. (As opposed to a computer designed specifically to control a manufacturing process.) A typical general-purpose digital computer consists of four subsystems: (1) Input/output, which permits communication with the outside world; (2) memory, which stores data and instructions; (3) central processing unit (CPU), or arithmetic unit, which performs the arithmetic and data processing operations; and (4) control, which ties all of the subsystems together so that they operate in a fully automated way.

general-purpose relay Device which can perform a wide variety of functions. Name is applied to distinguish them from relays that have a specific function.

generator 1. Rotating machine that converts mechanical energy into electrical energy. 2. Radio device that develops an alternating current voltage at a desired frequency and a desired shape when energized with direct current power. 3. Any device that generates electricity. 4. In computers, a routine designed to create specific routines from specific input parameters and skeletal coding. 5. A machine that converts mechanical energy into electrical energy. In its most common form, a large number of conductors are mounted on an armature that is rotated in a magnetic field produced by field coils. Also called *dynamo*. 6. A vacuum-tube oscillator or any other non-rotating device that generates an alternating voltage at a desired frequency when

energized with dc power or low-frequency ac power. Such generators are used to produce large amounts of high frequency power, such as that used for high-frequency heating and ultrasonic cleaning. 7. A circuit that generates a desired repetitive or nonrepetitive waveform such as a pulse generator.

genetic Pertaining to generation.

genetics Study of heredity and its variations.

genitourinary Relating to reproduction and to urination. Denoting the organs concerned in these functions.

geriatric Pertaining to old age.

geriatrics The study of disease among the elderly.

germanium (Ge) A brittle, grayish-white metallic element having semiconductor properties. Widely used in transistors and crystal diodes. Atomic number is 32.

germanium diode A semiconductor rectifier which uses germanium as the rectifying element.

gerontology The study of aging.

gestation The interval during which the fetus normally matures in the uterus, ending with labor and birth.

gestation sac The fetus with its enveloping membranes, decidua, etc. The contents of a pregnant uterus.

getter Alkali metal introduced into a vacuum tube during manufacture and fired after the tube has been evacuated to react chemically with any gases which may have been left in manufacture. The silvery deposit on the inside of the glass envelope of a tube, usually near the tube base, is the result of getting firing.

G.I. Abbreviation for gastrointestinal.

giga A prefix used to denote one billion (10^9).

gigahertz A contraction of the prefix *giga* meaning 10^9 and *hertz* meaning cycles per second. Replaces the more cumbersome term kilomegacycle.

gigawatt (GW) One thousand million watts (10^9 W).

gigo An acronym for garbage in, garbage out, used to describe an incorrect output of a computer. See also *kiss*.

gigohm 10^9 ohms. One billion ohms.

gilbert The unit of magnetomotive force in the centimeter-gram-second electromagnetic system. The value of the magnetomotive force in gilberts in any magnetic circuit is equal to the line integral around the circuit of the magnetic intensity in oersteds, with length being in centimeters. One gilbert is equivalent to 0.7956 ampere-turn.

gilberts per centimeter Practical centimeter-gram-second unit of magnetic intensity. Gilberts per centimeter are the same as oersteds.

glass electrode An electrode used to determine the potential of a particular solution, with respect to a reference electrode, in electronic pH determination.

glaucoma Eye disease characterized by an increase in the internal pressure in the eye, caused by alteration of the intraocular fluid flow, resulting in visual impairment and, if untreated, blindness.

-glia Suffix meaning glue.

globulin A specialized class of proteins distributed throughout the body with numerous specialized functions. One group, the gammaglobulins, are antibodies.

glomeruli (plural of glomerulus) The meshed network of capillaries and tubules within the kidneys, through which soluble wastes and water are extracted from the blood.

glossal Relating to the tongue.

glossopharyngeal nerve The ninth cranial nerve, arising in the medulla oblongata. This paired nerve conveys the impulses of taste from the rear third of the tongue and gives sensation to the rear part of the mouth. See also *cranial nerves*.

glottis The vocal apparatus of the larynx.

glow discharge Luminous discharge of electricity through a gas, without sparks. In a phototube, a glow discharge indicates excessive ionization and excessive current.

glow lamp A small lamp having two electrodes and filled with a gas which ionizes easily. When energized, the negative electrode is covered with a glow, orange-red if the gas is neon, and blue if the gas is argon. Widely used as indicator lamps.

glow potential Voltage at which a glow discharge begins in a gas-filled electronic tube as the voltage is gradually increased.

glucose Dextrose. A sugar containing six carbon atoms which is widely distributed in nature in disaccharides like sucrose and lactose, and in polysaccharides such as starch. Glucose is the principal source of energy production in metabolism.

gluteus maximus, gluteus medius, gluteus minimus The three large muscles of the buttock.

glycosuria The presence of glucose in the urine.

gold-bonded diode A semiconductor diode in which a preformed whisker of gold is brought into contact with an N germanium substrate and the junction is formed by electrical pulses of millisecond duration. The pulses may be either ac or dc.

go/no-go testing Testing designed to show only whether the device under test passed or failed, with no indication of how it failed or why.

gradient The difference between values of a function measured in different locations. Examples: There is a gradient of pressure upstream and downstream from a stenotic aortic valve. There is a gradient of oxygen tension between alveolar air and the arterial blood.

graft To join or to induce union between tissues. Tissues, and in some cases, organs, may be transferred in the same individual (autograft), or from one individual (donor) to another (recipient) of the same species (homograft), or of a different species (heterograft).

-gram Suffix meaning scratch, write, or record.

grand mal Major epilepsy.

granulation tissue Newly formed vascular connective tissue. It is typically formed at the surface of wounds as part of the process of healing.

granuloma A tumor composed of vascular connective tissue.

-graph Suffix meaning scratch, write, or record.

graphics A method of transmitting visual intelligence via telecommunications. A broader term than facsimile.

grass In echographic studies, low-amplitude echo traces not of diagnostic significance. They can be eliminated from an A-mode or M-mode display with the reject control of the echograph.

gray code Modified binary code used for analog to digital conversion, such as changing an angle to a binary value. Also called cyclic binary code and reflective code.

grid The control element in a vacuum tube. The voltage between the grid and the cathode determines the amount of current which flows between the cathode and the plate. Relatively small changes in grid voltage result in substantial changes in plate current.

grid bias A negative voltage applied to the control grid of an electron tube with respect to the cathode potential. It establishes operation at the desired point on the characteristic curve of the tube.

grid controlled rectifier Triode mercury vapor rectifier tube in which the grid determines the instant at which plate current starts to flow during each cycle, but does not determine how much current will flow.

grid radiotherapy The application of ionizing radiation to a patient through a sieve-like screen of lead, to improve tissue tolerance in palliative radiotherapy.

groin The lower anterior portion of the abdominal wall.

ground 1. Point in a circuit used as a common reference or datum point in measuring voltages. 2. Conducting connection, whether intentional or accidental, between an electric circuit or equipment and earth, or to some conducting body which serves in place of earth. Also called *earth*.

ground current Current in the earth or grounding connection, usually indicative of a fault.

ground detector Device that indicates ground faults in electrical circuits.

grounded Connected to the earth, or to a rod or pipe which makes a good electrical connection with the earth.

grounded-collector amplifier Transistor circuit in which the collector electrode is common to both the input and output circuits. The collector need not be directly connected to circuit ground. Also called *common-collector connection.*

grounded-emitter amplifier Transistor circuit in which the emitter electrode is common to both the input and output circuits. The emitter need not be directly connected to circuit ground. Also called *common-emitter connection.*

grounded-gate amplifier Field-effect transistor amplifier in which the gate electrode is connected to ground. The input signal is fed to the source electrode and the output is obtained from the drain electrode.

grounded parts Parts which are so connected that, when the installation is complete, they are substantially of the same potential as the earth.

grounded system An electrical system in which one conductor or a neutral point is intentionally grounded, either solidly or through a grounding device.

ground fault An unintentional electrical path between a part operating normally at some potential to ground, and ground.

ground fault current A fault current which flows to ground rather than between conductors.

ground fault interrupter See *current limiter.*

grounding The act of connecting an electric circuit or equipment to earth, or to a conductor connected to earth. A safety measure to avoid electric shock.

grounding conductor A conductor which, under normal conditions, carries no current, but serves to connect exposed metal surfaces to an earth ground, to prevent hazards in case of breakdown between current-carrying parts and the exposed surfaces. The conductor, if insulated, is colored green with or without a yellow stripe.

grounding device An impedance through which an electrical system is connected to ground, and which serves to limit ground fault currents.

grounding electrode A conductor imbedded in the earth for maintaining ground potential on conductors connected to it.

grounding outlet An alternating current receptacle which has a third contact connected to ground. Used with three-wire plugs and cords to safely ground housings of portable electric tools and appliances.

grounding plate An electrically grounded metal plate on which a person stands in order to discharge any static electricity that may be picked up by his body.

grounding transformer Transformer intended primarily for the purpose of providing a neutral point for grounding purposes.

ground loop Potentially detrimental loop formed when two or more points in an electrical system that are nominally at ground potential are connected by a conducting path. The term is usually employed when, by improper design or by accident, unwanted noisy signals are generated in the common return of relatively low-level signal circuits by the return currents or by magnetic fields from relatively high-powered circuits or components.

ground potential Having the same (zero) potential as the earth.

ground resistance The ohmic resistance between a grounding electrode and a remote or reference grounding electrode so spaced that their mutual resistance is essentially zero.

ground-return circuit A circuit using the earth as one side of the complete circuit.

ground wire A heavy copper conductor,

usually insulated, which is used to connect protectors or other equipment to a ground rod or cold water pipe.

grove cell Primary cell, having a platinum electrode in an electrolyte of nitric acid within a porous cup, outside of which is a zinc electrode in an electrolyte of sulfuric acid. This cell normally operates on a closed circuit.

guard See *guard shield*.

guarded input An input of an amplifier that has a third terminal which is maintained at a potential near the input-terminal potential for a single-ended input, or near the mean input potential for the differential input. It is used to shield the entire input circuit.

guard shield An internal floating shield which surrounds the entire input section of an amplifier. Effective shielding is achieved only when the absolute potential of the guard is stabilized with respect to the incoming signal. Also called *guard*.

gymnosperm The XX-bearing male gamete, which, when united with the female ovum, will produce a female offspring.

gynecology The management and treatment of disorders affecting the female generative organs (e.g., ovaries, uterus, vagina).

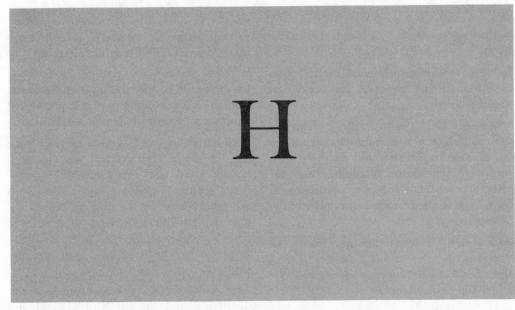

H

half-add In a computer, an operation that is performed first in doing a two-step binary addition. It adds corresponding bits in two binary numbers, ignoring any carry information. See also *exclusive OR*.

half adder A solid-state logic device which adds binary digits. It consists of two AND gates, an OR gate, and an inverter. It has two binary inputs, a sum output and a carry output. Since it has no carry input, it can be used only to add the units digit. To add the tens, hundreds, etc., digits, a full adder having a carry input is required.

half-duplex 1. A communication system in which information can be transmitted in either direction, but only in one direction at a time. 2. In communications, pertaining to an alternate independent transmission, made in one direction at a time.

half-life The time in which the total radiation emitted by a radioactive substance is reduced by decay to half its original value. It is a constant for each isotope and is independent of the quantity.

half shift register A logic device which is equivalent to one-half of a full master-slave flip-flop.

half-track tape Magnetic tape on which half the width of the tape is used for one sound path. Also known as two-track tape. Such a tape provides stero in one direction of tape travel, or mono sound in both directions.

half-wave rectifier Nonlinear device used to rectify alternating current into direct current. Only one-half of the input cycle is rectified, the output wave being a pulsating direct current.

half-word A continuous sequence of bits or characters which comprises half a computer word and is capable of being addressed as a unit.

Hall effect The generation of a voltage across a current-carrying flat conductor when it is placed in a magnetic field perpendicular to the flat surface. The Hall effect voltage is perpendicular to both the magnetic field and the carried current.

hallucination A sensation not based upon reality.

hallucinogen Drug causing hallucinations.

halogenated hydrocarbon aneshetics Volatile liquids that are used for inhalation anesthesia.

halogens The nonmetallic elements: fluorine, chlorine, bromine, iodine. They are anionic in solution and combine with metals to form salts. Numerous uses for elemental and salt forms of the halogens have been devised in medicine, principally in asepsis.

hamartoma Tumors arising from the overgrowth of developing tissues. The term describes benign lesions such as vascular naevi and neurofibromas.

hamming code A specific error-correcting code. See *error-correcting code*.

hamstrings The tendons traversing the popliteal region.

handshaking Synchronous transmission scheme, used to describe the process by which predetermined configurations of characters are exchanged by the receiving and transmitting equipment to establish synchronization.

hard copy Typewritten or printed characters on paper produced by a computer at the same time information is copied or converted into machine language that is not easily read by a human.

hard-copy printer An automatic device, sometimes resembling a typewriter, which produces intelligible symbols in a permanent form.

hardness That quality which determines the penetrating ability of X-rays. The shorter the wavelength, the harder and more penetrating are the rays.

hard tube An electron tube which has been evacuated to such a high degree of vacuum that the residual gases which remain do not affect the characteristics of the tube.

hardware The physical equipment or devices forming a computer and peripheral equipment. Contrasted with software.

hard-wired logic A group of solid-state logic modules mounted on one or more circuit boards and interconnected by electrical wiring. The logic control functions are determined by the way in which the modules are interconnected (as contrasted with a programmable controller or microprocessor in which the logic is in program form).

hard X-rays Highly penetrating X-rays as distinguished from soft X-rays, which are less penetrating.

harmartoma A term used to classify tumors arising from the overgrowth of developing tissues. The original Greek meant "missing the target with the javelin." Missing the target in development gives rise to benign lesions such as vascular naevi and neurofibromas.

harmonic A harmonic of a fundamental frequency is the frequency of its exact multiple. Thus twice the fundamental frequency is the second harmonic (one octave up) and so on. If the fundamental frequency is 1,000 Hz, the second harmonic would be 2kHz, the third 3kHz, and the fourth harmonic 4kHz (two octaves up from the fundamental).

harmonic analyzer A measuring device having tuneable circuits which can identify the frequency of each of the components of a complex wave, and with a meter measure their amplitudes separately.

harmonic distortion 1. The ratio, expressed in decibels, of the power at the fundamental frequency, to the power of a harmonic of that fundamental. 2. A form of nonlinear distortion in which harmonics of a fundamental frequency are generated by the nonlinearity of a circuit. 3. The sum of all signals in an output which are multiples of the input signal frequencies (harmonics). Their intensities are expressed as a percentage of the total output intensity. See also *distortion*.

Hartely In computers, a unit of information content equal to one of ten possible and equally likely values or states of anything used to store or convey information.

Hartley oscillator Oscillator in which a parallel-tuned tank circuit is connected between grid and plate of an electron tube or between base and collector of a junction transistor, the inductive element of the tank having an intermediate tap at cathode or emitter potential.

head A device that reads, records, or erases data on a storage medium; for example, a small electromagnet used to read, write, or erase data on a magnetic drum or tape, or the set of perforating reading or marking devices used for punching, reading, or printing on paper tape.

head gap Space or gap intentionally inserted into the magnetic circuit of the tape recorder head to force or direct the recording flux into the recording medium.

headphones Small sound reproducers, superficially resembling miniature loudspeakers, set in a suitable frame for wearing about the head and listening by close coupling to the ears. Recent headphones, improved greatly in fidelity, have become increasingly popular for private listening without disturbing others, as well as to prevent outside noises from interfering with the listening.

headset Small portable telephone receivers, usually in pairs, with a connecting clamp to support the phones against the ears, for operators of receiving equipment.

hearing aid A self-contained microphone, amplifying device, and transducer, designed to increase the acoustical intensity of sounds for application to the user's ear. The object is to compensate for hearing loss. Most modern hearing aids employ low-power, battery-operated, solid-state circuitry with compensatory filtering. These are small enough to fit within the ear.

heart A natural or artificial pumping device that maintains the circulation of blood within two systems: the first bearing oxygen-enriched blood to cellular structures; the second, carrying carbon dioxide-laden blood away from the structures to the lungs, for expiration and exchange of CO_2 for oxygen with the environmental air. 1. The natural heart is an organ created by specialization of the arteriovenous system. Lined by endothelial tissue, folds of which create the cardiac valves that make blood flow unidirectional, the heart is encased in an ordered structure of myocardial muscle, modified to conduct activating impulses arising in its own specialized pacemaker (the sinoatrial and atrioventricular nodes). Externally, the heart is enclosed within a membranous sac (the pericardium). 2. The artificial heart is a pumping mechanism that duplicates the output, rate, and blood pressure of the natural heart. The heart can be intracorporeal, extracorporeal, or paracorporeal, and can replace the function of the entire natural heart or only that of a portion of the natural heart. Intracorporeal hearts may be classified as being one of the following types: solenoid-drive, electro/motor-driven (including roller and pendulum types), hydraulically driven, piezoelectric, or gas-driven. Further classification of intracorporeal hearts can be made by determining whether the natural heart is totally replaced, partially replaced, or essentially intact with the addition of a booster heart or auxiliary ventricle. Extracorporeal hearts usually perform both a pumping function and an oxygenating function, with the blood by-passing both the heart and the lungs. In this case, the artificial heart can be called a pump oxygenator. Paracorporeal hearts are pumps worn at the side of the body and they may replace or augment the function of a portion or all of the natural heart.

heart block A disturbance of cardiac conduction somewhere between the sino-atrial node and the ventricles. Complete heart block results in the Stokes-Adams syndrome. Lesser degrees of heart block lead to limited conduction and some irregularity of cardiac rate or rhythm. An electronic pacemaker may be implanted to provide stimulus pulses to tissues beyond the breakpoint, thereby maintaining rhythmic contraction and cardiac function.

heart-lung machine Machine used in cardiac surgery to oxygenate the blood. See *pump oxygenator*.

heart murmurs Adventitious sounds detectable by auscultation of the heart, resulting from altered hemodynamics, in which the flow of blood through the heart becomes turbulent. Often, the murmurous sounds are created by flows which exceed a certain critical velocity. These conditions often occur when there is narrowing or stenosis of valves, abnormal communications between chambers of the heart, or where hyperthyroidism and severe anaemia create hyperdynamic circulation.

heartpacer A pulse generator used to stimulate the heart muscle in order to start or maintain a proper rhythm. See *pacemaker*.

heart-rate module A component of a pa-

tient monitor providing a tachometer circuit for detection and counting of cardiac contractions per minute. The usual input is the patient's ECG potentials, obtained from chest electrodes.

heart sounds The normal heart sounds are verbally represented by the words *lub-dup*. The first heart sound *lub* is due to the closure of the right and left atrioventricular valves which occurs at the beginning of the ventricular contraction (systole). The second sound *dup* is due to the closure of the aortic and pulmonary valves. Very occasionally, a third sound is heard due to rapid filling of the atria. Abnormal heart sounds may be due to exaggeration or distortion of the normal sounds, e.g., splitting of the components of the first or second sounds or the production of adventitious new sounds. See *heart murmurs*.

heart valve A structure in an outflow from an atrium, or a ventricle, that prevents the reflux of blood back into the heart chamber. The aortic valve is the valve at outflow from left ventricle to aorta. An artificial valve is a man-made heart valve. The ball-type valve (Starr-Edwards) is an artificial valve containing a small ball that moves freely within a cage. With diastole, the ball is in such a position that blood cannot reflux. The mitral valve is the valve at outflow from left atrium. The pulmonary valve is the valve at outflow from right ventricle. A semilunar valve is the valve at outflow from left atrium. The pulmonary valve is the valve at outflow from right atrium.

heart ventricle One of the two lower ventricles of the heart. The right ventricle receives blood from the right atrium and pumps it out to the lungs via the pulmonary arteries. The left ventricle receives blood from the left atrium and forces it out to the systemic circulation through the aorta.

heater The electric heating element in the center of an indirectly heated cathode of an electron tube. Permits using alternating current to heat the cathode without causing ac hum.

heat exhaustion Condition characterized by rapid pulse, dyspnea, and abdominal cramps, due to excessive sweating and loss of sodium chloride. The source of the adverse patient reaction is great heat.

heat loss The loss of useful power in an electrical device due to its being converted to unwanted heat, caused by resistance in wires and hysteresis in iron cores.

heat pipe A sealed tube lined with a capillary pumping structure, or wick, which is saturated with a working fluid. As heat is added at one end of the tube, the working fluid is vaporized and moves down the tube until it condenses at a cooler side and reverts to the liquid state. As liquid is returned to the heat source, via capillary action in the wick, heat is continuously transferred from one region to another through the nearly isothermal process of evaporation and condensation.

heat scanning An electronic technique that involves the rhythmic movement of a bolometer over an object for purposes of detecting infrared waves. See *thermography*.

heat sink 1. A mass of metal, often with fins, mounted on or under a circuit component which produces heat, such as a silicon rectifier or an electron tube. Absorbs and radiates the heat to maintain a safe working temperature. 2. A mounting base for semiconductor devices, usually metallic, which serves to dissipate, carry away, or radiate into the surrounding atmosphere heat which is generated within a device. The device's package itself often serves as a heat sink for the semiconductor chip, but, for higher power devices, a separate heat sink on which one or more packages are mounted is required to prevent overheating, and subsequent destruction, of the semiconductor junction.

heat-stress meter Instrument that measures humidity, heat, and wind for determining effect on humans, especially during exercise.

heatstroke Hyperpyrexia due to failure of temperature-regulating mechanisms of the body.

heat-writing recorder A type of strip-chart recorder employing a heated stylus to write on a chemically treated strip of paper. The paper is discolored or a coating is melted away by the heat, making visible the path followed by the stylus over the surface of the paper. The heat-writing recorder eliminates the mess and inconvenience of the pen-and-ink type of recorder.

helium-tight See *hermetic*.

hemagglutinin Antibodies present in the blood which combine with red blood corpuscles of a different blood group and cause agglutination. See *blood grouping*.

hematemesis The vomiting of blood.

hematinometer Hemoglobinmeter, an instrument for measuring hemoglobin in blood.

hematocele A swelling or cyst containing blood.

hematology The study of the blood, its nature, functions, and diseases.

hematology counter Also called hematometer. An instrument used to estimate the properties or constituents of blood.

hematoma A swelling composed of blood.

A bruise. In its usual form, a solid clot mass which readily becomes encapsulated by connective tissue. May constitute a visible, tumor-like swelling or false aneurism.

hematometer See *hematology counter.*

hematuria The presence of blood in the urine.

hemiparesis A condition in which one side of the body becomes paralyzed.

hemiplegia Paralysis of one side of the body.

hemisphere Half of any spherical or roughly spherical structure or organ, as demarcated by dividing it into approximately equal portions.

hemo 1. Prefix pertaining to blood. 2. Tetrapyrrollic ring containing an atom of ferrous iron. When combined with protein globin, forms hemoglobin.

hemochromatosis An inherited defect in the metabolism of iron with deposition of iron in tissues. This foreign presence interferes with their function. Diabetes mellitus, hyperpigmentation of the skin, and cirrhosis of the liver are symptomatic.

hemodialysis Passage of circulating blood through an artificial kidney, to restore normal balance of blood chemical constituents.

hemodilution Increase in the fluid content of the blood, with a decrease in its concentration.

hemodynamic Pertaining to the flow or to the pumping of blood, usually through the systemic circulation.

hemoglobin 1. The substance in the red blood cell which carries oxygen. 2. An iron-bearing pigment in red blood cells within a complex protein (globin). In combination it has the property of forming a reversible combination with oxygen and carbon dioxide.

hemolysis The separation of hemoglobin from the red blood cells and its appearance in the plasma, often a problem in prolonged use of assisted circulation techniques.

hemolytic Having the power to destroy red blood cells.

hemolytic anaemia A form of anaemia resulting from destruction of red cells as in forms of poisoning, or by the action of antibodies.

hemolytic disease of the newborn Jaundice in a Rhesus–positive infant caused by red cell destruction by anti–Rhesus antibodies generated in the Rhesus–negative mother's circulation during pregnancy. See *blood grouping.*

hemometer Apparatus for electronically measuring hemoglobin and volume of erythrocytes in blood.

hemophilia A congenital failure of the clotting factor in the blood. It occurs only in males, but is transmitted genetically, through the females of the family. The tendency to hemorrage is quite pronounced in the person afflicted by this disease.

hemophiliac A person suffering from hemophilia.

hemopoiesis The formation of the blood cells, particularly the red blood corpuscles. In the developed fetus, hemopoiesis occurs in the spleen and liver; in the adult, in the bone marrow.

hemoptysis The spitting up of blood.

hemorrhage Bleeding, ranging from small to large blood flows. The character of a hemorrhage is determined by its source: *venous hemorrhage* results in a steady flow of darkish blood; *arterial* *hemorrhage* is bright red in color and occurs in spurts, under the impetus of systolic pressure in the arterial system; *capillary hemorrhage* results in a slow oozing of blood, usually from a wound having relatively large surface area. The blood loss may be visible or concealed, depending upon the size of the injury. *Concealed hemorrhage* usually results in pallor, quick respiratory rate, rapid, weak pulse, below-normal temperature, restlessness, a sense of coldness and accompanying perspiration, and, in extreme cases, collapse. Most often, and where a wound is visible, hemorrhage will bring about its symptoms at the time of injury. This is called *primary hemorrhage*. In cases where symptoms are delayed up to 24 hours, it is termed *reactionary* hemorrhage. Where sepsis occurs within 7 to 10 days, it is called *secondary* hemorrhage.

hemorrhoidectomy Surgical removal of hemorrhoids.

hemostasis 1. The prevention of hemorrhage or the measures taken for its arrest. 2. Arresting a flow of blood, as by occluding an artery or vein with external pressure.

hemostat A clamp-like surgical instrument, used to stop blood flow through a vessel. Its opposing jaws apply pressure which collapses and occludes the vessel walls. A locking mechanism maintains the occlusal force.

hemostatic 1. The arresting of blood flow. 2. An agent used to arrest the flow of blood.

hemotherapy Treatment of disease, as by means of blood or blood fractions.

hemothorax A collection of blood in the thoracic cavity.

henry (h) Centimeter-gram-second elec-tromagnetic unit of inductance or mutual inductance. The inductance of a circuit is 1 henry when a current variation of 1 ampere per second induces 1 volt. Basic unit of inductance. In high-frequency electronics, smaller units are used, such as *millihenry,* which is one thousandth of a henry, and *microhenry,* which is one millionth of a henry.

heparin An acid which is produced by body tissues, principally, the liver and lungs. It is a complex anticoagulant, the principal constituent being a mucopolysaccharide comprising D-glucuronic acid and D-glucosamine, both of which are sulfated. Used with a serum protein cofactor, heparin prevents agglutination of platelets, thus acting to prevent thrombus formation. Heparin solutions are often used as postoperative anticoagulants to prevent thrombo-embolic difficulties. Also, in cardiac and vascular catheterization, to maintain potency of the catheter lumen.

heparinemia Presence of heparin in the blood as it is circulating within the body.

heparinized Said of a patient to whom heparin has been administered.

hepatic Relating to the liver.

hepatic coma A condition occurring in severe liver disease, or when blood from the gut bypasses the liver, so that detoxification does not take place. The coma results from ammonia poisoning of the brain cells. Ordinarily, ammonia formed by commensal organism of the intestine is reduced to harmless compounds by subsequent activity of the liver. When this does not occur, or when the liver is bypassed, hepatic coma is a constant danger.

hepaticostomy A surgical procedure to make a fistula in the hepatic duct.

hepatic portal system System of veins

which carry the blood from the intestine to the liver so that, with the exception of neutral fats, all the materials absorbed from the gut go straight to the liver.

hepatitis   Liver inflammation, caused by infection or toxic substances. It is characterized by jaundice (yellow coloration of skin and membranes, especially of the eye) and is usually accompanied by fever. Significant disturbances of enzyme production occur, adversely affecting blood clotting, protein utilization, and other vital functions.

hepatomegaly   Enlargement of the liver.

hepatosplenomegaly   Enlargement of the liver and the spleen.

hermaphroditic connector   A connector design which utilizes pin and socket contacts in a balanced arrangement such that both mating connectors are identical. The contacts may also be hermaphroditic and arranged as male and female contacts as for pins and sockets. Hermaphroditic contacts may also be used in a manner such that one-half of each contact mating surface protrudes beyond the connector interface and both mating connectors are identical.

hermetic   Permanently sealed by fusion, soldering, or other means, to prevent the transmission of air, moisture vapor, and all other gases. Also called *helium-tight, leak-tight* and *vacuum-tight*.

hermetic seal   An airtight seal allowing no detectable leakage.

herniotomy   A surgical procedure which divides the constricting band of a strangulated hernia and returns the protruding part to its usual location.

hertz (Hz)   International standard unit of frequency. Replaces, and is identical to, the older unit cycles per second.

heterodyne   When two wave motions of slightly different frequencies come together they combine to form a new frequency equal to their frequency difference. Thus, if two sound waves of 500 hertz and 510 hertz per second are sounded at the same time, a third frequency of 10 hertz per second is heard. This is the heterodyne or beat frequency. The heterodyne principle is also used in the reception of radio signals. In radio telegraphy, it operates when continuous wave signals are used. A local variable frequency oscillator circuit is provided at the receiver. This beat frequency oscillator (BFO) can be adjusted so that the frequency between it and the received radio frequency signals is an audible frequency, e.g., the frequency of the received signals is 400 kilohertz per second and the BFO frequency is adjusted to 399 to give a 1,000-hertz audible note in the receiver. In the superhet type of domestic radio receiver, the intermediate frequency (IF) which is generated by the mixing of the received signals and the local oscillator is usually 465 kHz, hence the name *supersonic* (above sound) *heterodyne* which is abbreviated to *superhet*. Because all received signals are changed to this frequency, it is possible to design economically a very high gain receiver using tuned-coupled stages with a minimum of controls which is also very selective.

heuristic   Empirical. Referring to knowledge or procedures determined by experience, but difficult to prove.

heuristic program   A set of computer instructions that simulate the behavior of human operators in approaching similar problems.

hiatal hernia   Hernia of the stomach through the diaphragm at the esophageal opening.

hiatus   An opening or space.

higher level language A programming language that is independent of the computer. It usually resembles natural languages and requires a compiler for translation into machine language. FORTRAN and ALGOL are examples.

high level In digital logic, the more positive of the two levels in a binary system. See also *low level logic, negative logic,* and *positive logic.*

high-pass filter A filter which passes, without appreciable attenuation or distortion, all frequencies above a specified cut-off frequency while attenuating all frequencies below the cutoff frequency.

high-resistance voltmeter Voltmeter having a resistance considerably higher than 1000 ohms per volt, so that it draws very little current from the circuit in which a measurement is made.

high-speed bus See *memory register.*

high-speed circuit breaker A device which starts to reduce the current in the main circuit in 0.01 second or less, after the occurrence of overcurrent or an excessive rate of current rise.

high-speed storage See *rapid storage.*

high threshold logic (HTL) High noise margin logic used primarily for industrial applications. Closely resembling DTL, the basic difference is that HTL uses a reverse-biased emitter junction as a threshold element, operating as a zener diode. Typical noise margin is 6 volts with a 15-volt supply.

hilum Site at which the pedicel of an organ is attached.

hindbrain See *brain.*

histamine Poisonous substance injected into that part of the body where there is damaged tissue or tissue exposed as a result of surgical operation. Its effect is to reduce blood pressure and in the latter case causes surgical shock. During recent years, antihistamine drugs have been prepared to reduce or eliminate this condition.

histogram 1. A description of one (or all) parameters, showing distribution, standard deviation, mean value failure limits, and sample-lot size for all samples within the lot. 2. A graphic presentation in which events occurring within sequential time units (i.e., minutes, hours, or days) are represented by a symbol or a stack of symbols which show the frequency of occurrence in that time unit. The resulting presentation shows the distribution of events over a time period and thus provides some trend information. 3. A graphical representation of a frequency distribution by a series of rectangles which have for one dimension a distance proportional to a definite range of frequencies and for the other dimension a distance proportional to the number of frequencies appearing within the range.

histology Branch of biology devoted to the study of tissues. Morbid histology is concerned with the study of diseased tissues and the changes which take place in them.

hold 1. To retain a display of physiologic waves (i.e., the ECG) on a cathode-ray tube, by storing in a shift register and periodically refreshing the display. 2. To maintain storage elements in charge storage tubes at equilibrium potentials by electron bombardment. 3. In computers, to retain information contained in one storage device after copying it into a second storage device. Opposite of clear.

hold mode In integrators or other charge-storage circuits, a condition or time interval in which input(s) are removed and the circuit is commanded (or expected) to maintain constant output.

hold-off voltage The maximum voltage an electronic flash tube will stand without self-flashing. Normal hold-off voltage is reduced at the end of lamp life, and in the presence of high temperatures or high voltage fields.

hole A mobile vacancy in the valance structure of a semiconductor crystal which acts like a positive charge and may be filled by an electron. When the vacancy is filled by an electron from an adjacent atom, the hole, in effect, moves. In P-type semiconducting material, current is carried primarily by holes.

hole conduction Conduction occurring in a semiconductor when electrons move into holes under the influence of an applied voltage, and thereby create new holes. The apparent movement of such holes is toward the more negative terminal, and is hence equivalent to a flow of positive charges in that direction.

hole current In a semiconductor, that current associated with apparent positively charged carriers designated as holes. Holes are created by the ejection of an electron from a covalent bond. When an electron from a neighboring bond transfers to fill the hole, a new hole is created, and the effect is the same as if a positive charge had been carried from the first hole position to the second. This process is repeated in random fashion. When an electric field is applied, the random drift of electrons and holes is transformed into a directed drift of the carriers in opposite directions. The directed drift of the holes constitutes the hole current.

hole-electron pair A positive (hole) and a negative (electron) charge carrier, considered together as an entity.

hollerith 1. A widely used system of encoding alphanumeric information onto cards; hence, the term hollerith cards is synonymous with punch cards. Such cards were first used in 1890 for the U.S. Census and were named after Herman Hollerith, their originator. 2. A particular type of code or punched card utilizing 12 rows per column and usually 80 columns per card.

hologram An interference pattern recorded on photographic film or similar media. This pattern is created by directing two beams of coherent light into the film. One, called the reference beam, strikes the film directly. The second, called the object beam, bounces off, or passes through the test specimen, then strikes the film. The interaction of these two beams make up the interference pattern called a hologram. To decode the swirls and dots of the pattern and create a visible image, a coherent light beam is directed onto the hologram. The image thus created has three-dimensional properties, unlike the two-dimensional character of a common photograph.

holography The recording of an object wave (usually optical) in such a way that an identical wave can subsequently be reconstructed. Whereas a conventional photograph records only the intensity of the light incident on it, a hologram records both the amplitude and phase. The additional phase information is contained in an interference pattern which is formed from the object wave and a reference wave.

homeostasis 1. That state of dynamic physiologic equilibrium that must be maintained in good health. 2. A control of the bodily processes through many feedback mechanisms, which leads to the stability of the internal environment of the organism. For example, in exercise, the level of oxygen in blood is sensed by the body and respiration is automatically increased, thereby maintaining the oxygen content in the blood at a near-normal level.

homeothermic Maintaining constant body temperature.

homogeneity The quality or state of being homogeneous.

homogeneous Of uniform structure or composition.

homosexuality Sexual attraction or relationship between members of the same sex.

hook transistor A transistor having four alternating P-type and N-type layers, with one layer floating between the base layer and the collector layer. This arrangement gives high emitter-input current gains. A PNPN transistor has a P-type floating layer, while a NPNP transistor has an N-type floating layer.

hormone The secretion of a ductless gland or other endocrine tissue.

horsepower (hp) A unit of mechanical power equivalent to 550 foot-pounds per second, or to 745.7 watts.

hot 1. Connected, electrically alive, energized. Pertains to terminal or any ungrounded conductor. 2. Not grounded. 3. Capable of imparting heat energy to surrounding media or objects in contact through radiation, conduction, convection, or a combination of these. 4. A strongly radioactive material. 5. Excited to a relatively high energy level.

hot cathode Cathode in which electron emission is produced by heat.

housekeeping In computers, that portion of a program that involves the setting up of constants and variables to be used in the program. The housekeeping must be done before any productive work is done by the computer.

HTL Abbreviation for high threshold logic.

hum Noise from the power line, either its actual frequency or harmonics of it, that intrudes into the reproduced sound and mars listening quality.

human engineering The science and art of designing machines for human use, taking into account the abilities, limitations, habits, and preferences of the human operator.

humidity transducer A layer of hygroscopic (moisture-absorbing) substance deposited between two metal electrodes. These electrodes establish electrical contact with the hygroscopic chemical, which serves as a resistance element. Since the chemical coating tends to absorb moisture from the surrounding air, its resistance decreases as humidity increases. In this manner, humidity variations are converted to resistance variations.

hum-loop A condition arising from the connection of two or more grounds to an amplifier system whereby circulating currents of low value at power line frequency and harmonics are added to the program signals, causing hum to occur in the background.

hum modulation Modulation of an RF signal or detected AF signal by hum. This type of hum is heard in a radio receiver only when a station is tuned in.

hunting 1. A control system's continuous, cyclical search for a desired or ideal value. Very rapid hunting is usually termed *oscillation,* slower cycles called *bird-dogging.* 2. Continuous attempt of an automatically controlled system to attain a stable condition.

HV Abbreviation for high voltage.

hybrid 1. An electronic circuit having both vacuum tubes and transistors. 2. A mixture of thin-film and discrete integrated circuits. 3. A computer having both analog and digital capability. 4. A blend or composite of two different technologies. The context defines the technology. 5. Transformer or combination of transformers or resistors affording paths to three branches, circuits A, B and C, so arranged that A can send to C, B can receive from C, but A and B are effectively isolated. 6. A hybrid junction.

hybrid circuit 1. A combination of passive and active subminiature devices on an insulating substrate to perform a complete circuit function. 2. A combination of one or more integrated circuits with one or more discrete components. 3. The combination of more than one type of integrated circuit into a single package. See *integrated circuit.*

hybrid computer A computer that combines both analog and digital equipment for purposes of solving problems that cannot be adequately or economically handled by either type of computer operating independently. The term hybrid computer does not denote the use of some analog equipment to preprocess data that is then converted to digital form and subsequently entered into a conventional digital computer. Rather, there is usually a continual flow of data in both directions between analog and digital equipment.

hybrid integrated circuit An arrangement consisting of one or more integrated circuits in combination with one or more discrete devices. Alternatively, the combination of more than one type of integrated circuit into a single integrated component.

hybrid junction A transformer, resistor, or device that has four pairs of terminals so arranged that a signal entering at one terminal pair will divide and emerge from the two adjacent terminal pairs, but will be unable to reach the opposite terminal pair. Also called *bridge hybrid* and *hybrid.*

hybrid microcircuit Microcircuit in which thin–film or diffusion techniques are combined with separately attached semiconductor chips to form the circuit.

hybrid thin-film circuit Microcircuit formed by attaching discrete components and semiconductor devices to networks of passive components and conductors that have been vacuum deposited on glazed ceramic, sapphire or glass substrates.

hybrid-type circuit See *multichip circuit.*

hydration The act of combining with water; the state of having adequate body fluids.

hydraulic Pertaining to the action of liquids.

hydrodynamics That branch of physics concerned with the movement of fluids. The term can be applied to both the design of submarines and the flow of blood through arteries, with the basic laws of hydrodynamics holding in both cases.

hydrometer An instrument for determining the specific gravity of liquids, especially of storage battery electrolyte. It consists of a weighted glass float having a graduated stem which sinks into the liquid to a point determined by the specific gravity of the liquid. The float is usually contained in a glass and rubber syringe which can be used to withdraw a sample of the liquid. See also *specific gravity.*

hygrostat A device for closing a pair of contacts when the air humidity reaches a predetermined level.

hyper A prefix denoting excessive, above, or increased.

hyperalgesia Excessive sensitivity to pain.

hyperbaric A pressure greater than atmospheric pressure.

hyperbaric chamber Pressurized enclosure used to deliver oxygen and nitrogen at greater than room pressure.

hypercalcemia An excessive amount of calcium in the blood.

hypercapnia Excess carbon dioxide in arterial blood resulting in extra stimulation of the respiratory center. $PaCO_2$ is above 45 mm Hg. Also called *hypercarbia*.

hypercarbia See *hypercapnia*.

hypercholesterolemia A decrease or deficiency in blood cholesterol content. Cholesterol (C_{27} H_{45} OH) is a solid monohydric alcohol; a consituent of all animal fats and oils. It is important in the metabolism and may be activated to form a vitamin D. Total serum cholesterol in normal blood ranges from 150 to 250 mg per 100 cc.

hyperemia Excessive blood in any part of the body.

hyperglycemia An increased concentration of glucose in the blood.

hyperkalemia An excessive amount of potassium in the blood.

hypermotility Increased motor activity.

hypernatremia An excessive amount of sodium in the blood.

hyperplasia An abnormal increase in the number of cells in a tissue or organ.

hyperpnea Abnormal increase in depth and rate of respiratory movements.

hyperpolarization An increase in the membrane potential.

hyperpolarizing current An applied electrical current in the inward direction through the cell membrane causing hyperpolarization.

hypertension An abnormally high arterial blood pressure; that is, above its classified normal limits of a systolic pressure of 140 and a diastolic pressure of 95. Resting blood pressure hypertension may be primary or essential, i.e., cause unknown or secondary to arterial obstruction, renal disease, endocrine disturbances, and other factors.

hyperthermia An abnormally high body temperature; fever.

hyperthermia apparatus Equipment for raising patient's body temperature by means of electric or circulated-water blankets.

hyperthymia An overactive state of mind with a tendency to perform impulsive actions.

hypertrophy An abnormal increase in the size of an organ or tissue as a result of an increase in the size of the cells, such as heart or liver.

hyperventilation Increased pulmonary ventilation beyond that needed to maintain the blood gases within normal range. Overbreathing.

hypnosis 1. A state of increased receptivity to suggestion and direction, initially induced by the influence of another person. The degree may vary from mild suggestibility to a trance state so profound as to be used in surgical operations. 2. Artificially induced sleep which is sometimes used by psychoanalysts and psychotherapists to treat patients suffering from disorders seated in the unconscious. Most normal people are suitable subjects for hypnosis, so much so, in fact, that it has been exploited by some for theatrical acts.

Although effective in curing such minor mental disorders as sleeplessness, stammering, claustrophobia, and the like, it can have damaging results on the mental harmony of a person if loosely and not medically controlled.

hypnotic barbiturates Sedatives and sleeping pills.

hypo Prefix meaning under, less, or below.

hypocalcemia A decrease in the amount of serum calcium.

hypocapnia Deficient carbon dioxide in arterial blood. $PaCO_2$ is below 35 mm Hg. Also called *hypocarbia*.

hypocarbia See *hypocapnia*.

hypochloremia A reduced concentration of chloride in the blood.

hypochondria A neurotic state of anxiety about personal health, in which the patient believes himself to be afflicted by any of countless illnesses. Though symptoms are usually imaginary, the suffering induced by the neurosis is quite real.

hypochondriasis Overconcern with the state of physical or emotional health, accompanied by various bodily complaints without demonstrable organic pathology.

hypochondrium Surface anatomy nomenclature relating to the region of the anterior abdominal wall beneath the ribs.

hypodermic Below the skin, as in the use of a hypodermic syringe to subcutaneously administer medication, or to inoculate a patient for preventive, curative, or experimental purposes.

hypodermoclysis The introduction of fluids into the subcutaneous tissues.

hypoglossal nerve The twelfth cranial nerve, arising in the medulla oblongata. It innervates the tongue musculature.

hypoglycemia A reduction in the normal amount of glucose in the blood.

hypokalemia A reduction in the normal amount of potassium in the blood.

hyponatremia A decrease in the normal amount of sodium in the blood.

hypostasis Deposit, passive congestion.

hypotension An abnormally low blood pressure.

hypothermia 1. An abnormally low body temperature. 2. A technique for reducing the normal metabolic requirements of the body's cells, especially brain cells, by decreasing temperature. In major heart surgery, this is achieved by cooling the blood through a heat exchange device, so that body temperature is decreased below 29°C. At this temperature, the metabolic rate is slowed at the cellular level so that minimal oxygen is required to sustain life processes. This makes it possible to stop circulation, if need be, so that surgical procedures can be carried out under asystalic conditions. The period during which circulation can be stopped is prolonged, but nevertheless, limited, by use of the hypothermia technique.

hypothermic blanket A blanket enclosing tubing through which cooling fluid is circulated, to carry away body heat from the patient and induce hypothermia.

hypothyroidism A morbid condition due to severe deficiency of thyroid hormone. In its advanced form, it is called *myxedema* or *cretinism*. In the less severe form, it is a condition associated with a basal metabolic rate about 20 percent below normal, and a milder degree of the symptoms associated with myxedema. Loss of function due to surgical removal of thyroid tissue, atrophy, or infection is called *primary hypothyroidism*. Reduction in function due to inadequate stimulation is called *secondary hypothyroidism*.

hypoventilation Respiratory minute volume inadequate for metabolic needs with resultant increase in $PaCO_2$.

hypovolemia Blood volume inadequate for capacity of the vascular bed. The problem may be a low volume or a large capacity.

hypoxemia Oxygen deficiency in arterial blood.

hypoxia A decrease in the oxygen content of blood and tissues.

hysteresis The dependence of a measurement upon the conditions to which a system has been recently exposed (its history). Examples: 1. The difference in response of a unit to an increasing signal and to a decreasing signal. 2. The amount of lag (due to molecular friction) between the actual magnetization of a ferrous material and the magnetizing force. 3. Oscillator behavior typified by the correspondence of multiple values of the output power and/or frequency to given values of an operating parameter. 4. A lag.

hysteresis curve A graph showing the amount of magnetism imparted to a magnetizable material as the result of a varying magnetic field. This coincides with the variations of the applied field only through a relatively narrow range between zero magnetism and saturation, but the addition of a bias field allows an audio signal to be recorded on a magnetic tape within this linear range for minimum distortion.

hysteresis loop 1. A curve (usually with rectangular coordinates) which shows, for a magnetic material in a cyclically magnetized condition, for each value of the magnetizing force, two values of the magnetic induction; one when the magnetizing force is increasing, the other when it is decreasing. 2. For any magnetic material, a graphical plot of magnetizing current versus magnetic flux density will appear as a loop, with one value of flux when the current is increasing and a second value when the current is decreasing. The area within the loop is proportional to the power loss due to hysteresis.

hysteresis loss The power loss in a magnetic core, such as in a transformer energized by an alternating current, which is due to hysteresis.

hysterical personality A personality type characterized by shifting emotional feelings, susceptibility to suggestion, impulsive behavior, attention seeking, immaturity, and self-absorption; not necessarily disabling.

hysterosalpingography X-ray examination of uterus and fallopian tubes after the injection of radiopaque material.

Hz Abbreviation for Hertz, meaning cycles per second (of any periodic phenomenon).

-ia Suffix meaning condition or state of.

-iasis Suffix meaning a process.

iatro Prefix meaning a physician.

iatrogenic 1. An abnormal state or condition which is induced by the actions, instructions, medications, or manner of a physician. 2. A disorder resulting from treatment.

-ible Suffix meaning capable of.

-ic Suffix meaning pertaining to or connected with.

IC An integrated circuit. A very small electronic device containing a silicon chip on which hundreds to thousands of electronic components are fabricated. This is a monolithic IC, as opposed to a hybrid IC. Sometimes the chip itself is referred to as an IC but, strictly speaking, an IC is the packaged chip with leads from the chip brought out through the package.

ichor A serous discharge from a wound or an ulcer. It is usually an acrid, thin fluid.

-ician Suffix meaning one who practices.

iconoscope A camera tube in which a beam of high-velocity electrons scans a photoemissive mosaic that is capable of storing an electrical charge pattern.

-icus Suffix meaning pertaining to or connected with.

idiopathic An occurrence without apparent cause.

IEEE Institute of Electrical and Electronics Engineers, Inc., a professional organization of scientists and engineers whose purpose is the advancement of electrical engineering, electronics, and allied branches of engineering and science. (The IEEE resulted from the merger of the IRE and AIEE.)

ignore A punched tape code which indicates that no action should be taken.

illumination Light flux which is incident on a unit projected area. The photometric counterpart of irradiance, expressed in foot-candles.

I.M. Abbreviation for intramuscular.

image converter 1. An electron tube that

creates a visual replica of an electromagnetic image projected onto its cathode. Electrons ejected from the photosensitive cathode by the incident electromagnetic radiation are accelerated to and focused upon a fluorescent phosphor screen, thus forming the visual replica. Image converters can be used in the infrared, ultraviolet, and X-ray range, as well as in the visible. An example of an infrared-sensitive image converter is the snooperscope. 2. Solid-state optoelectronic device capable of changing the spectral characteristics of a radiant image. Examples of such changes are infrared-to-visible and X-ray-to-visible.

image-enhancing equipment An elaborate device, often involving a computer, in which a photograph is scanned by a point of light, the amplitude of the electrical signal being modified electronically before being rerecorded on another film.

image iconoscope Iconoscope in which greater sensitivity is obtained by separating the function of charge storage from that of photoelectric emission. An optical image is projected on a continuous photosensitive screen and the electron emission from the back of this screen is focused electromagnetically onto a mosaic screen that is scanned by an electron beam as in the original emitron cathode-ray tube. British term is *super-emitron*.

image intensifier 1. Device used in X-ray to obtain a brighter fluoroscopic image (up to several hundred times) and thus to reduce radiation to which the patient is exposed. 2. An electronic tube equipped with a light-sensitive electron emitter at one end and a phosphor screen at the other end; an electron lens inside the tube relays the image. These devices are used in astronomy for photographing very faint celestial objects. 3. A system for increasing sensor response to a radiation pattern

or image by interposing active elements between the sensor and the image, and supplying power to the active element. This is normally done by focusing the scene to be imaged on the photocathode of the tube, giving rise to a photoelectron pattern corresponding to the optical image. This pattern is accelerated and focused onto a phosphor which emits light to reproduce a visual image of the scene.

image transfer constant See *transfer constant*.

immediate access 1. In computers, the ability to obtain data from or place data in a storage device without serial delay due to other units of data and usually in a relatively short period of time. 2. Ability of a computer to put data in (or remove it from) storage without delay.

immediate access store A computer store whose access time is negligible in comparison with other operating times.

immune globulins A protein derived from human blood that, when injected into persons exposed to certain infectious diseases, confers temporary immunity.

immunoelectrophoresis Analysis of antigens using a combination of agargel electrophoresis and an antigen-antibody reaction in the gel, to obtain separation of immunologically different proteins. See also *electrophoresis*.

immunofluorescence A method of demonstrating the presence of antigens and antibodies by use of a fluorescent dye technique and microscopic examination. In the technique, a fluorescein-labeled antibody is used to identify a bacterial, viral, or other antigenic material specific for the labeled antibody. The binding of the antibody is determinable by the production of a characteristic visible light when the preparation is exposed to ultraviolet light.

impaction The condition of being firmly lodged or wedged.

impedance Total opposition offered to the flow of an alternating current at a particular frequency. It may consist of any combination of resistance (R), inductive reactance, and capacitive reactance (X) all expressed in ohms. Impedance $(Z) = (R^2 + X^2)^{1/2}$. It is also expressed as the vector sum of resistance and reactance $(R + j X)$, or as a vector of magnitude Z at an angle theta $(Z \Theta)$.

impedance compensator Electric network designed to be associated with another network or a line with the purpose of giving the impedance of the combination a desired characteristic with frequency over a desired frequency range.

impedance coupling The coupling that exists between two circuits through a mutual impedance.

impedance ground An earth connection made through an impedance of predetermined value usually chosen to limit the power short-circuit ground current.

impedance match The condition existing between two circuits when electrical waves passing between them do not suffer any reflection loss. This condition is satisfied when the impedances of the two circuits are alike, both in magnitude and angle, including sign.

impedance matching Making the load impedance equal to the source impedance in order to obtain maximum power transfer. Implies that the impedances must be equal in magnitude, but not necessarily equal or opposite in sign. See also *conjugate impedances.*

impedance phlebograph Apparatus utilizing an electrical circuit to measure variations in venous volume and to detect deep vein thrombosis.

impedance plethysmograph An instrument for detecting the increased volume of blood in the body tissues during each contraction of the heart. See electrical-impedance cephalography and finger plethysmograph.

impedance pneumograph An instrument for measurement of respiration by detecting the change in trans-thoracic impedance accompanying each inspiration and expiration.

implantable pacemaker A miniaturized pulse generator surgically implanted beneath the skin. Its output leads are connected directly to the heart muscle. The electrodes may contact either the outside or inside wall of the heart. Electrodes that contact the outer wall of the myocardium (heart muscle) are known as *myocardial electrodes;* electrodes that contact the inner surface of the heart chamber are known as *endocardial electrodes.*

implant devices Plastic or metal, sometimes radioactive, prosthetic inserts.

implied AND See *wired AND.*

implied OR See *wired OR.*

impregnated Said of a porous material which has all voids filled with an insulating oil, varnish, or wax.

impulse A sequence of electrical and chemical events in a neuron, serving as a unit signal in the nervous system.

impulse excitation Method of producing oscillatory current in an electric circuit in which the duration of the impressed voltage is relatively short compared with the duration of the current produced. Also called *shock excitation.*

impulse timer A timing device electrically powered by a synchronous motor, featuring a mechanical stepping device which enables it to advance a predetermined number of degrees within a predetermined time interval controlling a multiple number of circuits. Said circuits being controlled by individual cams which program their activity.

impurity An element such as boron, phosphorus, or arsenic which is introduced into a semiconductor in minute quantity to change its resistivity and convert it from one type of material to another, such as converting N-type to P-type. (In N-type material, current conduction is by negative charges; and in P-type material, conduction is by positive charges.)

incandescence The generation of light caused by passing an electric current through a wire filament. The resistance of the filament to the current causes the filament to heat and emit radiant energy, some of which is in the visible range.

inclusive OR In computers, the logical operator which has the property that A or B is true if A is true, or both A and B are true.

incomplete antibody Antibody which will not by itself form precipitate *in vitro* with corresponding antigen. See also *antibody*.

incontinence Inability to restrain a natural discharge, as of urine or feces.

incus A small anvil-shaped bone of the middle ear.

indifferent electrode An electrode in the unipolar ECG lead systems devised by Wilson. It is placed at the mid-thorax and joined through 5,000-ohm resistors to right arm, left arm, and left leg electrodes. In this way, the algebraic sum of all cardioelectric potentials appearing at this common electrode connection is made to be practically zero. An exploring electrode is placed at various sites on the body and the ECG potential difference between it and the indifferent electrode is recorded. The unipolar precordial leads V1 through V6 and the extremity leads VR, VL, and VF are the most common leads using the indifferent electrode. Also called *Wilson center terminal.*

indirect address An address, in a computer instruction, that indicates a location where the address of the referenced operand is to be found. In some computers, the indicated machine address can itself be indirect. Such multiple levels of addressing are terminated either by prior control or by a termination symbol synonymous with second-level address.

indirect addressing A method of cross reference in a computer when one memory location indicates where the correct address of the main fact can be found.

indirectly heated cathode In an electron tube, a cathode which is a coated tube which encloses a heating element. Prevents noise and bias potential from the heater supply voltage.

indirectly heated thermistor A thermistor which incorporates, as part of its composite structure, an electrical heater. A thermistor whose body temperature in use is significantly higher than the temperature of its surrounding medium, as a result of current passing through its heater.

indirect material Semiconductor material in which electrons do not drop directly from the conduction band to the valence band, but rather drop in steps due to the trapping levels in the forbidden gap.

induced charge An electrostatic charge produced on a conducting body when it is brought near to or connected to another body which bears an electric charge.

induced current Current which flows through a conductor when it is moved through a magnetic field at right angles to the direction of the field; or when it is subjected to a magnetic field of varying intensity.

induced voltage The voltage produced in a coil of wire either (a) when the coil moves through a magnetic field or (b) when the magnetic flux through the coil is varied.

inductance When electric current flows through a coil it generates a corresponding magnetic field. Changes in current through the coil result in changes in the magnetic field. The changes in the field induce a voltage which opposes the change in current. Thus, a coil induces a voltage which reacts against changes in current through the coil. This effect is called *inductive reactance* or *inductance*.

induction 1. When an electrically charged body is brought near another body it produces an electric charge in the neighboring body. A similar effect can be obtained by bringing a magnet close to a piece of metal which has magnetic poles induced in it. In both cases, the phenomenon is known as induction. 2. The process by which a change in current in one circuit causes a corresponding change in an adjacent circuit due to magnetic coupling, or by which a voltage on one conductor causes the opposite voltage to appear on another conductor with which there is electrostatic coupling. See also *self–induction*.

induction coil A device for changing direct current into high-voltage alternating current. Its primary coil contains relatively few turns of heavy wire, and its secondary coil, wound over the primary, contains many turns of fine wire. Interruption of the direct current in the primary by a vibrating-contact arrangement induces a high voltage in the secondary.

inductive coupling 1. The coupling that

exists between two circuits through a mutual inductance, such as that in a transformer. 2. Coupling between two circuits through an inductance that is common to the two circuits. Direct inductive coupling.

inductive interference Effect arising from the characteristics and inductive relations of electric supply and communication systems of such character and magnitude as would prevent the communication circuit from rendering service satisfactorily and economically if methods of inductive co-ordination were not applied.

inductive reactance Reactance which is due to inductance (possibly from a coil of wire) in a circuit. In a circuit having net inductive reactance, the current wave lags behind the voltage wave.

inductive transducer A transducer which receives stimulus information by means of changes in inductance.

inductor 1. A coil inserted in a circuit to supply inductance. Large values of inductance are used to smooth ripples in direct current. Smaller values are used to pass dc and low frequencies, while preventing the passage of high frequencies. 2. A choke. 3. A passive fluidic element which, because of fluid inertness, has a pressure drop that leads flow by essentially 90 degrees. See *coil*.

indwelling catheter A thin tube communicating to the body surface, inserted into the vascular system and positioned to permit pressure measurements or blood sampling over a relatively long time.

inevitable hemorrhage Bleeding due to placenta praevia.

infarct 1. An area of necrosis in a tissue or organ resulting from obstruction of circulation by a thrombus or embolus. 2. The death of tissue resulting from the blockage of the nourishing arterial blood flow to the affected part. Typically, a loss of part of the functional myocardial tissue, resulting from atherogenesis and ultimate occlusion of a coronary artery.

infarction   See *myocardial infarct*.

infection   The invasion of the body by disease-producing microorganisms.

inferior   Located or directed below.

inferior vena cava   A major vein emptying into the right atrium of the heart. It receives and combines the venous blood flows from several lesser veins serving the tissues and organs of the lower trunk and the legs.

infinite resolution   Capable of stepless adjustment. This phase is used as a synonym for cursive, continuous, or analog. (Note that this term really has nothing to do with resolution.)

inflammation   A series of reactions in tissues produced by microorganisms or other irritants, marked by redness of the affected area. There is an influx of erythrocytes with exudation of plasma and leukocytes.

information   That property of a signal or message whereby it conveys something meaningful and unpredictable to the recipient. Usually measured in bits.

information bits   In telecommunications, those bits which are generated by the data source and which are not used (for error control) by the data transmission system.

information retrieval   The recovery of data from a collection for the purpose of obtaining information. Retrieval includes all the procedures used to identify, search, find, and remove specific information or data stored. It includes both the creation and the use of the data.

information retrieval system   A system for locating and selecting on demand certain documents or other graphic records relevant to a given information requirement from a file of such material.

information theory   The analysis of information content in signals or the effort to discover the most efficient system for transmitting signals over a channel that has only limited capacity (bandwidth) and a finite amount of noise. Information theory has been used in studying neurological data processing.

infra   Prefix meaning below, beneath, or less than.

infrared   That portion of the electromagnetic frequency spectrum just below the visible light portion (which extends from red to violet). The portion just above the visible light spectrum is ultraviolet. Both infrared and ultraviolet rays are invisible to the human eye. Infrared is generated by thermal agitation; therefore, the higher the temperature, the greater the radiation. Masers and lasers operate in the infrared portion of the spectrum as well as in the visible portion.

infrared detector   A transducer which is sensitive to invisible infrared radiation (wavelength between 0.75 and 1000 microns) usually using a semiconductor a thermocouple bolometer or pneumatic (pressure) device to detect the radiation.

infrared light   Light in which the rays lie just below the red end of the visible spectrum, extending in wavelength from about 0.75 to 1000 microns.

infrared optics   Lenses, prisms, and other optical elements for use with infrared radiation (radiation with wavelength between 0.75 and 1000 microns).

infrared radiation (IR)   Radiation in the wavelength range between 7,500 angstroms (red) and about 10,000,000 angstroms (microwaves); invisible to the human eye.

infrared sources Emitters of radiation with wavelength between 0.75 and 1000 microns.

infrasonic Having a frequency below the audible range (less than 20 Hz). Frequencies above the audible range are called ultrasonic or supersonic.

infundibulum 1. An orifice or funnel-shaped passage. 2. That portion of the brain formed by the growth of its floor, especially that portion forming the pituitary gland.

inguinal That anatomical portion of the body within the groin.

inguinal canal That portion of the anatomy which in the male is occupied by the spermatic cord (or ductus deferens), and in females by the round ligament. In both, it is a short, narrow passage between the abdominal ring and the inguinal ring, within the groin.

inhalation The inspiration of air or other gases into the lungs.

inhalation therapy That branch of therapeutic service devoted to the restoration of pulmonary function.

inhibit 1. To hold in check. 2. To prevent from operating. 3. A signal applied to a logic gate to prevent passage of another signal to a following stage.

inhibit-gate Gate circuit whose output is energized only when certain signals are present and other signals are not present at the inputs.

inhibiting input An input, as to a gate circuit, which prevents operation of a circuit.

inhibition 1. Interference with or restriction of activities; the result of an uncon-

scious defense against forbidden instinctual drives. 2. Literally, restraint. The term is used to imply the prevention of some activity; thus psychological inhibition, enzymatic inhibition, nerve inhibition, and in describing controlling signals in logic circuitry.

inhibition gate See *inhibitor*.

inhibitor 1. In a digital computer, a logic circuit that clamps a specified output to the zero level when energized. Also called inhibition gate. 2. A chemical agent which blocks a reaction or prevents a neurological response.

inhibitor gate A circuit that prevents information or data pulses from passing when the circuit is triggered by an external source.

inhibit pulse A current pulse applied to an inhibit winding on a magnetic memory core which prevents the core from being affected by a write pulse.

inhomogeneities See *inhomogeneity*.

inhomogeneity Something which is not homogeneous.

initialize To set counters, switches, and addresses to zero or other starting values at the beginning of, or at prescribed points in, a computer routine.

initializing In computers and logic equipment, setting flip-flops or memory elements to known states prior to commencing operation.

injection laser A solid-state semiconductor device with at least one PN junction capable of emitting coherent or stimulated radiation under specified conditions. Incorporates a resonant optical cavity. See also *laser*.

innervation The supply of nerves or the conveyance of nervous impulses to or from a part.

innominate artery The large artery which arises from the arch of the aorta and divides into the right common carotid and right subclavian arteries.

inorganic Mineral as opposed to living material or its products.

input 1. Information that is received by a device. For example, information read into a computer is its input, a physiologic parameter received by a patient monitoring system is also an input, and visual information is the input to the eye. 2. The signal which is fed into a circuit. 3. The terminals which receive the input signal. 4. The power which energizes an electrical device. 5. Data to be processed. 6. State or sequence of states occurring on a specified input channel. 7. Channel for impressing a state on a device or logic element. 3. Device or collective set of devices used for bringing data into another device. 9. Process of transferring data from an external storage to an internal storage of a computer. 10. Signals fed into data processing equipment in the appropriate form for storage, processing or control.

input block 1. A section of internal storage of a computer reserved for the receiving and processing of input information. 2. An input buffer. 3. A block of computer words considered as a unit and intended or destined to be transferred from an external source of storage medium to the internal storage of the computer.

input channel A channel for impressing a state on a device or logic element.

input device The device or collective set of devices used for conveying data into another device.

input impedance The impedance which is seen looking into the input terminals of a line or electrical device when the input signal source has been disconnected.

input offset current The difference in the currents into the two input terminals of an operational amplifier when the output is at zero.

input offset voltage The voltage that must be applied between the two input terminals of an operational amplifier to obtain zero output voltage.

input/output (I/O) 1. The process of transmitting information from an external source to the computer or from the computer to an external source. 2. General term for the equipment used to communicate with a computer and the data involved in the computer.

input/output (I/O) devices These are devices which enable people to communicate with the computer. They include teletypes, computer controlled typewriters, line printers, video terminals, analog-to-digital converters, digital-to analog converters, and others.

input register The register of internal computer storage able to accept information from outside the computer at one speed, and supply the information to the computer calculating unit at another, usually much greater, speed.

input transformer A transformer used to transform the impedance of a signal source to the impedance of the device of which it is a part.

input voltage offset The dc potential difference between the two inputs of a differential amplifier when the potential between the output terminals is zero.

insertion gain Gain resulting from the insertion of a transducer in a transmission system is the ratio of the power delivered to that part of the system following the transducer to the power delivered to that same part before insertion. If more than one component is involved in the input or output, the particular component used must be specified. This ratio is usually expressed in decibels.

insertion loss   The loss at a particular frequency caused by the insertion of apparatus or a network into a communication system. It is the ratio, expressed in dB, of the power at that frequency delivered beyond the point of insertion before and after the insertion.

insight   Self-understanding. A major goal of psychotherapy. The extent of the individual's understanding of the origin, nature, and mechanisms of his attitudes and behavior.

in situ   Located in place; also, in the normal place.

inspiration   The act of taking air into the lungs.

inspiratory capacity   Commencing from the resting expiratory level, the maximum volume of gas which can be drawn into the lungs.

inspiratory reserve volume   The maximum volume of gas which can be inspired, commencing from the end-inspiratory level.

inspissated   Thickened by evaporation.

instillation   The dropping of a liquid into a cavity such as the ear.

instruction   In computers, information which tells a machine where to obtain the operands, what operations to perform, what to do with the result, and sometimes, where to obtain the next instruction. See also *instruction code*.

instruction code   An artificial language for expressing or describing the instructions which can be carried out by the computer. Usually, a list of symbols, names,

and definitions of the instructions, which are understandable by a given computing system.

instruction counter   See *program-address counter*.

instruction register   Register that temporarily stores the instruction currently being used in a computer. As such, it serves as a scratch-pad memory, retaining data only while needed in an operation, then clearing for the insertion of new data to be temporarily held.

instrumentation   The introduction and use of devices of any kind into a process for the purpose of augmenting the observer's sensory, analytic, or memory capacity, or to obtain more precise data (improving quality), or for securing information not otherwise discernible. Typically accomplished by generating, transmitting, displaying, controlling, correlating, and recording signals.

instrument shunt   Particular type of resistor connected in parallel with the measuring device to extend the current range beyond some particular value for which the instrument is already complete.

instrument transformer   A highly accurate low-power transformer having negligible leakage flux which can be used to transform high voltages or high currents to low ranges so they may be measured on low-voltage or low-current meters. See *voltage transformer* and *current transformer*.

insufflator   A type of instrument which is employed to blow air, gas, vapor, or a powder into a body cavity.

insulated   A condition occurring when the substance between two conductors, circuits, or devices cannot conduct current at operating voltage and beyond (to the limit of breakdown voltage). The substance may be air as in the case of a busbar, liquid (such as oil) as in a power transformer, or a solid as in the substrate of a printed circuit board. See *isolated*.

insulated-substrate monolithic circuit Integrated circuit which may be either an all-diffused device or a compatible structure, so constructed that the components within the silicon substrate are insulated from one another by a layer of silicon dioxide, instead of reverse-biased PN junctions used for isolation in other techniques.

insulating strength Measure of the ability of an insulating material to withstand electric stress without breakdown. It is defined as the voltage per unit thickness necessary to initiate a disruptive discharge. Usually measured in volts per centimeter. See *dielectric strength* and *electric strength*.

insulation 1. Nonconducting material used to prevent the leakage of electricity from a conductor and to provide mechanical spacing or support to protect against accidental contact. 2. Use of material in which current flow is negligible to surround or separate a conductor to prevent loss of current.

insulation resistance The resistance offered by an insulating material to the flow of current resulting from an impressed voltage. It includes both the volume and surface resistances. Industrial specifications usually call for a certain minimum value (usually in the hundreds of megohms), determined with a specific applied voltage.

insulator 1. Material of such low conductivity that the flow of current through it can usually be neglected. 2. Device having high electric resistance, used for supporting or separating conductors so as to prevent undesired flow of current from the conductors to other objects. 3. A material in which

the outer electrons are tightly bound to the atom and no electrons are free to move. Thus no current can flow when a voltage is applied. Resistivity is above 10^8 ohm-cm and generally decreases with temperature rise.

integrated circuit 1. A small chip of solid material (generally a semiconductor) upon which, by various techniques, an array of active and/or passive components have been fabricated and interconnected to form a functioning circuit. Integrated circuits, which are generally encapsulated with only input, output, power supply, and control terminals accessible. Offer great advantages in terms of small size, economy, and reliability. 2. The physical realization of a number of electrical circuit elements inseparably associated on or within a continuous body of semiconductor material to perform the function of a circuit. 3. An electronic device containing several elements active or passive, which perform all or part of a circuit function. 4. An interconnected array of conventional components such as transistors, diodes, capacitors, and resistors fabricated in situ within and on a single crystal of semiconductor material with the capability of performing a complete electronic circuit function. Also called functional device or hybrid circuit. 5. An electronic circuit containing transistors, diodes, resistors, and perhaps capacitors along with interconnecting electrical conductors processed and contained entirely within a single chip of silicon.

integrated electronic circuits See *integrated circuit*.

integrated electronics That portion of electronic art and technology in which the interdependence of material, device, circuit, and system-design considerations is especially significant; more specifically, that portion of the art dealing with integrated circuits.

integrated electronic systems See *integrated circuit.*

integrated microcircuits See *integrated circuit.*

integrator A device whose output is proportional to the integral with respect to the input variable.

intelligence 1. The intellectual capacity of a person. 2. The degree of sensor/processing circuitry provided in a control system. 3. The information product resulting from the collection, evaluation, analysis, integration, and interpretation of all available data. 4. Data, information, or messages that are to be transmitted.

intensive care unit Specialized hospital ward where acute illnesses are treated by a dedicated staff who constantly monitor the patient's condition and institute corrective action, as necessary.

intensifier electrode Electrode provided in some types of electrostatic cathode-ray tubes which permits additional acceleration of the electron beam after it has been deflected. This electrode permits greater intensity of the trace without materially reducing the deflection sensitivity of the tube. Also called *postaccelerating electrode.*

intensity 1. Relative brightness of a light source, or the light produced by fluorescence of a phosphor (as in a CRT) by the impact of a stimulating electron beam. 2. In radiology, the amount of energy per unit time passing through a unit area perpendicular to the line of propagation at the point in question. Often this term is used incorrectly in the sense of dose rate.

intensity modulation The process and/or effect of varying the electron beam current

in a cathode-ray tube resulting in varying brightness or luminance of the trace.

intensity of radiation The radiant energy emitted in a specified direction per unit time, per unit area of surface, per unit solid angle.

inter A Latin prefix meaning *between* that is used with many medical terms.

intercellular message The stimulation of body cells ultrasonically sometimes known as micromessage.

intercostal Between the ribs.

intercostal space The spaces between the ribs, designated numerically from topmost, down. These spaces are palpable and used for diverse reference purposes, including determining the correct placement of monitoring ECG electrodes, and for situating the crystal probe of an echocardiograph.

interelectrode capacitance The capacitance existing between the electrodes in an electron tube.

interface 1. The junction or point of interconnection between two systems or equipments having different characteristics. They may differ with respect to voltage, frequency, speed of operation, type of signal, and/or type of information coding. 2. To interconnect two different systems with care to resolve their incompatibilities. 3. The hardware through which a computer communicates with a peripheral device (mass storage or input-output). 4. A common boundary between two or more items. May be mechanical, electrical, physiological, functional, or contractual. 5. A common aspect at the boundary between two systems involving intersystem communication, e.g., the interaction between research and development, basic and applied science, or engineering and systems development. 6. The physical and space boundary surrounding the system, subsystem, equipment, or component, through which all environmental and operational stimuli essential to the device or affecting its proper operation must propagate or interact with other related devices or structures. 7. The hardware for linking two units of electronic equipment; for example, a hardware component to link a computer with its input (or output) device. 8. The means of connection between two logic elements, often elements that belong to two different families. 9. A point, device, or tissue structural boundary where a transition is made between media, power levels, modes of operation, etc. 10. The two surfaces on the contact side of mating connectors or plug-in component (e.g., relay) and receptacles, which face each other when mated.

interface circuit   Circuit used to link one type of logic family with another, or with analog circuitry. In effect, it translates the logic voltage swing of one into the logic swing of the other.

interface equipment   Equipment used between two other equipments which would otherwise be incompatible. It converts the terminal voltage, power level, impedance, or type of signal of one equipment to match those of the second equipment. Examples are transformers, amplifiers, pads, and converters.

interface unit   A device which translates incoming signals that are incompatible with the electrical characteristics of the computer without changing the information content. Also translates outgoing signals for the benefit of associated equipment that is designed to different electrical standards.

interference   Any electrical or electromag-

netic disturbance, phenomenon, signal, or emission, man-made or natural, which causes or can cause undesired response, artifact, malfunctioning, or degradation of electrical performance of electrical and electronic equipment.

interlace   1. In a computer, to assign successive storage location numbers to physically separated storage locations in order to reduce access time. 2. In video, the overlaying of two successive fields of lines to create a frame of pictorial information.

interleave   In a computer, to insert segments of one program into another program in such a way that the two programs can be in essence executed simultaneously (e.g., a technique used in multiprogramming).

intermediate storage   The portion of the computer storage facilities that usually stores information in the processing stage.

intermittent fevers   Those fevers in which there are regular pauses between the attacks.

intermodulation   The production in a nonlinear circuit element, of frequencies corresponding to the sums and differences of frequencies which are transmitted through the circuit element. Usually, this action is undesired and creates a peculiar form of distortion.

intermodulation distortion   See *distortion*.

internal arithmetic   Any computations performed by the arithmetic unit of a computer, as distinguished from those performed by the peripheral equipment.

internally stored program   A sequence of instructions (program) stored inside a computer in the same storage facilities as the computer data, as opposed to being stored externally on punched paper tape, pin there are regular pauses between the boards, etc.

internal memory   See *internal storage.*

internal pneumocardiograph   A pneumocardiograph that detects pressure changes in the respiratory passages.

internal resistance   The resistance of a voltage source, such as a generator, which acts to reduce the terminal voltage of the source as current is drawn.

internal storage   1. The storage of data on a device which is an integral part of a computer. 2. The storage facilities forming an integral physical part of the computer and directly controlled by the computer. In such facilities all data are automatically accessible to the computer, for example, magnetic core, magnetic tape on-line. Also called *internal memory.* See also *external storage.*

interphase   The interval occurring in a cell cycle between two successive cell divisions.

interpreter   1. A punch-card machine which will read the information conveyed by holes punched in a card, and print its translation in characters arranged in specified rows and columns on the card. 2. A computer program for translating a higher level language one line at a time and executing that line as soon as it is translated. Basic is usually translated by an interpreter. An interpreter is especially convenient for short tasks where immediate feedback to the operator is desirable.

interpreter routine   An executive computer routine, which as the computation progresses, translates a stored program expressed in some machine-like pseudocode into machine code and performs the indicated operations by means of subroutines, as they are translated.

interpretive programming   The writing of computer programs in a *pretend* machine language, which the computer precisely converts into actual machine language instructions before performing them.

interrupt   1. In a computer, a break in the normal flow of a system or routine such that the flow can be resumed from that point at a later time. An interrupt is usually caused by a signal from an external source. 2. Cessation of power, communication, or signal flow, resulting from switching or a failure.

interstitial   Between the cells of tissue.

interstitial edema   Accumulation of fluid in the interstitial spaces of the lung.

interstitial spaces   Those voids between parts, i.e., in connective tissue.

intra   Prefix meaning within.

intracardiac   An adjective meaning inside the heart which is used to describe instruments whose pickup element is inserted through a vessel directly into the heart chambers. The term is used in connection with measurements such as: intracardiac electrocardiography, intracardiac phonocardiography, and intracardiac oximetry.

intracardiac ECG   1. The cardioelectric potentials associated with the contraction of the heart musculature, as measured from a point within the heart. 2. The activating impulses flowing through the ECG (between P and QRS wave inscriptions), usually measured by means of a bipolar or multipolar electrode catheter within the heart.

intracardiac oximetry   A method for measurement of oxygen content of the blood within the heart, in which the sensory agent is a catheter.

intracardiac phonocardiogram   A method

for detecting, analyzing, and recording the sounds associated with heart action by means of a catheter-placed sensitive microphone and suitable external equipment.

intracardiac phonocardiography A variation of phonocardiography in which a minute microphone is inserted directly into the heart. The technique reveals abnormalities not detectable by surface auscultation.

intracellular Being or occurring within a protoplasmic cell.

intracorporeal See *heart*.

intradermal Within the dermis.

intramuscular Within the muscle tissue.

intraocular fluids The fluid contents of the eye (the aqueous humor and the vitreous humor). The aqueous humor is constantly formed by the ciliary body and is drained through the canal of Schlemm. If this circulation is impaired the intraocular fluid pressure may rise. See *glaucoma*. The normal intraocular pressure is about 25 mm Hg.

intraosseous Within the bone.

intraspinal See *intrathecal*.

intrathecal Located within the spinal canal. Also called *intraspinal*.

intravascular An adjective meaning, *inside the vessels*. Intravascular catheterization is carried out to detect and measure flows, pressures, and other characteristics not ascertainable from the body surface.

intravenous 1. Within a vein. 2. Pertaining to the lumen of a vein. Abbreviated I.V.

intravenous infusion The therapeutic introduction of a fluid into a vein.

intraventricular system The structure which separates the right and left ventricles of the heart.

intrinsic 1. Inherent. 2. Peculiar to a part.

intrinsically safe Equipment and wiring incapable of releasing sufficient electrical energy under normal or abnormal conditions to cause ignition of a specific hazardous atmospheric mixture. Abnormal conditions include accidental damage to any part of the equipment or wiring insulation, or other failure of electrical components application of overvoltage, adjustment, and maintenance operations and other similar conditions.

intrinsic characteristics Characteristics of a material which are due to the material itself and do not depend on or are caused by impurities.

intrinsic material Semiconductor material that has an equal number of holes and electrons; i.e., no impurities.

intrinsic semiconductor A semiconductor which is a pure crystal without donor impurities.

introitus Opening into a viscus. Usually refers to the external opening of the vagina.

intubation Insertion of a tube into a passage or organ, especially tracheal intubation.

intussusception Condition in which part of the intestine is drawn into a more distal part.

invasive techniques Methods of measurement (i.e., catheterization) which require that the patient's body be entered surgically for purposes of making the measurement.

inverse feedback See *negative feedback*.

inverse peak voltage 1. Peak value of the instantaneous voltage across a rectifier tube during the half of the cycle that is not conducting. 2. Maximum voltage between cathode and plate of a vacuum tube when it is not conducting.

inverse voltage   The effective value of the alternating voltage that exists across a device which conducts current in only one direction during the half-cycle when current is not flowing.

inversion   1. The changing of direct current to alternating current. 2. Changing a positive pulse to a negative pulse, and vice versa. 3. Changing the phase of an alternating current by 180 electrical degrees. 4. A turning inward.

inverter   1. An electronic or electromechanical device which changes dc to ac (or vice versa). 2. A circuit used to reverse the polarity of an electrical signal or pulse. Often used in computer logic circuits. 3. A device that accepts an input that is a function of maximum voltage and changes it into an output that is a function of both maximum voltage and time. 4. A circuit with one input and one output and its function is to invert the input. When the input is high, the output is low, and vice versa. The inverter is sometimes called a NOT circuit, since it produces the reverse of the input.

inverter circuit   See *NOT circuit.*

in vitro   Literally in glass, i.e., in the test-tube as opposed to in life.

in vivo   In the living body.

-ion   Suffix meaning act or state of.

ion   An electrically charged atom or group of atoms. A postively charged ion is an atom or group of atoms with a deficiency of electrons; a negatively charged ion is an atom or group of atoms with an added electron.

ion-exchange electrolyte cell   Fuel cell which operates on hydrogen and oxygen in the air, similar to the standard hydrogen-oxygen fuel cell with the exception that the liquid electrolyte is replaced by an ion-exchange membrane. Operation is at atmospheric pressure and room temperature.

ion implantation   A method for doping semiconductors where impurities which are ionized and accelerated to high velocity penetrate the semiconductor surface and deposit in the interior.

ionization   The process of giving net charge to a neutral atom or molecule by adding or subtracting an electron. Can be accomplished by radiation, or by creation of a strong electric field.

ionization effect   Effects of ionizing radiation which leads to bond rupture, establishing free radicals, coloration, luminiscence, etc. in solids. In metals it appears as heat and temperature rise.

ionization gauge   A type of radiation detector that depends on the ionization produced in a gas by the passage of a charged particle through it. One of the best known is the Geiger-Mueller counter, although cloud chambers and spark chambers can also be included in this category.

ionization potential   Potential at which ionization begins within a gas-filled tube. This potential is slightly lower than the firing or striking potential at which complete ionization takes place.

ionizing radiation   Photons or particles possessing sufficient energy to produce ions in substances through which they pass. High-energy radiation of this type is capable of nuclear interaction.

ionization time   Of a gas tube, the time interval between the initiation of conditions for and the establishment of conduction at some stated value of tube voltage drop.

ionization transducer A transducer in which the displacement of the force-summing member is sensed by means of induced charges in differential ion conductivity.

ionize To make an atom or molecule of an element lose an electron, as by X-ray bombardment, and thus be converted into a positive ion. The freed electron is a negative ion, or it may attach itself to a neutral atom or molecule to form a negative ion.

ionizing radiation 1. Particles or photons of sufficient energy to produce ionization in their passage through air. 2. Particles that are capable of nuclear interactions with the release of sufficient energy to produce ionization in air. 3. Any electromagnetic or particle radiation capable of producing ions, directly or indirectly, in its passage through matter.

ion migration Movement of ions produced in an electrolyte, semiconductor, etc., by the application of an electric potential between electrodes.

iontophoresis The introduction, by means of an electric current of ions, of soluable salts into the tissues of the body for therapeutic purposes. Certain pharmaceutical agents that ionize easily can be placed on the skin to which an electrode is applied. In many instances, effective levels of the agent, such as an antibiotic, can be found as deep as 1 cm within the body tissues.

iontophoresis apparatus Device for using electric current to analyze a solution as its ions separate in an electric field, also to produce ions of soluble salts in tissues of body.

I²R Power in watts expressed in terms of the current I and resistance R.

I²R loss 1. The power which is lost in electrical conductors due to their resistance. The lost power in watts is equal to the square of the current in amperes times the resistance in ohms. 2. Power loss in transformers, generators, connecting wires, and other parts of a circuit due to current flow I through the resistance R of the conductors. Also called *copper loss*.

IR drop The voltage drop that exists across a resistance of R ohms when a current of I amperes is flowing through it.

iridectomy The formation of an artificial pupil for the eye through the surgical removal of a portion of the iris.

iridocele A rupture of the cornea through which a portion of the iris protrudes.

iris The pigmented membrane behind the cornea of the eye.

iron loss Power loss occurring in iron cores of electric machines, coils, transformers, etc., due to hysteresis and eddy currents.

irradiation The exposure of tissue or substance to the effects of alpha, beta, gamma rays, neutrons, or other atomic particles, or to electromagnetic waves, including infrared, ultraviolet, or visible light.

irritability The ability of an organism to respond to an outside influence (stimulus). A stimulus can take many forms, such as a change in temperature, an electric current, or intrusion of a foreign object or harmful bacteria. The response to a stimulus can also take many forms, such as contraction, expansion, discharge, movement, or chemical and physical changes.

irritant An agent applied to the skin to produce a reaction.

ischemia 1. The effect of constricting the flow of oxygenated blood to tissue. It is seen in cases of frostbite and angina pectoris, and in Raynaud's disease. The cause of the diminution of blood supply may be vasoconstriction, due to vascular spasm, or local obstruction, due to an embolus or atherosclerotic deposit. 2. Diminished supply of blood to a part.

isodose curves The description of doses of radiation by means of diagrams depicting contours through body tissue that receives equal doses. This problem is often quite complex, especially if two or more radiation sources are rotated about the patient, and computer solutions are required.

isoelectric That which has a uniform electrical charge throughout. Since no potential difference exists between any two points on an isoelectric body, no current flow is possible.

isolation 1. The condition which exists between two circuits, which prevents their interacting with each other. Isolation can be achieved in several nonequivalent ways: (a) by increasing the loss between two circuits, (b) by inserting a transformer or capacitors to prevent the flow of longitudinal current, (c) by applying decoupling to joint branches of the circuits, or (d) by inserting a one-way amplifier between two circuits, expressed in decibels.

isolation amplifier Amplifier employed to minimize the effects of a following circuit on the preceding circuit.

isolation transformer A transformer used to separate two sections of a circuit, often to prevent longitudinal currents by separating a grounded section from an ungrounded section.

isomers Substances having the same chemical composition but differing in structure and properties, owing to a difference in the relative positions of atoms within their molecules.

isometric Of the same dimensions. As applied to muscle tissue, a muscle which applies force without changing length is said to be acting isometrically.

isoniazid The most important drug in the treatment and prophylaxis of tuberculosis.

isostatic Under equal pressure from every side.

isothermal See *isothermic.*

isothermic Having the same temperature.

isotone One of several different nuclides having the same number of neutrons in their nuclei.

isotonic 1. A uniformity in tension, such as occurs in a muscle in a state of full contraction. 2. A solution having an osmotic pressure equivalent to normal physiological saline (0.9 percent sodium chloride); hence, one which has the same fluid phase as a cell or tissue. 3. Having the same tone.

istonic solutions These have the same osmotic pressure as normal or physiological saline; 0.9 percent NaCl.

isotope One of several nuclides having the same number of protons in their nuclei, and hence having the same atomic number but differing in the number of neutrons and therefore in the mass number. Almost identical chemical properties exist between isotopes of a particular element. Elements with the same atomic number but with different atomic masses. See *radioisotope.*

isotropic The property of having the same appearance, dimensions, or characteristics, regardless which axis is observed or measured.

isotropic material A material having the same magnetic characteristics along any axis.

isotropy See *isotropic.*

-ist Suffix meaning one who practices.

isthmus The neck or constricted part of an organ.

iterations per second In computers, the number of approximations per second in iterative division; the number of times an operational cycle can be repeated in one second.

iterative Recurring an infinite number of times. Said of a network with an infinite number of identical sections, or of the impedance looking into such a network.

iterative array In a computer, an array of a large number of interconnected identical processing modules, used with appropriate drive and control circuits to permit a large number of simultaneous parallel operations.

iterative division In computers, a method of dividing by use of the operations of addition, subtraction, and multiplication. A quotient of specified precision is obtained by a series of successively better approximations.

iterative impedance The impedance which will terminate the output of a network in such a way that the impedance then measured at the input of the network will be equal to the (iterative) terminating impedance.

-itis Suffix meaning inflamed, e.g., dermatitis, inflammation of the skin.

I.V. Abbreviation for intravenous.

-ive Suffix meaning having the power to.

J

jack Connecting device to which a wire or wires of a circuit may be attached and which is arranged for the insertion of a plug.

jaundice A yellow discoloration of the skin, mucous membrane, and deeper tissues due to bilirubin, a bile pigment.

jejunum That portion of the small intestine which lies between the duodenum and the ileum.

J–FET See *junction field–effect transistor.*

J–K flip-flop A flip-flop that has two conditioning inputs (designated J and K) and one clock input. If both conditioning inputs are disabled prior to a clock pulse, the flip-flop will remain in its present condition when a clock pulse occurs. If the J input is enabled (HI) and the K input is disabled (LO) the flip-flop will go to the 1 condition on a clock pulse. If the K input is enabled, and the J input is disabled, the flip-flop will go to the 0 condition on a clock pulse. If both the J and K inputs are enabled prior to a clock pulse, the flip-flop will complement or go to the opposite state on a clock pulse.

JK inputs See *synchronous inputs.*

Johnson counter A counter comprised of an N-state shift register with the complement of the last stage returned to the input. It normally has 2N states through which it cycles. It has the distinguishing characteristic that only one stage changes state at each count. Also called *mobius counter* or *twisted ring counter.*

joule Unit of work and named after James Prescott Joule (1818–89) who investigated the relationship between heat and mechanical work; he also determined the mechanical equivalent of heat. The unit is the measure of the work done in 1 second by a current of 1 ampere flowing through a resistance of 1 ohm; it also equals 10,000,000 ergs, or 0.738 foot-pounds.

joule effect 1. Heating effect produced by the flow of current through a resistance. 2. Change in the dimensions of a ferromagnetic object when placed in a magnetic field. Also called *magnetostriction.*

Joule's law Gives the heat generated when a current passes through a conductor. Heat in calories = I^2Rt where I = current, R = resistance, and t = time.

jugular Relating to the neck.

juice 1. Electric current (slang). 2. To feed electric current to a circuit.

jump A departure from the normal sequence of executing instructions, synonymous with transfer.

junction A region of transition between P- and N-type semiconductor material. The controllable resultant asymmetrical properties are exploited in semiconductor devices. There are diffused, alloy, grown, and electrochemical junctions.

junction diode 1. A two-terminal single junction semiconductor device which exhibits different conduction characteristics depending on the polarity of applied voltage. A convential PN diode is said to be forward biased when the P region is positive with respect to N region. A for-

ward biased diode has a low impedance. Reversing the polarity results in a reverse bias diode and a high diode impedance. 2. A semiconductor diode in which rectification occurs at a junction between N-type and P-type materials, rather than at a point contact.

junction field-effect transistor (J–FET) 1. An FET whose gate element is a region of semiconductor material (ordinarily, the substrate) insulated by a PN junction from the channel, which is material of opposite polarity. All junction-FETs are depletion-type (normally turned on). 2. A transistor that consists of a gate region diffused into a channel region. When a control voltage is applied to the gate, the channel between the source and drain is depleted or enhanced by enlargement of the PN junction. Current cannot flow when the channel is pinched off.

junction transistor A transistor without point contacts having three layers of semiconductor material, either NPN or PNP.

K

K Stands for kilowords or kilobytes. These terms are usually used in describing the size of a computer or the amount of memory required to hold a particular body of information. By convention, K stands for 1024 rather than 1000. Thus, a computer with 16K words of memory actually contains 16,384 words.

k Abbreviation for kilo (10³).

katharometer Electrical apparatus for determining basal metabolic rates; instrument for determining gas mixture composition by variations in thermal conductivity.

kc Abbreviation for kilocycle; an obsolete term, now replaced by kilohertz (kHz).

keloid Overgrowth of connective tissue arising in scars.

Kelvin (celsius) temperature scale A scale of temperatures which uses Centigrade degrees, but dropped so that absolute zero is 0 degrees Kelvin. Water freezes at 273 degrees K and boils at 373 degrees K.

keratectomy Surgical removal of part of the cornea.

keV or kev Abbreviation for kiloelectron-volt.

keyboard entry 1. An element of information inserted manually, usually via a set of switches or marked punch levers, called keys, into an automatic data processing system. 2. A medium, as in 1, for achieving access to or entrance into an automatic data processing system.

keypunch A machine controlled by a typewriter-like keyboard which enables an operator to punch data in Hollerith code into data cards.

kHz Abbreviation for kilohertz, 1000 hertz.

kick-sorter See *pulse-height analyzer*.

kidney Either of two organs, situated in the posterior portion of the upper abdominal viscera, behind the peritoneum, involved in the elimination of water and soluble waste products from the body. Each kidney is about 12 cm in length, weights about five ounces, and is bean shaped. Both function as components of the urinary system. Each kidney contains at least a million filtering units called nephrons, and each nephron is surrounded by a capillary vessel network meshed with tubules, called the glomeruli. Blood enters this network from the renal artery serving each kidney, and the cells within the tubules selectively extract water and soluble wastes from the blood. These are converted to urine, which flows to the bladder from each kidney by way of a connecting tube called the ureter.

kidney failure The inability of the kidney(s) to sustain a level of waste removal which is consistent with a functional state of health. The causes of kidney failure may be renal (within the kidney) or extrarenal (outside the kidneys). Acute inflammation of the nephrons (nephritis) and several other diseases of kidney tissues cause renal failures. A severe drop in blood pressure, as in shock, hemorrhage, or dehydration, reduces blood flow to the kidneys, bringing on extrarenal failure.

kilo A prefix meaning one thousand.

kiloampere (kA or ka) One thousand amperes.

kilobit One thousand bits.

kilocycle (kc) One thousand cycles. Generally interpreted a meaning one thousand cycles per second. Obsolete term, replaced by kilohertz (kHz).

kiloelectron-volt (keV or kev) One thousand electron-volts. The energy acquired by an electron that has been accelerated through a voltage difference of 1,000 volts.

kilogram (kg) 1. A unit of weight equal to 1000 grams or approximately 2.2 pounds. 2. Unit of mass. The mass of a particular cylinder of platinum-iridium alloy, called the international prototype kilogram, which is preserved in a vault at Sevres, France, by the International Bureau of Weights and Measures.

kilohertz (kHz) 1. One thousand hertz. 2. One thousand cycles per second.

kilohm (kΩ) One thousand ohms.

kilomega (kM) Obsolete prefix for giga (G), representing 10^9, or 1,000,000,000.

kilosecond One thousand seconds.

kilovolt One thousand volts.

kilovoltage A voltage of the order of a thousand volts.

kilovolt-ampere (kva) One thousand amperes.

kilowatt (kw) One thousand watts.

kilowatt hour (kwhr) A measure of the energy supplied by 1,000 watts in a period of one hour.

kinaesthesis The sense of muscular movement.

kinematics The study of motion.

kinescope Cathode-ray tube used for image presentation.

kinesiology The science of motion.

kinesis Movement. Usually applied to cells migrating in response to a stimulus.

kinesthesiometer Device for making quantitative measurements of kinesthesia (the means of sensing how well a person perceives his own weight, limb position, etc.).

kinetocardiography See *vibrocardiography*.

Kirchoff's Laws 1. Sum of the current flowing to a given point in a circuit is equal to the sum of the current flowing away from that point. 2. Algebraic sum of the voltage drops in any closed path in a circuit is equal to the algebraic sum of the electromotive forces in that path.

kit A prepared package of parts, with instructions, for assembling and/or wiring a component or chassis. Also, a small accessory item, such as a tape-cleaning kit.

Korotkoff sounds Detectable sounds produced by the pulsation of arterial blood through a partially occluded artery, as heard through a stethoscope applied over the artery, during blood pressure measurement with a sphygmomanometer.

KUB Abbreviation for kidneys, ureters, and bladder.

kussmaul breathing Deep, rapid respiration of metabolic acidosis.

kV or kv Abbreviation for kilovolt.

kW or kw Abbreviation for kilowatt.

labium A lip or lip-shaped structure, such as the labium minus of the female reproductive system.

laceration A wound having torn edges, of irregular shape, as compared to a wound consisting of a clean cut.

lacrimal Pertaining to the tears or tear-producing apparatus.

lag 1. An interval of time by which a particular phase of one waveform follows the corresponding phase of another, or the vectorial notation for this interval in electrical degrees. 2. A synonym for delay—the time lapse between cause and effect (stimulus and response).

lagging current In an alternating current circuit, a current wave which lags in phase behind the voltage wave which produced it. The current lags in a circuit in which the net reactance is inductive.

lambert A unit of luminance (photometric brightness) equal to $1/\pi$ candela per square centimeter and, therefore, is equal to the uniform luminance of a perfectly diffusing surface emitting or reflecting light at the rate of one lumen per square centimeter. The lambert also is the average luminance of any surface emitting or reflecting light at the rate of one lumen per square centimeter. For the general case, the average must take count of variation of luminance with angle of observation. Also, of its variation from point to point on the surface considered.

Lambert's law of absorption Each layer of equal thickness absorbs an equal fraction of the light which traverses it.

Lambert's law of illumination The illumination of a surface on which the light falls normally from a point source is inversely proportional to the square of the distance of the surface from the source. If the normal to the surface makes an angle with the direction of the rays, the illumination is proportional to the cosine of that angle.

lamella A thin sheet of tissue; e.g., bone. Plural: lamellae.

lamina A thin layer of tissue or fluid.

lamination Any of the multiple stamped

pieces of sheet iron which are interleaved to make a magnetic core. They are individually insulated with varnish to reduce eddy current losses.

land The conductive area of a printed circuit board to which components or separate circuits are attached, usually surrounding a hole through the conductive pattern and the base material. Also called *boss, pad, terminal point, blivet, tab, spot,* or *donut.*

language 1. The association of symbols or the techniques used to represent coded information. 2. System, in electronic computers, consisting of: (a) A well defined, usually finite, set of characters; (b) Rules for combining characters with one another to form words or other expressions; and (c) A specific assignment of meaning to some of the words or expressions, usually for communicating information or data among a group of people, machines, etc. 3. System similar to the above but without any specific assignment of meanings. Such systems may be distinguished from 2, by referring to them as *formal* or *uninterpreted* languages. Although it is sometimes convenient to study a language independently of any meanings, in all practical cases at least one set of meanings is eventually assigned.

language converter A data processing device designed to change one form of data, i.e., microfilm, strip chart, into another (punch card, paper tape).

LAO Abbreviation for left anterior oblique.

laparoscope Instrumentation for the visual examination of the peritoneal cavity. Typically, an endoscope, used in combination with a sheath, obturator, biopsy forceps, sphygmomanometer bulb and tubing, scissors, and syringe.

laparotomy Surgical opening of the abdominal cavity, usually as an investigative procedure.

LAPO Abbreviation for left anteroposterior oblique.

large-scale integration (LSI) 1. The accumulation of a large number of circuits on a single chip of semiconductor material. 2. The simultaneous realization of large-area circuit chips and optimum component packaging density for the express purpose of reducing costs by maximizing the number of system interconnections made at the chip level. 3. Usually refers to monolithic digital ICs with a complexity of typically 100 or more gates or gate-equivalent circuits (e.g., MOS ROMs). Each manufacturer has his own number of gates per chip that he uses to define LSI; LSI sometimes describes hybrid ICs built with a number of MSI or LSI chips.

laryngectomy Surgical removal of the larynx.

laryngoscope A light source and optical system combined as an instrument for visual examination of the larynx. The instrument is adaptable to use in diagnosis, therapy, and surgery as well.

larynx The upper trachea, containing a complex of cartilages, including the vocal cords. Vibration of these structures by controlled airflow is the source of speech in man.

LASCR Light-activated silicon-controlled rectifier; a PNPN device with incident light taking the place of gate current; three of the four semiconductor regions are available for circuit connections; a photoswitch. See also *LASCS.*

LASCS Light-activated silicon-controlled switch. A four-terminal semiconductor device that can be triggered by light positive signals (at the gate terminal) and negative signals (at the anode gate terminal). See also *LASCR*.

laser Acronym formed from the words Light Amplification by Stimulated Emission of Radiation. The laser generates a beam of light which is nearly parallel and coherent. Coherent light is of the same wavelength or color as opposed to ordinary light, which is spherical and diffuses rapidly. Laser beams is that they can be modulated—that is, made to carry data such as communications signals. This quality is the basis for their use in the communications and computer industries.

There are three basic types of lasers: *optically pumped, gas,* and *injection.* Optically pumped lasers rely on a light source to raise the energy level in the lasing material to the point at which it emits light. The first laser, a ruby crystal laser, was optically pumped. Other materials used for this type are glass, liquids, gas, and plastic.

Gas lasers consist of a mixture of helium and neon and are operated by either a radio frequency or a direct current.

Injection lasers consist of a semiconductor diode made of gallium arsenide or gallium arsenidephosphide. The device emits light when an extremely high current is passed between the terminals of the semiconductor diode.

The three types of lasers have separate applications in most cases. Optically pumped devices are used in industry and medicine. Gas types, which emit a continuous wave, are used frequently in scientific work and have some applications in communications.

Biomedical applications of lasers include: tumor therapy, microsurgery, cauterization, diagnosis of deep pathological lesions, and retinal welding.

laser cavity An optically resonant, and hence, mode-selecting low-loss structure in which laser action occurs through the buildup of electromagnetic field intensity upon multiple reflection.

laser diode A PN junction semiconductor electron device which converts direct forward bias electrical input (pump power) directly into coherent optical output power via a process of stimulated emission in region near the junction. Also called *diode, injected laser, coherent electroluminescence device,* and *semiconductor laser.*

laser welding A method of welding in which material heating is accomplished by concentration of a beam of coherent light on the area until fusion of the materials takes place.

lasing In a laser, the generation of visible or infrared light waves that have very nearly a single frequency, by pumping or exciting electrons into high-energy states.

lat Abbreviation for lateral.

latch 1. Usually a feedback loop in a symmetrical digital circuit (such as a flip-flop) for retaining a state. 2. A simple logic storage element. The most basic form is two cross-coupled logic gates which store a pulse presented to one logic input until the other input is pulsed, thus storing the complementary information in the latch.

latch-up 1. A condition in which the collector voltage in a given circuit does not return to the supply voltage when a transistor is switched from saturation to cutoff. Instead, the collector finds a stable operating point in the avalanche region of the collector characteristics. 2. An unintended stable circuit mode which will not revert to a previous intended circuit mode after removal of a stimulus, such as a spurious signal or radiation. The effect is usually caused by parasitic circuit elements.

latency 1. In a serial storage computer

system, the time necessary for the desired storage location to appear under the drum heads. 2. In computers, the time required to establish communication with a specific storage location, not including transfer time, i.e., access time less word time. 3. A state of seeming inactivity, such as that occurring between the instant of stimulation and the beginning of response.

latent heat Quantity of heat required to change the state of a unit mass of matter.

latent homosexuality A condition characterized by unconscious homosexual desires.

lateral 1. On the side. 2. Pertaining to the side. 3. In the direction of the side. 4. To either side of the medial vertical plane.

lateral lemniscus Cartilage between femur and tibia on the outside of the knee.

lateral position Lying on the side. A position, favored by some obstetricians, which may be adopted by the mother during delivery.

lattice 1. A three-dimensional pattern of atoms which is repeated throughout a single crystal. 2. The term lattice is used to denote a regular array of points in space, as in the case of the sites of atoms in a crystal. Lattice designs form a class of experimental designs enabling a large number of unrelated treatments to be compared in randomized blocks of reasonable size. In lattice dynamics, the theory of solid state deals with the properties of the thermal vibrations of crystal lattices.

lavage Therapeutic washing out of an organ such as the stomach.

law of electromagnetic indication See *Faraday's Law*.

law of electromagnetic systems 1. Any two circuits carrying currents tend to dispose themselves in such a way that the flux of magnetic induction linking the two

will be a maximum. 2. Every electromagnetic system tends to change its configuration so that the flux of a magnetic induction will be a maximum.

LCD Abbreviation for liquid crystal display. A seven-segment (typically) display device consisting basically of a nematic liquid hermetically sealed between two thin-film metallized glass plates. The metallization pattern defines the segments of the display. One type of LCD (dynamic scattering) depends upon ambient light for its operation, while a second type depends upon a backlighting source. The readout is either dark characters on a dull white background or white on a dull black background. LCDs have very low power requirements.

LC ratio The ratio of inductance to capacitance, equal to the inductance in henrys divided by the capacitance in farads.

lead 1. An electrical conductor or wire used to connect a test point to an electronic device or instrument; a wire terminated by an electrode. 2. A particular connection scheme for the input leads in an electrocardiogram, e.g., voltage between right arm and left arm constitutes lead I. Number of leads taken during an electrocardiogram may vary between 8 and 12, normally designated I, II, III, AVR, AVL, AVE, V, CR, CL, and CE. 3. The particular oscillographic wave resulting from one of the connections above.

lead-acid storage cell A storage cell in which both plates are lead-antimony or lead-calcium grids filled with sponge lead for the negative plate and lead peroxide for the positive plate. On discharge, the material in both plates is converted to lead sulfide. The electrolyte is a solution of sulfuric acid having a specific gravity of 1.200. The cell voltage is nominally 2 volts, rising to 2.15 volts on float and dropping to 1.85 volts on discharge.

leading current   In an alternating current circuit, a current wave which precedes in phase the voltage wave which produces it. The current leads in a circuit in which the net reactance is capacitive.

leading edge   The first part of a pulse or electromagnetic wave, during which the current or voltage rises rapidly from zero to a relatively steady-state value.

leak   A condition of low insulation or high capacitance which permits a current to leak from its conductor.

leakage   1. Current which drains from a conductor through shunt resistance. 2. The shunt resistance (conductance) through which a leakage current drains. 3. Magnetic flux which bypasses a path on which it can do useful work. 4. Current resulting from capacitance between conductors in an ac system.

leakage rate   A laboratory procedure used to determine the amount and duration of resistance of an article to a specific set of destructive forces or conditions.

leakage reactance   That portion of the reactance of a transformer primary which is due only to leakage flux.

leakage resistance   The resistance of a path over which a leakage current flows. Normally of a high value, measured in megohms.

leaks   Condition in which current is shunted away from its destination, due to low resistance or high capacitance.

leak-tight   See *hermetic*.

leapfrog test   Computer check routine using a program that performs a series of arithmetical or logical operations on one section of memory locations, transfers it-self to another location, checks correctness of transfer, and repeats the series of operations. Eventually all storage positions will have been checked.

learning   The ability of a system to increase its effectiveness when environment and machine are stable.

learning machines   Systems that change their performance to produce an output that is a desired, but not completely specified, function of their input.

least significant bit (LSB)   The lowest weighted digit of a binary number.

least significant digit   In a binary number, that digit having the lowest place value; the right-most digit of the number.

LeClanche cell   A primary cell having a carbon positive electrode and a zinc negative electrode in an electrolyte of sal ammoniac and depolarizer. A common dry cell.

LED   Abbreviation for light-emitting diode. A P-N junction semiconductor device specifically designed to emit light when forward-biased. This light can be one of several colors—red, amber, yellow, or green —or it may be infrared and thus invisible. Electrically, an LED is similar to a conventional diode in that it has a relatively low forward voltage threshold. Once this threshold is exceeded, the junction has a low impedance and conducts current readily. This current must be limited by an external circuit, usually a resistor. The amount of light emitted by an LED is proportional to the forward current over a broad range, thus it is easily controlled, either linearly or by pulsing. The LED is extremely fast in its light output response after the application of forward current. Typically, the rise and fall times are measured in nanoseconds.

left-hand rule 1. For motors and generators, stretch the thumb and first finger of the left hand at right angles to each other in the same plane and the second finger at a 90° angle perpendicular to the plane of the thumb and first finger. For a conductor in a generator armature, when the thumb indicates the direction of motion the first finger indicates the direction of magnetic lines of force, and the second finger indicates the direction of electron flow. For a motor, the right hand is used. 2. For a current-carrying wire, if the fingers of the left hand are closed around the wire so that the thumb points in the direction of electron flow, the fingers will be pointing in the direction of the magnetic field.

left heart bypass The flow of blood from the pulmonary veins to the aorta, avoiding the left atrium and the left ventricle.

Lenz's law 1. The current induced in a circuit due to a change in the magnetic flux through it or to its motion in a magnetic field is so directed as to oppose the change in flux or to exert a mechanical force opposing the motion. 2. If a constant current flows in a primary circuit A, and if by motion of A or the secondary circuit B, a current is induced in B, the direction of induced current will be such that, by its electromagnetic action on A, it tends to oppose the relative motion of the circuits.

-lepsy Suffix meaning seizure.

lesion 1. Any pathological or traumatic destruction of an area or part of the body. 2. Any change in function or structure of tissue due to disease or injury.

lethargy Inaction.

letters patent See *patent*.

leukemia A disease of blood-forming organs, characterized by increase of white corpuscles of the blood.

leukocyte A white (actually, colorless) ameboid cell of the blood having a nucleus and either granular or nongranular cytoplasm. Some are phagocytes, acting to engulf and destroy invading pathogens within the blood stream. See also *blood corpuscles*.

leukocytolysis The destruction of leukocytes, usually as an effort by the body to restore normal white cell count after an infection has been brought under control.

leukocytosis A sharp increase in the number of white blood cells (leukocytes) found in a blood sample, beyond the number which is considered normal. Typically, a healthy adult will show some 7,000 leukocytes per cubic millimeter of blood. In cases of severe infection, this count may rise to greater than 35,000 leukocytes per cubic millimeter.

leukoerythroblastic anaemia Descriptive term applied to the appearance of nucleated red cells and primitive white cells in the circulation. It is due to neoplastic infiltration of bone marrow.

leukopenia Reduction in the number of white cells in the blood to a value below the usual norm (approximately 7,000 leukocytes per cubic millimeter in normal adults).

leukopoiesis Formation of white blood cells in the bone marrow and in other places within the body.

leukosis Too many white blood cells.

leukotomy Transection of nerve fibers passing to and from a lobe of the brain. Usually, an operation to relieve mental disorder in which the prefrontal lobes are surgically isolated from the rest of the brain. Also called *lobotomy*.

levator A muscle which raises up a part.

levator ani A muscle of the pelvis which

plays an important part in keeping pelvic viscera in position.

levator palpebrae superioris Muscle which raises the upper eyelid.

level 1. The number of bits per character in a code, such as an eight-level ASCII code, or a twelve-level Hollerith code. 2. The difference between a quantity, such as voltage, power, or sound volume, and a specified reference quantity. This difference is usually specified in decibels. 3. An expression of the relative signal strength at a point in a communication circuit, compared to a standard level such as zero dBm or to the level at a reference point on the same circuit such as the zero transmission level point. 4. The measurable quantity of a substance within another substance. 5. The quality of being parallel to the horizontal plane, or perpendicular to the force of gravity.

level translator A circuit which accepts digital input signals having one pair of voltage levels and whose output signals are of a different pair of voltage levels.

lever A rigid bar which revolves around a fixed point.

Leyden jar Early form of capacitor used in many of the original experiments in electrostatics. It consists of a glass jar with an interior and exterior coating of tinfoil, the former being connected to a metal knob fixed on a rod passing through the insulating stopper. The tinfoil forms the plates of the capacitor and the glass jar is the dielectric.

library 1. In computer programming, an ordered collection of proven routines and subroutines by which problems and parts of problems may be solved. These instructions are usually written and stored in some form of relative addressing. 2. A group of standard, proven routines which may be incorporated into larger routines in a computer.

LIC Abbreviation for linear integrated circuit.

life test Any test conducted to evaluate the longevity of an article or change in characteristics as a function of time of operation.

ligament 1. A band of fibrous tissue at the joints which acts to join together two or more bones. 2. A tough band of tissues connecting the extremeties of bones.

ligation The application of a ligature, or thread, to connect or close tissues or vessels.

ligatures Threads of silk, wire, catgut, fascia, nylon, etc., used to tie arteries, stitch tissue, etc.

light As defined by the I.E.S. (Illuminating Engineering Society), visually evaluated radiant energy; visible spectrum is about 3800 to 7800 angstroms.

light-activated silicon-controlled rectifier (LASCR) Silicon controlled rectifier with the ability to be triggered by current to the gate terminal, or by a light flashed on a light-sensitive part of the semiconductor pellet.

light-activated silicon-controlled switch (LASCS) A semiconductor device that combines the LASCR and the PSPS. Having four terminals, the LASCS can be triggered by light positive signals (at the gate terminal) and negative signals (at the anode gate terminal).

light activated switch Semiconductor diode which is triggered into conduction by light-irradiation of a light sensitive part of the semiconductor pellet.

light-beam cathode-ray tube recorder Recorder using an electron beam to make multiple traces on CRT screen. Traces are reflected from a fixed plane mirror onto moving photosensitive paper via an optical system.

light-beam galvanometer A modified form of the D'Arsonval meter movement used for projecting a beam of light into a photographic strip-chart recorder. A small mirror is cemented to a moving coil mounted in the field of a permanent magnet. As the signal current flows through the coil, a magnetic field is established and the coil is deflected angularly. The mirror reflects a beam of light onto a moving strip of photographic paper. When developed, the paper chart shows the waveform of the signal current through the coil.

light beam oscillograph Recorder using a mirror on a galvanometer to achieve recording response to 5 kHz.

light-emitting diode A PN junction that emits light when biased in the forward direction. See *LED*.

light flux See *luminous flux*.

lightning rod A pointed metal rod carried above the highest point of a pole or building, and connected to earth by a heavy copper conductor. Intended to carry a direct lightning discharge directly to earth without damage to the protected structure.

light pen 1. A light-sensitive device used with a computer-operated CRT display for selecting a portion of the display for action by the computer. 2. A photosensor placed in the end of a pen-like probe. It is used in conjunction with a CRT (cathode-ray tube) display for drawing, erasing, or locating characters. Operation is by comparison of the time it senses a light pulse to the scanning time of the display. 3. A hand–held data-entry device used only with refresh displays. It consists of an optical lens and photocell, with associated circuitry, mounted in a wand.

light pipe 1. A bundle of transparent fibers which can transmit light around corners with small losses. Each fiber transmits a portion of the image through its length, reflection being caused by the lower refractive index of the surrounding material, usually air. Used in laparoscopes, endoscopes, and cystoscopes. 2. Transparent matter that usually is drawn into a cylindrical or conical shape through which light is channeled from one end to the other by total internal reflections. Optical fibers are examples of light pipes.

light relay See *photoelectric relay*.

limbus Literally a border; applied to the junction between the sclera and cornea of the eye.

limiter 1. Device in which some characteristic of the output is automatically prevented from exceeding a predetermined value. 2. Transducer in which the output amplitude is substantially linear with regard to the input up to a predetermined value and substantially constant thereafter.

linear 1. A ratio in which a change in one of two related quantities is accompanied by an exactly proportional change in the other. 2. Having an output that varies in direct proportion to the input.

linear amplifier An amplifier that operates on the linear portion of its forward transfer characteristic so that its output signal is always an amplified replica of the input signal.

linear circuit 1. A circuit whose output is an accurate, amplified replication, or a predetermined variation, of its input signal. 2. A circuit whose output voltage is approximately directly proportional to the input voltage. This relationship generally holds only over a limited signal voltage range and often for a limited frequency range.

linear control Rheostat or potentiometer having uniform distribution of graduated resistance along the entire length of its resistance element.

linear device See *linear IC*.

linear differential transformer A type of electromechanical transducer that converts physical motion into an output voltage whose phase and amplitude are proportional to position. See also *linear motion transducer*.

linear feedback control system Feedback control system in which the relationship between the pertinent measures of the system signals are linear.

linear IC 1. Also called linear device. An IC in which there is a proportional relationship between the output and input signals. For example, in a linear amplifier, the output signal is an amplified replica of the input signal. 2. Typically, an IC having a linear relationship between its input and output. (An amplifier, for example, has an output which is an amplified replica of the input.)

linear integrated circuit (LIC) An amplifier whose output remains proportional to the input level. Generally taken to mean an analog IC, including voltage regulators, comparators, sense amplifiers, drivers, etc., as well as linear amplifiers. The circuit can be made to operate nonlinearly by connecting the basic linear amplifier to circuit elements that have thresholds or other nonlinear characteristics.

linearity The *straightlineness* of the transfer curve relating an input to an output; that condition prevailing when output is directly proportional to input. See *nonlinearity*.

linear motion transducer An instrumentation component that translates straight line (linear) mechanical motion into an ac analog which is usable as a feedback signal for control or display. A transformer-type device in which a moveable magnetic core is displaced axially by the moving component being monitored. When the core is moved in one direction, from the center of its stroke, output voltage is in phase with the excitation voltage and when the core is moved in the opposite direction from center, output voltage is 180° out of phase. At the center, output voltage is virtually zero. In either direction from center, voltage increases as a precise linear function of probe displacement. Thus, the output signal has two basic analog components; phase relationship with the excitation voltage, indicating direction of travel, and voltage amplitude, indicating length of travel.

linear network A network having constants and transmission characteristics that do not vary with the magnitude of the voltage impressed or the current flowing through the network.

linear transducer Transducer for which the pertinent measures of all the waves concerned are related by a linear function, e.g., a linear algebraic differential or integral equation.

line driver A buffer circuit with special output characteristics (i.e., high current and/or low impedance) suitable for driving logic lines of longer than normal interconnection length (greater than a few feet). It may have complementary (push-pull) outputs to work with the differential inputs of a line receiver.

line drop A voltage loss occurring

between any two points in a power or transmission line. Such loss, or drop, is due to the resistance, reactance, or leakage of the line. An example is the voltage drop between a power source and load when the line supplying the power has excessive resistance to the amount of current.

line filter   A filter associated with a transmission line, such as a filter used to separate speech frequencies from carrier frequencies.

line printer   A device capable of printing one line of characters across a page; i.e., 100 or more characters simultaneously as continuous paper advances line by line in one direction past type bars or a type cylinder that contains all characters in all positions.

lines of force   In electric and magnetic fields, the electric and magnetic forces of repulsion or attraction which are taken to follow certain imaginary lines radiating from the electric charge or the magnetic pole. (It is assumed that any unit electric charge or unit magnetic pole placed in the appropriate field will be acted upon so as to move in the direction of these imaginary lines.)

line spectrum   The spectrum of a periodic, discrete signal consisting of one or more frequencies. For example, a square wave is characterized by a fundamental and odd-order harmonics.

lingual   Relating to the tongue.

lipemia   Presence of excess of fat in the blood.

liquid crystal display   See *LCD*.

liquid crystals   Nematic liquids which are doubly-refracting and which display interference patterns in polarized light.

lissajous figure   The pattern appearing on a cathode-ray oscilloscope when the horizontal plates are fed one alternating

frequency, and the vertical plates are fed another alternating frequency.

lithium   A drug found effective in the treatment of manicdepressive psychoses.

lithium chloride sensor   A hygroscopic element that has fast response, high accuracy, and good long-term stability, and whose resistance is a function of relative humidity. Also called *Dunmore cell*.

lithotomy   Operation of cutting into the bladder to remove a stone.

live   1. Energized. 2. Connected to a source of an electrical voltage. 3. Charged to an electrical potential different from that of the earth. 4. Reverberant, as a room in which there are reflections of sound. 5. Display the vital signs of brain activity, heart action, pluse, and respiration consistent with the living state.

load   1. A circuit component that consumes electric power. 2. A device that receives the useful output of an electric generator, oscillator, or other signal source. 3. The amount of electric power that is drawn from the source. 4. The burden acted upon by a muscle in contraction; also, the preload (resting state) and afterload (muscle in fully contracted state). 5. To enter information into the storage or memory portion of a computer.

load cell   1. Transducer which measures an applied load by a change in its properties, such as a change in resistance (strain-gauge load cell), pressure (hydraulic load cell), etc. 2. A device which produces an output signal proportional to the applied weight or force.

load characteristic   Relation between the instantaneous values of a pair of variables such as an electrode voltage and an electrode current, when all direct electrode supply voltages are maintained constant. Also called *dynamic characteristic*.

loader   A program used for loading other programs.

loading error   A decrease in output from a signal source by the amount of voltage dropped across its internal impedance when signal current flows.

load life   Ability of a device to withstand its full power rating over an extended period of time, usually expressed in hours.

lobe   A somewhat rounded projection or division of a bodily organ or part.

lobotomy   See *leukotomy.*

localized edema   See *edema.*

loculated   Divided into many cavities.

logic   1. Science dealing with the criteria or formal principles of reasoning and thought. 2. Systematic scheme which defines the interactions of signals in the design of an automatic data processing system. 3. Basic principles and application of truth tables and interconnection between logical elements required for arithmetic computation in an automatic data processing system. 4. In computers and information-processing networks, the systematic method by which information is processed, usually with each successive step influencing the next one. 5. The systematic scheme which defines the interactions of signals in the design of an automatic data processing system. 6. A form of mathematics based upon two-state truth tables. Electronic logic uses two-state gates and flip-flops to perform decision making functions.

-logical   Suffix meaning study or science of.

logical decision   1. The choice or ability to choose between alternatives. Basically this amounts to an ability to answer yes, or no with respect to certain fundamental questions involving equality and relative magnitude; e.g., in an inventory applica-

tion, it is necessary to determine whether there has been an issue of a given stock item. 2. In a computer, the operation of selecting alternative paths of flow depending on intermediate program data. 3. Ability of a computer to choose between alternatives. Basically, the ability to answer yes or no to certain fundamental questions involving equality and relative magnitude.

logical diagram   1. In logic design, a diagram representing logical elements and their interconnections without necessarily expressing construction or engineering details. 2. A pictorial representation of interconnected logic elements using standard symbols to represent the detailed functioning of electronic logic circuits. The logic symbols in no way represent the types of electronic components used, but represent only their functions.

logical element   In a computer or data processing system, the smallest building blocks which can be represented by operators in an appropriate system of symbolic logic. Typical logical elements are the AND gate and the flip-flop, which can be represented as operators in a suitable symbolic logic.

logical flow chart   A detailed graphic presentation of work flow in its logic sequence —often, the built-in operations and characteristics of a specific machine, with symbols to indicate types of operations.

logic function   A combinational, storage, delay, or sequential function expressing a relationship between variable signal input(s) to a system or device and the resultant output(s).

logical operations   Those operations which are considered nonarithmetical in character such as selecting, searching, sorting, matching, or comparing.

logical sum   A boolean expression denoting the union of two classes or propositions; i.e., a + b implies a or b, or both. In computers, a logical sum is the resultant of an inclusive or operation on two arguments.

logic arrays An integrated device in which fifty or more circuits are integral to a single, silicon chip. In addition, the circuits are interconnected on the chip to form some electronic function at a higher level of organization than a single circuit. Logic arrays are constructed by the unit cell method in which a simple circuit (or function) is repeated many times on a slice. The interconnection pattern for converting groups of cells into large functions is determined after cell probe tests are completed. Each interconnection pattern may be unique to a single slice. In general, logic arrays are characterized by multiple levels of metallization to effect the large-scale function.

logic circuit An electronic circuit whose input-output relationship corresponds to a boolean algebra logic function.

logic design The specification of the working relations between the parts of a system in terms of symbolic logic and without primary regard for its hardware implementation.

logic device Digital components which perform logic functions, such as AND, OR/NOT, and flip-flop. They can gate or inhibit signal transmission with the application, removal, or other combinations of input signals.

logic diagram A picture representation for the logical functions of AND, OR, NAND, NOR, NOT.

logic function 1. Means of expressing a definite state or condition in magnetic amplifier, relay, and computer circuits. 2. The boolean algebra functions of AND,

OR, and NOT, or combinations of these.

logic instruction An instruction that executes an operation that is defined in symbolic logic, such as AND, OR, NOR, NAND.

logic levels Nominal voltages which represent binary conditions in a logic circuit.

logic swing The voltage difference between the two logic levels 1 and 0.

logic switch Diode matrix or other switching arrangement that is capable of directing an input signal to one of several outputs.

logic system A group of interconnected logic elements which act together to perform a relatively complex logic function.

-logy Suffix meaning study or science of.

long-nose pliers Pliers with long, narrow holding jaws suitable for wrapping wire around closely-spaced terminals.

long-persistence screen Term applied to the phosphor-coated screen of a cathode-ray tube which has been treated so that the screen fluoresces for a finite time, after being excited by the electron beam. The time of persistence varies with the electron beam accelerating voltages employed and the type of phosphor coating of the screen.

loop A sequence of instructions that is executed repeatedly until a terminal condition prevails.

loop gain 1. The increase in gain that is observed when the feedback path of an amplifier is opened, but with all circuit loads intact. 2. In an operational amplifier circuit, the product of the transfer characteristics of all of the elements (active or passive) encountered in a complete trip around the loop, starting at any point and returning to that point.

loose connection A contact between two

conductors which is not tight, and therefore has resistance to the flow of current. Current flow through that resistance produces heat and electrical noise.

loose coupling Coupling between two circuits such that little power is transferred from one to the other, and such that an impedance change in one circuit has little effect on the other circuit.

loss 1. Amount of electrical attenuation in a circuit, or the power consumed in a circuit component. 2. Energy dissipated without accomplishing useful work. Usually expressed in decibels.

lot A group of similar components which have been grouped in either of two ways: (a) All manufactured in a continuous production run from homogeneous raw materials under constant process parameters. (b) Assembled from more than one production run and submitted for random sampling and acceptance testing.

low-energy circuit A circuit application that functions at low voltage (i.e., approximately 10 volts or less) and low current (i.e., approximately 1 ma or less).

low level In digital logic, the more negative of the two logic levels in a binary system. See also *high-level logic, negative logic,* and *positive logic.*

low-level signal Very small amplitudes serving to convey information or other intelligence. Variations in signal amplitude are frequently expressed in microvolts.

low-pass filter 1. Filter network which passes all frequencies below a specified frequency with little or no loss but which discriminates strongly against higher frequencies. 2. Wave filter having a single transmission band extending from zero frequency up to some critical or cutoff frequency, not infinite. 3. Filter which passes all frequencies below a certain cutoff point and attenuates all frequencies above that point.

LPAO Abbreviation for left posteroanterior oblique.

LPO Abbreviation for left posterior oblique.

LSB See *least significant bit.*

LSI See *large-scale integration.*

lumen 1. The cavity of a tubular organ (the lumen of a blood vessel). 2. The bore of a tube or catheter.

luminance An attribute of light or color. Luminous intensity or brightness, as measured by a photometer instead of by the human eye. Measured in foot-lamberts. (One foot-candle of light, scattered by a perfectly diffusing white surface without loss, would produce one foot-lambert.)

luminescence A general term which is applied to the production of light, either visible or infrared, by the direct conversion of some other form of energy. The general term is then subdivided to denote the particular energy conversion involved (i.e., thermoluminence, cathodoluminence, photoluminence, and electroluminence).

luminous flux Also called light flux. 1. The time rate of flow of light (the total visible energy produced by a source per unit time). Usually measured in lumens. 2. Radiant power weighted at each wavelength in accordance with the ability of the eye to perceive it.

luminous intensity The luminous flux through a unit of solid angle and is usually measured in candelas (lumens/steradian).

lumped constant A resistance, inductance, or capacitance which is connected at a point, rather than being distributed uniformly throughout the circuit.

lux The unit of illumination, predominantly in Europe; equal to one lumen per square meter. Ten lux is approximately equal to one foot-candle.

luxation Dislocation of a joint.

lymph A clear, transparent, slightly yellowish alkaline fluid derived from the body tissues.

lymphatic Pertaining to or containing lymph.

lymphatic leukemia A disease in which large numbers of primitive lymphocytes appear in the blood.

lymphatic system Tiny tubes and a system of nodes which make up a circulatory drainage system interspaced throughout the tissues and the body structure with the capillaries and arteries of the blood circulatory network. The system operates through the lymphatic fluid, a plasma-like substance containing lymphocytes; it starts its travel in the interstitial, intracellular, or tissue spaces to which it has leaked from blood capillaries and where it has served to equalize pressure that would otherwise cause tissue swelling. From the interstitial spaces, the lymph passes through four structures in its journey before eventually arriving in the venous system: lymph capillaries, which pass into lymph vessels, narrow tubes which pass into lymph nodes, enlargements producing lymphocytes and in turn serving to filter out, engulf, and destroy bacteria, infectious organisms, and other debris that may have entered the body. Nodal output travels through larger and larger vessels to the lymph ducts, which pass the accumulated fluid into the venous system.

lymphocytes See *blood corpuscles.*

-lysis Suffix meaning solution, loosening, or dissolve.

lysis The gradual fall of an elevated temperature; the disintegration of a cell.

M

m Abbreviation for milli, 1/1,000 or 10^{-3}.

M Abbreviation for mega, 1,000,000 or 10^6. 2. Abbreviation for megohm, 1,000,000 ohms.

mA Abbreviation for milliampere, 10^{-3} ampere.

mach Unit of speed measurement equal to speed of sound. Sound travels 759 mph at sea level, it decreases about 2 percent for every thousand feet of altitude.

machine code An operation code that a machine is designed to recognize.

machine cycle The shortest complete process or action that is repeated in order. The minimum length of time in which the foregoing can be performed.

machine language 1. The symbols and rules used to express information to be handled by computers or data processing equipment. 2. A language designed for interpretation and use by a machine without translation. 3. A system for expressing information which is intelligible to a specific machine; e.g., a computer or class of computers. Such a language may include instructions which define and direct machine operations, and information to be recorded by or acted upon by these machine operations. 4. The set of instructions expressed in the number system basic to a computer, together with symbolic operation codes with absolute addresses, relative addresses, or symbolic addresses. 5. A programming language which is only one step removed from machine language. Assembly languages are referred to as machine-oriented because of the one-to-one relation between assembly instructions and machine instructions, and because of the degree to which the format of the machine instruction determines the format of the assembly instruction. 6. Information recorded in a form which may be used by a computer without prior translation. 7. A series of bits written as such to instruct computers. The first level of computer language. See *assembly language* (second level) and *compiler language* (third level).

macro assembler An assembly program of a computer which translates alphanumeric language into machine language.

macrocode In digital computer programming, a coding system which assembles groups of computer instructions into single code words. This requires interpretation or translation so that a computer can follow it.

macroelement An ordered set of two or more elements used as one data element with a single data-use identifier. For example, the macroelement data may be the ordered set of the data elements year, month, or day.

macro instruction 1. In a computer, an instruction in a source language that is equivalent to a specified sequence of machine instruction. 2. Computer programming instructions in which a single programmer language instruction (macroword) produces a series of machine instructions.

macroprogram A computer program written as a sequence of instructions in a source language.

macroprogramming In a computer, the process of writing machine procedure statements in terms of macro instructions.

macroscopic Large enough to be observed by the unaided eye.

macroshock Electric shock due to contacts applied to the exterior of the body. Here, currents in the range of 100 to 300 milliamperes can cause heart fibrillation. The lowest body resistance, with well-prepared electrodes, is in the range of 1000 to 1600 ohms, so that interelectrode voltages of the order of 75 to 120 volts (at commercial power frequencies) could be dangerous, if the voltage source is not current-limited. All hospital patients are exposed to macroshock from defective electric devices such as lamps and bed-adjusting motors—just as they might be outside the hospital. But, in addition, macroshock may result from a defective electrocardiograph machine in operation after well-prepared electrodes are applied to the patient's body.

magnetic amplifier An electrical device in which the flow of a heavy alternating current through a reactor can be controlled by a relative weak direct current which is used to saturate the core of the reactor and thus reduce its reactance.

magnetic circuit A closed path of magnetic flux, which is forced through the reluctance of a magnetic circuit by a magnetomotive force.

magnetic core 1. Ferromagnetic material placed inside of coils or transformer windings to provide a low reluctance path for magnetic flux, to increase the inductance of single winding coils and increase the coupling between transformer windings. 2. A device using magnetic remanence to store binary information, consisting of a small doughnut-shaped core of a magnetic material, having an open square hysteresis curve. Arranged in a matrix, each core is threaded with a vertical *write wire,* a horizontal write wire, and a *sense (read) wire.* Some types also have an *inhibit (cancel) wire.* Energizing the vertical and horizontal wires with a half-write current will switch the core at their intersection to the one state.

magnetic core memory A storage device in which binary data are represented by the direction of magnetization in each unit of an array of magnetic material, usually in the shape of toroidal ferromagnetic rings.

magnetic core storage A configuration of magnetic material that is, or is intended to be, placed in a rigid special relationship to current-carrying conductors and whose magnetic properties are essential to its use. For example, it may be used to concentrate an induced magnetic field as in a transformer, induction coil, or armature, to retain a magnetic polarization for the purpose of storing data, or for its nonlinear properties as in a logic element. It may be made of such material as iron, iron oxide, or ferrite in such shapes as wires, tapes, toroids, or thin film.

magnetic coupling Coupling between two circuits by way of a magnetic flux which links coils in both circuits.

magnetic delay line Delay line, used for the storage of data in a computer, consisting of a metallic medium along which the velocity of the propagation of magnetic energy is small compared to the speed of light. Storage is accomplished by the recirculation of wave patterns containing information, usually in binary form.

magnetic disc A flat circular plate with a magnetic surface on which data can be stored by selective magnetization of portions of the flat surfaces.

magnetic disc storage A storage device or system consisting of magnetically coated discs, on the surface of which information is stored in the form of magnetic spots arranged in a manner to represent binary data.

magnetic drum storage A precisely-made cylindrical metal drum plated with a ferromagnetic cobalt-nickel alloy, and made to rotate at a constant speed between 3000 and 12,000 rpm. Read/write heads are arranged along the length of the drum, each associated with a track of stored data. The writing heads store data as magnetized spots on the drum surface. Reading is nondestructive.

magnetic field 1. Any space or region in which magnetizing forces are of significant magnitude with respect to conditions under consideration. A magnetic field is produced by any current or permanent magnet and including the earth itself. 2. The area within which the forces of a magnet may be observed. 3. The region around a conductor or coil which is carrying a current within which moving electric charges will be acted upon by a magnetic force.

magnetic hysteresis loop A closed curve showing the relation between the magnetization force and the induction of magnetization in a magnetic substance when the magnetized field force is carried through a complete cycle.

magnetic ink character recognition The machine recognition of characters printed with magnetic ink.

magnetic integrated circuit An integrated component that utilizes magnetic elements to perform all or at least a major portion of its intended function.

magnetic material A mass of material consisting of one or more substances which exhibit magnetic properties and which may be a single crystal, a massive polycrystalline aggregate, powder particles, or compressed powder particles with or without a binder or insulator. The material may be in the form of a bar, rod, wire, sheet, strip, tape, thin film, shell, particular core shape, compressed powder, lamina, laminated structure, or the like. The term also includes (a) fine particle magnets, (b) magnetic alloys, or any composition of two or more metals or alloys intimately mixed to form a body of aggregate exhibiting magnetic properties, (c) intermetallic compounds, mixtures, or solid solutions, (d) compounds such as oxides or sulfides, which possess the structure that exhibits ferromagnetism, and (e) ferromagnetic iron oxide material containing ferrosoferric oxide, Fe_3O_4, or gamma ferric oxide, Fe_2O_3, or a mixture thereof.

magnetic memory plate Form of digital

computer magnetic memory consisting of a ferrite plate having a grid of small holes through which the readin and readout wires are threaded. Printed wiring may be applied directly to the plate in place of conventionally threaded wires, permitting mass production of plates having a high storage capacity.

magnetic moment A measure of the magnetic flux set up by the gyration of an electric charge in a magnetic field.

magnetic pole 1. One of the two centers or spots on a magnet where magnetic attraction is the strongest. These spots are where the magnetic lines of force enter and leave the magnet. 2. Either of the earth's two magnetic poles.

magnetic reed switch Essentially two overlapping flexible-metal strips surrounded by glass tubing. The ends of these strips click together when a sufficiently strong magnetic field is induced into the two reeds.

magnetic shield An enclosure made from high-permeability magnetic materials which can protect the circuits or equipment it encloses from the effects of external magnetic fields.

magnetic tape 1. A tape with a magnetic oxide surface on which data can be stored by selective polarization of portions of the surface. 2. A tape of magnetic surface used as the constituent in some forms of magnetic cores.

magnetic tape reader A data-processing device which can read at high speed the signals recorded on magnetic tape, and deliver these signals to a computer or other data processing equipment.

magnetic tape recorder A device which can record audio-frequency signals on a plastic tape coated with magnetizable iron oxide particles. Consists of a tape transport,

a recording head, a playback head, an erase head, and an amplifier.

magnetic tape storage Storage system utilizing magnetic spots (bits) on metal or oxide-coated plastic tape. The spots are arranged to read out in the desired code as the tape travels past the read-write head.

magnetism A property possessed by iron, alnico, and other magnetic materials, by which they can produce and maintain an external magnetic field which will attract magnetic materials, and will either attract or repel other magnets.

magnetize To arrange the molecules in a material in a definite polarized pattern and as a result produce magnetic lines of force.

magnetizing current The current in a transformer which is just sufficient to magnetize the core and supply the iron losses, but which will not provide any current to a secondary load.

magnetocardiogram The record of electromagnetic force changes accompanying cardiac contractions.

magnetocardiograph Instrument that generates electrical signals proportional to heart's magnetic pulses.

magnetoencephalograph (MEG) Instrument for recording magnetic signals emanating from electrical activity in the brain and proportional to EEG waves.

magnetohydrodynamics (MHD) The dynamics of ionized fluids under the influence of an applied magnetic field and a self-generated electric field.

magnetometer Device for measuring the strengths of magnetic fields.

magnetomotive force The force which tends to produce lines of force in a magnetic circuit. It has the same relation to a magnetic circuit as voltage has to an electrical circuit. Its practical unit is the ampere-turn, equal to 1.257 gilberts.

magnetoplasmadynamic generator Electricity generator which produces current by shooting an ionized gas (plasma) through a magnetic field.

magnetoresistance Change in resistance due to a change in the magnetic field in some materials.

magnetostriction The contraction of a ferromagnetic rod, particularly one of iron or nickel, when placed in a longitudinal magnetic field having a strength of over about 250 oersteds. When the magnetic field is removed, the rod returns to its original length. See also *joule effect*.

magnetostriction transducer A loudspeaker element used for underwater sound ranging. Multiple magnetostrictive elements in parallel are used to convert electrical energy into sound energy for radiating under water.

magnet wire Any coated conductor used to wind an electromagnetic coil in order to develop and maintain a magnetic field under prescribed conditions.

magnitude The size of a quantity irrespective of its sign. For example, $+10$ and -10 have the same magnitude.

main bang In echographic studies, the initial echo complex resulting from ringing in the crystal transducer during and immediately following transducer excitation.

main frame 1. The central processor of the computer system. It contains the main storage, arithmetic unit and special register groups. 2. All that portion of a computer exclusive of the input, output, peripheral and, in some instances, storage units.

main service panel The main electrical switch or circuit breaker and the circuit panel box which houses the circuit breakers or fuses for a branch circuit.

main storage Usually the fastest storage device of a computer and the one from which instructions are executed.

maintainability The ability to restore a system to its specified operational conditions within a specified total downtime, when maintenance action is initiated under stated conditions.

maintenance All actions necessary for retaining an item in, or restoring it to, a serviceable condition. Maintenance includes servicing, repair, modification, modernization, overhaul, inspection, condition determination, and initial provisioning of support items.

majority carrier In semiconductors, the type of carrier constituting more than half of the total number of carriers. The majority carriers may be either holes or electrons, depending on the construction of the semiconductor.

make The closing of a relay, key, or other contact to complete a circuit.

make-before-break The action or closing of a switching circuit before opening another associated circuit.

mal A prefix meaning bad, abnormal, inadequate, wrong, or irregular. 2. Seasickness (as in *mal de mer*). 3. A major epileptic seizure (*grand mal*). 4. A minor epileptic seizure (*petit mal*).

malaise An overall feeling of bodily weakness, discomfort, or distress, sometimes the first sign of infection.

malar Relating to the cheekbone.

mal de mer Seasickness.

malignant Virulent, fatal. A malignant tumor or growth is one which if not removed completely will spread and cause similar growths in other parts of the body until the patient dies (e.g., carcinoma, sarcoma).

mania   A suffix or combining form denoting a pathological preoccupation with some desire, idea, or activity; a morbid compulsion. Some frequently encountered manias are dispomania, compulsion to drink alcoholic beverages; egomania, pathological preoccupation with self; kleptomania, compulsion to steal; megalomania, pathological preoccupation with delusions of power or wealth; manomania, pathological preoccupation with one subject; necromania, pathological preoccupation with the dead; pyromania, morbid compulsion to set fires.

manic-depressive reaction   A group of psychiatric disorders marked by conspicuous mood swings, ranging from normality to elation or to depression, or alternating. Officially regarded as a psychosis but may also exist in milder form. 2. Depressed phase characterized by depression of mood with retardation and inhibition of thinking and physical activity. 3. Manic phase characterized by heightened excitability, acceleration of thought, speech, and bodily motion, and by elation of mood, and irritability.

manipulated variable   The one variable (value, condition, quantity, etc.) of a process that is being controlled. By manipulating this variable, the process can be controlled.

manometer   An instrument for measuring the pressure of gases and liquids (e.g., the sphygmomanometer, which indirectly measures blood pressure).

manual input   1. The entry of data by hand into a system at the time of processing. 2. Direct computer entry through manual intervention or drum entry of manual data through card machines.

marrow   The soft spongy substance which fills the medullary canal of a long bone and the small spaces in cancellous bone. The red cells of the blood are formed in the red bone marrow.

mAs   Abbreviation for milliampere second or milliamperage second.

maser   Acronym for Microwave Amplification by Stimulated Emission of Radiation. 1. A means of focusing a stream of particles, which concentrates only on the high energy particles. These are passed into a resonator which is resonating as the radiation frequency of the particles. The particles are in this state raised to a strong oscillation, and can be used for control purposes. By reducing the flow of particles to the resonator, to maintain oscillations, it can be used as an amplifier. (There are many other applications.) 2. Devices for amplifying a microwave frequency signal by stimulated emission of radiation, i.e., the weak microwave signal causes electrons in an atom to change orbit in such a manner as to emit an amplified signal of the same frequency as the weak signal.

mask   1. A fabric covering designed to cover the mouth and nostrils of health care personnel for the purpose of preventing infection. 2. An implement, usually a thin sheet of metal containing an open pattern, which shields selected portions of a base during a deposition process. 3. An implement used to shield selected portions of photosensitive material during photo processing.

masochism   Pleasure derived from undergoing physical or psychological pain inflicted by oneself or by others. It may be consciously sought or unconsciously arranged or invited. Present to some degree in all human relations and to greater degrees in all psychiatric disorders. It is the converse of sadism, in which pain is inflicted on another, and the two tend to coexist in the same individual.

mass spectrometer Analytical instrument which identifies a substance by sorting a stream of electrified particles (ions) according to their mass. When the stream of charged particles enters a magnetic field, it is deflected into a semicircular path, ultimately striking a photographic plate or photomultiplier tube sensor.

mass storage devices Auxiliary memories, generally containing magnetic media, which are used to supplement the main memory of a computer. It is less expensive than core memory per unit of information stored, but slower to access. Examples include magnetic tape, magnetic disc, and flexible disc.

master clock 1. The primary timing or synchronizing element in a device or system using timed signals. 2. In computers, the source of standard signals required for sequencing computer operation.

master routine See *executive routine.*

master-slave A binary element containing two independent storage stages with a definite separation of the clock function to enter information to the master and to transfer it to the slave.

master/slave flip-flop A flip-flop circuit which actually contains two flip-flops: a master flip-flop and a slave flip-flop. A master flip-flop receives its information on the leading edge of a clock pulse and the slave or output flip-flop receives its information on the trailing edge of the pulse.

master/slave operation See *parallel operation.*

mastication The act of chewing food.

matched impedance Condition which exists when two coupled circuits are adjusted so that the impedance of one circuit equals the impedance of the other.

matching Connecting two circuits or parts together in such a way that the maximum transfer of energy occurs between the two circuits and the impedance of either circuit will be terminated in its image.

matching impedance Impedance value that must be connected to the terminals of a signal-voltage source for proper matching.

mathematical check A check which uses mathematical identities or other properties occasionally with some degree of discrepancy being acceptable; e.g., checking multiplication by verifying that $A \times B = B \times A$. Synonymous with arithmetic check.

mathematical logic Exact reasoning concerning nonnumerical relations by using symbols that are efficient in calculation. Also called *symbolic logic.*

mathematical model 1. A mathematical representation that simulates the behavior of a process, device, or concept. 2. A consistent description of system behavior. The description may be symbolic in the form of conventional mathematical notation or a computer program, or analogic in the form of a topological representation.

matrix 1. An array of quantities in a prescribed form; in mathematics, usually capable of being subject to a mathematical operation by means of an operator or another matrix according to prescribed rules. 2. An array of coupled circuit elements; e.g., diodes, wires, magnetic cores, and relays, which are capable of performing a specific function; such as, the conversion from one numerical system to another. The elements are usually arranged in rows and columns. Thus a matrix is a particular type of encoder or decoder. Clarified by encoder and decoder. 3. A network which converts a group of individual inputs into a combination of outputs.

maxilla   The upper jawbone.

maximal breathing capacity   The volume of air which a subject can breathe with a voluntarily great effort, integrated over a given time interval.

maxwell   Centimeter-gram-second electromagnetic unit of magnetic flux. Equal to 1 gauss per square centimeter or to one magnetic line of force.

maxwell bridge   Four-arm ac bridge used to measure inductance (or capacitance) in terms of resistance and capacitance (or inductance). Bridge balance is independent of frequency.

Mc   Abbrevation for megacycle, an obsolete term superseded by MHz (megahertz).

mean time before failure (MTBF)   The average time between successive failures of a system. It is the reciprocal of the sum of the failure rates of every component and connection in the system.

measurand   A physical quantity, property, or condition which is measured.

meatus   The opening of the terminus of a canal, as in the meatus urethrae, the orifice of the urethra.

medial   Pertaining to the middle; situated toward the midline.

median   A point below which there are as many instances as there are above.

medical amplifier   Amplifier designed for receiving medical and biological signals (EEG, ECG, etc.) with the purpose of increasing their magnitude, separating component frequencies, or counting recurring events.

medical diathermy   The production of heat in body tissues for therapeutic purposes by high-frequency currents that are insufficiently intense to destroy tissues or to impair their vitality. Diathermy has been used in treating chronic arthritis, bursitis, fractures, gynecologic diseases, sinusitis, and other conditions.

medical electronics   The application of the toóls, techniques, and methods of electronic technology to the problems of medicine.

medical X-ray apparatus   Equipment for producing penetrating X-radiation (radiation with wavelength between 10^7 and 10^9 cm), directing the radiation through objects to be examined, and detecting the transmitted radiation by fluorescence, photography, or ionization effects.

medical sonic applicator   An electromechanical transducer designed for the local application of sound for therapeutic purposes; for example, in the treatment of muscular ailments.

medio   Prefix meaning middle.

medium scale integration (MSI)   Integrated circuits which form a self-contained logic system, such as a decade counter, or a five-bit shift register. These IC chips often contain up to 100 gates.

medulla   1. The central portion of a structure, organ, or tissue, as in the bone marrow. 2. The medulla oblongata, which see.

medulla oblongata   The lowest part of the brain where it passes through the foramen magnum and becomes the spinal cord. It contains the centers which govern circulation and respiration. See *brain*.

medullary   Relating to the marrow.

MEG   Abbreviation for magnetoencephalograph.

meg (MΩ)   Abbreviation for megohm.

mega (M)   A prefix meaning one million or 10^6.

megabit   One million binary digits.

megacycle (Mc)   One million cycles. Obsolete term replaced by megahertz (MHz).

megahertz (MHz)   One million hertz.

megavolt (MV)   One million volts.

megawatt (MW)   One million watts.

megohm   One million ohms. That resistance which limits current flow to one microampere per volt of applied potential. The unit used when measuring insulation resistance.

megohm-microfarad   A term used to indicate the insulation resistance of capacitors. For larger HV capacitors, megohm-farads are used.

mel   Unit of pitch; a simple tone of frequency 1000 hertz, 40 decibels above a listener's threshold, produces a pitch of 1000 mels. The pitch of any sound that is judged by the listener to be n times that of the 1-mel tone is n mels.

melanin   A black or brown pigment formed by the polymerization of quinonoid molecules. It is widely distributed in nature and is the skin pigment of man.

melanosis   Black spots in the tissues.

membrane   A thin tissue layer covering a surface, a space, or an organ or acting to divide two surfaces, spaces, or organs.

membrane action potential   An action potential recorded from a space-clamped axon or a small, uniformly responding patch of membrane.

membrane capacitance   The electrical capacitance of the cell membrane. The values found experimentally vary from 1 microfarad per square centimeter (squid axon) to 20 (muscle fiber of crab).

membrane potential   The electric potential that exists across the two sides of a cell membrane. See also *resting potential*.

membrane resistance   The electrical resistance of a square centimeter of cell membrane. In the squid axon, for example, it is about two thousand ohms. During the action potential, the membrane resistance falls rapidly to a relatively low value reflecting the enhanced permeability of ions through the membrane.

memory   1. A general term referring to the equipment and media that are used to hold information in machine language in electrical or magnetic form. The word memory usually means storage within a control system, while storage refers to magnetic drums, discs, cores, tapes, punched cards, etc., external to the control system. See *memory storage*. 2. That portion of the computer in which instructions and data are stored. 3. Any device or circuit capable of storing a digital word or words. 4. The component of a computer, control system, or other instrument, designed to provide ready access to data or instruction previously recorded so as to make them bear upon an immediate problem. 5. The faculty of the brain and nervous system to store accessible information. 6. Tendency of a material to return to its original shape after deformation.

memory buffer register   A computer register wherein a word is stored as it comes from memory (reading) or just prior to its entering memory (writing).

memory cell   Single storage element of a memory, such as a magnetic core, together with associated circuits for storing and reading out one bit of information.

memory cycle   A computer operation

consisting of reading from and writing into memory.

memory device See *memory storage.*

memory dump A process of writing the contents of memory consecutively in such a form that it can be examined for computer or program errors. See also *dump.*

memory fill Placing a pattern of characters in the memory registers not in use in a particular problem, to stop the computer if the program, through error, seeks instructions taken from forbidden registers.

memory hierarchy A set of computer memories with differing sizes and speeds and usually having different cost-performance ratios. A hierarchy might consist of a very-high-speed, small semiconductor memory, a medium-speed core memory, and a large, slow-speed core.

memory printout See *dump.*

memory register A register is used in all transfers of data and instructions between the memory, the arithmetic unit, and the control register in some computers. Also called *high-speed bus, distributor,* or *exchange register.*

memory storage The act of storing information. Any device in which information can be stored, sometimes called a memory device. In a computer, a section used primarily for storing information, called *memory* or *store.*

memory timer A timing device wherein the cycle duration is infinitely variable within the specified overall cycle time and having the ability, once the cycle is selected, to repeat this cycle repeatedly by a simple mechanical actuation of the timer shaft.

memory unit That part of a digital computer that stores information in machine language, using electrical or magnetic techniques.

memory word A group of bits read into or out of the memory, as a unit.

meninges The membranes surrounding and covering the brain and spinal cord. They are, from without; the *dura mater,* the *arachnoid,* the *pia mater.*

meningitis Inflammation of the enveloping membranes of the brain or spinal cord, caused by virus, bacteria, yeasts, fungi, or protozoa. It is a serious disease and may be a complication of another bodily infection.

meningocele Protrusion of meninges from a bony defect usually in the spine.

meniscectomy Removal of a semilunar cartilage from the knee joint.

meniscus 1. A semilunar cartilage. 2. A lens. 3. The crescent-like surface of a liquid in a narrow tube or vessel.

meperidine A powerful pain-relieving drug.

mercuric-oxide-cadmium cell A primary-cell electrochemical system, its chief advantage is its long shelf life in the fully charged condition and its operation at low temperatures, far below 0°C. Nominal cell voltage is about 0.9 volt.

mercury cell A primary cell with a zinc anode, a mercuric oxide cathode, in a potassium hydroxide electrolyte. This cell offers a higher capacity than the alkaline or zinc-carbon cells and a much flatter voltage-discharge characteristic. Delivers a steady 1.35 volts until end of life.

mercury switch An evacuated glass tube having sealed-in contacts and a pool of mercury which bridges the contacts when the tube is tilted.

mes Prefix meaning middle.

mesarteritis Inflammation of the middle coat of an artery.

mesa transistor A transistor that is formed by chemically etching away a transistor chip formed by either a double-diffused or diffused alloy process. When the etching process is complete, the base and emitter regions appear as plateaus above the collector region. One result of the mesa construction is a reduction in collector base capacitance as a result of lowering the junction area.

meta Prefix meaning beyond or over.

metabolic rate The rate at which a living organism ingests energy-containing substances, converts them into life-sustaining forms, and excretes the residue.

metabolism The total of the physical and chemical processes occurring in the living organism by which its substance is produced, maintained, and exchanged with transformation of energy; this energy itself provides fuel for all body functions and heat production. The chemical processes taking place in living cells can be divided into constructive processes (*anabolism*) and destructive processes (*catabolism*).

metacarpals The five bones of the hand joining the fingers to the wrist.

metal-base transistor A transistor with a base of a thin metal film sandwiched between two N-type semiconductors with the emitter doped more heavily than the base to give it a high electron-current to hole-current ratio.

metal film resistor An electronic component in which the resistive element is an extremely thin layer of metal alloy vacuum-deposited on a substrate.

metallized resistor A fixed resistor, in which the resistance element is a thin film of metal deposited on the surface of a glass or ceramic substrate.

metallizing Applying a thin coating of metal to a nonmetallic surface. This may be done by chemical deposition or by exposing the surface to vaporized metal in a vacuum chamber.

metal oxide resistor A type of film resistor in which the material deposited on the substrate is tin oxide, which provides good stability.

metal oxide semiconductor (MOS) 1. A field effect transistor whose gate is isolated from its channel through an oxide film. Also a capacitor formed by using similar techniques; semiconductor material forms one plate, metal the other plate, with oxide forming the dielectric. 2. A process which results in a structure of metal over silicon oxide over silicon. By appropriate topology, this generates field-effect transistors, capacitors, or resistors. 3. A circuit in which the active region is a metal-oxide-semiconductor sandwich. The oxide acts as the dielectric insulator between the metal and the semiconductor. 4. A field-effect transistor (MOSFET) that has a metal gate insulated by an oxide layer from the semiconductor channel. A MOSFET is either enhancement-type (normally turned off) or depletion-type (normally turned on). MOS also refers to integrated circuits that use MOSFETs (virtually all enhancement-type). 5. Technology that employs field-effect transistors having a metal or conductive electrode which is insulated from the semiconductor material by an oxide layer of the substrate material. Whereas bipolar devices permit current to flow in only one direction, MOS devices permit bidirectional current flow.

metal-oxide semiconductor field-effect transistor (MOSFET) The basic element of a

MOSIC. Consists of diffused source and drain regions on either side of a P- or N-channel region, plus a gate electrode insulated from the channel by silicon oxide. Application of a control voltage to the gate converts the channel to the same type of semiconductor as the source and drain. This eliminates part of the PN junction and allows current to flow between source and drain. Functionally the chief difference between a MOSFET and a bipolar transistor is that the MOSFET is "double-ended"; that is, the source and drain are interchangeable (unlike the bipolar emitter and collector) since the direction of current flow depends upon the relative potential of the electrodes.

metal-oxide semiconductor transistor (MOST)   An active semiconductor device in which conduction is controlled in a region between two electrodes by a voltage applied to an insulated electrode over the region.

metastasis   Transfer of a disease (usually cancer) or the causative agent of a disease (e.g., bacteria or cancer cells) from one organ or part of the body to another, through the blood or lymph vessels.

metatarsalgia   Pain in the fore part of the foot.

meter   1. Basic unit of linear metric measurement, abbreviated m. Originally based upon distance from equator to North Pole = 10^5 meters. 2. An electrical measuring instrument which may be calibrated in such units as volts, amperes, ohms, and watts. The typical movement is a D'Arsonnal milliammeter.

meter-kilogram-second system   Absolute system of units, based on the meter, kilogram, and second as fundamental units and extended to electrical effects by the measurement of potential difference by the power per unit current. The measure of current by its magnetic effects is included. The mechanical units of the system are based on unity as the proportionality factor in the recognized equations of mechanics, plus insertion of a new unit of force, the newton. Each of the electrical units of the meter-kilogram-second system has the same name and the same value as the corresponding unit of the centimeter-gram-second practical system of electrical units.

methicillin   The first penicillin effective against staph (staphylococcus) infections that are resistant to ordinary penicillin.

MeV   Abbreviation for million electron-volts.

MHD   Abbreviation for magnetohydrodynamics.

mho   The unit of conductance or admittance. It is the reciprocal of the resistance or impedance. Note that mho is ohm spelled backwards.

MHz   Abbreviation for megahertz.

mica   A naturally occurring crystalline form of aluminum-potassium-silicate which can be split into thin somewhat flexible sheets, translucent or transparent. Heat resistant, and an excellent insulator.

micro (μ)   1. A prefix used to denote one one-millionth (10^{-6}). A microvolt is one one-millionth of a volt. 2. Prefix meaning small.

microammeter   A meter calibrated to read microamperes.

microampere (μA)   One-millionth of an ampere.

microcircuit   1. Electronic circuit that may be constructed of integrated circuits, thin-film circuits, hybrid microcircuits, and similar miniature circuits. 2. Another name for integrated circuit.

microcircuit module An assembly of microcircuits or an assembly of microcircuits and discrete parts, designed to perform one or more electronic circuit functions and constructed such that for the purposes of specification, testing, commerce, and maintenance, it is considered to be indivisible.

microcircuitry A small circuit having a high equivalent circuit element density which is considered as a single part and which is a combination of interconnection elements inseparably associated on or within a single substrate to perform an electronic circuit function. This excludes printed wiring boards, circuit card assemblies, and modules composed exclusively of discrete electronic parts.

microcircuit wafer A microwafer carrying one or more circuit functions such as a flip-flop or gate. Integrated-circuit chips may be bonded to deposited conductors.

microcode 1. A system of computer coding that includes suboperations such as multiplication and division that are not ordinarily accessible in programming. 2. A list of very small program steps.

microcomponents Those components smaller than existing discrete components by several orders of magnitude.

microcomputer 1. A computer whose major sections, CPU, control, timing, and memory, are each contained on a single, integrated-circuit chip, or, at most, a few chips. 2. A general purpose computer composed of standard LSI components built around a central processing unit (CPU). The CPU (or microprocessor) is program controller featuring arithmetic and logical instructions, and general purpose parallel I/O bus. The CPU is contained on a single

chip or a small number of chips, usually not greater than four. Generally intended for dedicated applications, the microcomputer also includes any number of ROMs and RAMs (for instruction and data storage) and in some cases, one or more I/O devices. The simplest microcomputer consists of one CPU chip and one ROM.

microcurie (μCi) One-millionth of one curie.

microdensitometer An instrument used in spectroscopy, to measure lines in a spectrum by light transmission measurement.

microelectrode An electrode of extremely small dimension. Microelectrodes small enough to contact a single biological cell are available.

microelectronic circuit Discrete electrical components assembled and connected in extremely small and compact form.

microelectronic device An alternate term for integrated circuit.

microelectronics The term covering all techniques for fabricating very small electronic circuits, generally including all types of silicon integrated circuits, thin-film circuits, and thick-film circuits.

microfarad (μF) A unit of capacitance equal to one-millionth of one farad. Most capacitances used in medical electronics are of sizes which can be conveniently stated in microfarads.

micromessage See *intercellular message.*

micromho (μmho) One-millionth of one mho.

microminiaturization The technique of packaging a microminiature part or assembly composed of elements radically different in shape and form factor. Electronic parts are replaced by active and passive elements, through use of fabrication processes such as screening, vapor deposition diffusion, and photoetching.

micromodule Cube-shaped, plug-in, miniature circuit composed of potted microelements. Each microelement can consist of a resistor, capacitor, transistor, or other element, or a combination of elements.

micron 1. A unit of length equal to 10^{-6} meter. 2. A unit used in the measurement of very low pressures, equal to 0.001 mm of mercury at 32°F.

microohm ($\mu\Omega$) One-millionth of an ohm.

microorganism An organism which can be seen only by means of a microscope.

microphonism 1. Production of noise as a result of magnetic shock or vibration. 2. Quasiperiodic voltage output of a tube produced by mechanical resonance of its elements as a result of mechanical impulse excitation. 3. Periodic voltage output of a tube produced by mechanical resonances of its elements as a result of sustained mechanical excitation. 4. Output voltage of a tube acting as an electrical transducer of mechanical energy.

microphonograph A device which amplifies and records weak sounds, used in training deaf to speak.

microphonoscope A binaural stethoscope, using a membrane in the chest piece to accentuate the sound.

microprocessor 1. The basic building block of a microcomputer system; specifically the digital processor on a chip which performs arithmetic logic and control logic. 2. A semiconductor device that can perform arithmetic logic and decision-making operations under the control of a set of instructions stored in a memory device. It can also communicate with a set of peripheral devices via some defined input/output structure. 3. An IC that provides the basic logic and computing functions of a small central processing unit on one single chip. When combined with memory, miscellaneous support and input/output devices, one can create a microcomputer. 4. A complete processing unit on one large-scale integrated circuit. A particular collection of logic in IC form or otherwise, that is controlled by a microprogram.

microprogram 1. In a computer, a sub-element of a conventional program built up of a sequence of even smaller operations called microinstructions. Each microinstruction is further subdivided into a collection of micro-operations carried out in one basic machine cycle. (For example, the computer program consists of a sequence of instructions that are carried out in a specific order. Each instruction consists of a routine of one or more steps. This sequence of computer machine cycles necessary to execute a single instruction is called a microprogram.) 2. A special-purpose program, stored in a fixed memory, that is initiated by a single instruction in a system's main program. For example, one instruction in the main program may initiate a stored microprogram of 6 or 7 instructions needed to execute the single main program instruction. 3. A computer program written in the most basic instructions or subcommands that the computer is capable of executing. It is often stored in a read-only memory. See also *firmware*.

microprogramming 1. In a computer, a machine language coding in which the coder builds his own machine instructions from the primitive basic instructions built into the hardware. 2. A computer technique for using setup bits in the instructions, working with small, fast, read-only memories to directly control gates in the computer. 3. A method of operating the control part of a computer where each instruction is broken into several small steps (microsteps) that form part of a microprogram. 4. A method of organizing a general-purpose computer to perform desired functions, using instructions stored in a control array.

microradiometer   Thermosensitive detector of radiant power in which a thermopile is supported on, and connected directly to, the moving coil of a galvanometer.

microscopic   Small enough to be visible only with the aid of a microscope.

microsecond (μs)   One-millionth of one second.

microshock   Literally, an electrical shock of very small magnitude. It is a hazard to every patient who has a lead or electrical conductor from the interior of the heart extending out through the body's surface; but the number of patients who are thus internally wired to pacemakers or cardiac catheters is very small indeed. Currents in the range of 100 microamperes and higher, caused by accidental voltages, can produce ventricular fibrillation of the human heart under these conditions. See *patient isolation*.

microtome   An instrument for cutting fine sections for microscopic examination.

microvolt (μV)   One-millionth of one volt.

microwatt (μW)   One-millionth of one watt.

microwave communication system   Wireless radio system for hospital-hospital, hospital-university, hospital-clinic communications.

microwave integrated circuit   The physical realization of an electronic circuit operating at frequencies above one gigahertz and fabricated by microelectronic techniques. Either hybrid or monolithic

integrated circuit technology may be utilized.

micturition   The act of passing urine.

midbrain   See *brain*.

migration   In the case of some plated metals, such as gold plated over silver, the silver ions can migrate or move through the gold plating to the surface of the gold and oxidize. (Can be a problem when silver is used as an underplate for gold plating.)

mil   1. One-thousandth of an inch. 2. An angle, such that it is subtended by a 1-foot arc at a distance of 1000 feet. One mil equals 0.0575 degree, and 17.4 mils equal one degree.

milli (m)   A prefix meaning one-thousandth of (10^{-3}).

milliammeter   A meter which measures current in thousandths of an ampere.

milliampere (ma)   One-thousandth of one ampere. A small current.

millihenry (mH)   One-thousandth of one henry. A useful sized unit of inductance. If the current through a one millihenry inductance varies at the rate of one ampere per second, there will be one-thousandth of a volt potential across the inductance.

millisecond (ms)   One-thousandth of one second.

millivolt (mV)   One-thousandth of one volt.

millivoltmeter   An indicating meter that measures electrical potential drop in millivolts, used in checking small dc flow and also, with an instrument shunt for measuring much larger currents.

milliwatt (mW)   One-thousandth of one watt.

miniature lamps   Very small tungsten incandescent or neon lamps that are used where space is limited. Typical applications include indicator, alarm, and on/off visual annunciation.

minicomputer 1. A loosely-used term for describing any general-purpose digital computer in the low-to-moderate price range. The approximate ceiling price used to define a minicomputer is subject to wide interpretation. 2. A true digital computer similar to its larger predecessors but reduced in size by modern electronic packaging by eliminating certain optional features, and by setting a limit on the amount of high-speed memory available. Physically, it weighs approximately 50 pounds and has about the same dimensions as a stereo radio. It has no special operating requirements (voltage, air conditioning), being able to operate in the office, home, or factory. The computational speed is equivalent to computers of higher cost—a character can be retrieved from memory in one microsecond and the addition of two numbers takes three microseconds. Data can be transmitted in and out of the computer at a million characters per second. The major limitations of a minicomputer are the maximum memory size of 64,000 characters (bytes) as compared to several hundred thousand characters for larger computers, and the speed with which simultaneous data transfers can be made between high-speed data storage devices and the minicomputer. These limitations restrict the minicomputer system from applications involving data bases with hundreds of millions of characters.

minimum access routine In a computer with a serial memory, a routine coded with judicious arrangements of data and instructions in such a way that actual waiting time for information from the memory is much less than the expected random access waiting time.

minor cycle In a digital computer using serial transmission, the time required for the transmission of one machine word, including the space between words.

minute volume The sum of the air volumes either inspired or expired over a precisely timed 60-second interval.

mirror galvanometer Galvanometer having a small mirror attached to the moving element to permit use of a beam of light as an indicating pointer.

mirror scale Meter scale with a mirror arc used to align the eye perpendicular to the scale and pointer when taking a reading. By eliminating parallax (human error in reading), accuracy can be improved by half.

mismatch A condition whereby the coupled impedances inhibit optimum power transference or the source or output signal fails to match the amplifier sensitivity or the load's (i.e., speaker) power capacity.

missed beat The absence of a detectable heartbeat in the time between two detected beats separated by two complete R-R intervals.

mitochondria Small granules or rod-shaped structures found in differential staining in the cytoplasm of cells.

mitral stenosis A hardening of the mitral valve leaflets which interferes with the flow of blood between the left atrium and the left ventricle.

mitral valve Valve of the heart between the left atrium and the left ventricle. The function of this valve is to prevent backflow of blood into the left atrium during ventricular systole. See also *bicuspid valve* and *heart valve*.

mix 1. To combine two or more input signals in a transducer so as to produce a single output. If the transducer is linear, the output consists of a superposition of the input signals. If the transducer is nonlinear, the output consists of the heterodyne products of the input signals. 2. To combine physically, as with a mortar and pestle.

mixer 1. A device for combining two or more input signals while exercising individual control over the volume of each. 2. A device for blending of two or more signals for special effects. 3. A circuit which accepts two input frequencies and generates output frequencies equal to the sum and difference of the two input frequencies.

mmHg Millimeters of mercury. A measure of absolute pressure, being the height of a column of mercury that the air or other gas will support. Standard atmospheric pressure will support a mercury column 760 millimeters high (760 mmHg). Any value less than this represents some degree of vacuum.

M-mode In echocardiographic studies, a chart record which presents a sectional representation of moving tissue structures intersected by the ultrasound beam. Horizontal paper motion is at a constant speed. A high-speed top-to-bottom chart sweep is originated each time the crystal transducer is pulsed. Echo amplitude controls the intensity of the recording along the vertical sweep axis. Thus, the M-mode record is formed by successive vertical sweeps of an intensity-modulated writing beam. Since structures are in motion, the positions of beam intensification change on successive sweeps, gradually building up the M-mode echogram. The M-mode display is also possible on a high-persistence CRT, or on a storage CRT.

mnemonic Pertaining to the assisting, or intending to assist, human memory; thus, a mnemonic term, usually an abbreviation, that is easy to remember; e.g., mpy for multiply and acc for accumulator.

mnemonic code 1. Instructions for a computer written in a form which is easy for the programmer to remember, but which must later be converted into machine language. 2. A memory jogger.

mnemonic language Any programming language based on easy-to-remember symbols and which can be assembled by the computer into machine language.

mnemonic symbol A symbol chosen to assist the human memory, e.g., an abbreviation such as mpy for multiply.

MNOS Metal nitride oxide semiconductor. A semiconductor manufacturing process wherein the dielectric between metal and semiconductor is fabricated from silicon nitride.

mode 1. A method of operation. 2. The most frequent value in a frequency distribution. 3. A pattern in which waves vibrate. 4. One of several configurations in which energy propagates through a waveguide. 5. One of the modes of vibration in a piezoelectric material, either the extensional, flexural, or shear mode.

modem Acronym for *mo*dulator/*dem*odulator. 1. A device that transforms a typical two-level (binary) computer signal into a signal suited to the telephone network, for example, converting a two-level signal into a two-frequency sequence of signals. 2. A device that modulates and demodulates signals transmitted over communication facilities. 3. A device that performs modulation in the form of signal conversion, interfacing computers or computer peripheral equipment to the telephone line. Instead of trying to send a logic 0 or 1 dc voltage level over phone lines where voltages transients or noise pulses could be interpreted as false signals at the other end of the line and where transformer coupling is used, the modem changes the logic 0 or 1 into pulsed audio tones. The tones travel over the phone line and enter a companion modem at the other end of the line and are converted back into ones and zeros to properly interface and communicate with a computer or computer peripheral equipment.

mode number 1. The number of whole cycles that a mean-speed electron remains in the drift space of a reflex klystron. 2. The number of radians of phase shift in going once around the anode of a magnetron divided by 2π. Thus, n can have integral values 1, 2, 3 ... N/2 where N is the number of anode segments.

modification The physical alteration of a system, subsystem, or equipment to change its designed capabilities or characteristics. Any change or correction in equipment structures, arrangement, or accessories affecting capabilities or characteristics.

modulate To vary the amplitude, frequency, or phase of a high-frequency wave, called a carrier, in step with the amplitude variations of another wave, called the modulating wave. The carrier is usually a sine wave, but the modulating wave is often a complex voice-frequency wave or biophysical signal.

modulation 1. The process whereby the amplitude or frequency or phase of a single-frequency wave (called the carrier wave) is varied in step with the instantaneous value of, or samples of, a complex wave (called the modulating wave). 2. The intelligence which is carried on a modulated carrier wave. 3. A varying of the characteristics of a wave by controlling it with another wave. 4. The process by which some characteristic of one wave is varied in accordance with another wave or signal. 5. The controlled variation of frequency,

phase, and/or amplitude of a carrier wave of any frequency in order to transmit a message.

modulator A device for applying the information contained in a complex (modulating) wave to a higher frequency sinusoidal (carrier) wave, so as to enable transmission of the information at the higher carrier frequency. Device for tuning, regulating, and adjusting electrical frequency, acoustic volume, etc.

module 1. A combination of electronic components, contained in one package, or so arranged that they are common to one mounting, which contributes a complete function (or functions) to a system and/or subsystem in which it operates. 2. A basic unit of circuit packaging, which may be plugged in or wired in. 3. An incremental block of storage capacity for expanding a computer's memory.

module N counter See *programmable counter*.

molecular electronics The study and technique of the molecular rearrangement of materials in such a manner as to allow single blocks of matter to perform the function of complete electronic circuits.

molecular integrated circuit An integrated circuit in which the identity and location of specific electrical elements is indeterminable even upon microscopic disassembly of the material of which the circuit is formed. As opposed to convential microelectronic circuitry, the molecular integrated circuit can be defined only by function. The function can be described only by mathematical models and incremental circuit representation.

molecule 1. An aggregate of two or more atoms of a substance that exists as a unit. 2. A combination of two or more atoms of different elements which form a new substance (e.g., an atom of sodium united with an atom of chlorine produces a molecule of sodium chloride).

monaural See *monophonic*.

monitor 1. An electronic instrument which can detect one or more vital signs, for purposes of continuously determining patient condition. 2. An oscilloscopic instrument on which patient physiologic traces appear. 3. A video display unit in a closed-circuit TV system. 4. The specific program which schedules and controls the operation of several related or unrelated routines, performs overlapped 1/0, and allocates resources so that the computer's time is efficiently used.

monitoring 1. Automatic recording of physiological parameters, e.g., pulse, blood pressure. 2. Study, by human or computer, of physiologic data (i.e., ECG waves) for purposes of ascertaining patient condition and to detect premonitory events which may signal the onset of an acute episode.

monitoring amplifier Device which, when bridged on a circuit, absorbs a negligible amount of energy, but amplifies that energy so that it can be heard in an earphone or measured.

monitor routine See *executive routine*.

monochromatic 1. Pertaining to or consisting of a single color. 2. Radiation of a single wavelength.

monocytes See *blood corpuscles*.

monocytic leukemia A form of leukemia characterized by the exhibition of monocytes.

monolithic 1. An undivided whole. 2. In electronics, a semiconductor integrated circuit which is complete on one chip or piece of substrate. 3. Opposed of hybrid.

monolithic integrated circuit An integrated circuit whose elements are formed in situ upon or within a semiconductor substrate with at least one of the elements formed within the substrate. See also *integrated circuit*. To further define the nature of a monolithic integrated circuit, additional modifiers may be prefixed. Examples are: (a) PN junction isolated monolithic integrated circuit. (b) Dielectric isolated monolithic integrated circuit. (c) Beam-lead monolithic integrated circuit.

monophonic Single-channel sound. Formerly, and incorrectly, called monaural, which really means single-eared.

monostable A term used to describe a circuit that has one permanently stable state and one quasi-stable state. An external trigger causes the circuit to undergo a rapid transition from the stable state to the quasi-stable state, where it remains for a time and then spontaneously returns to the stable state. See *multivibrator*.

monostable multivibrator A multivibrator having one stable and one semistable condition. A trigger is used to drive the unit into the semistable state where it remains for a predetermined time before returning to the stable condition.

morbus Latin for disease.

moribund In a dying state or condition.

morphine The principal narcotic derived from opium; it has analgesic and addictive properties.

morphology The shape of a wave (i.e., the ECG), which indicates a change in the muscle tissue originating the ECG. The effect may be useful in diagnosis.

MOS Abbreviation for metal-oxide semiconductor, referring to a field-effect transistor (MOSFET) that has a metal gate insulated by an oxide layer from the semiconductor channel. A MOSFET is either enhancement-type (normally turned off) or depletion-type (normally turned on). MOS also refers to integrated circuits that use MOSFETs (virtually all enhancement-type).

MOS device Semiconductor component (typically, in an IC) formed by metal-oxide semiconductor process. In an MOS device, current can flow in either of two directions, whereas in a bipolar semiconductor current flows in only one direction.

MOS insulated-gate field-effect transistor A semiconductor device that consists of two electrodes (source and grain) diffused into a silicon substrate. The source and drain are separated by a finite space, thus forming a majority-carrier conducting channel. A metal gate is placed above the channel and insulated from it.

MOS monolithic IC A single-chip IC, consisting largely of interconnected unipolar active-device elements or MOS field-effect devices. This class of circuits results in higher equivalent parts density.

MOS random-access memory For MOS storage cells, the address decoding and sensing circuits can be either MOS or bipolar. Usual sizes are 256 bits/chip. Access time is 240 ns or less with bipolar decoding and 0.5 to 1.5 μs with MOS. Because readout is nondestructive, access and cycle time are equal.

MOST Abbreviation for metal-oxide semiconductor transistor.

most significant bit (MSB) The highest weighted digit of a binary number.

most significant digit 1. In a binary number, that digit having the highest place value. 2. The leftmost nonzero digit.

mother board A relatively large piece of insulating material on which components, modules, or other electronic subassemblies are mounted and interconnections made by welding, soldering, or other means, using point-to-point or matrix wire, or circuitry fabricated integral with the board.

motile Capable of moving independently.

motility 1. Ability of an organism to move on its own without outside assistance. This action enables the organism to warm itself, secure food, avoid threatening situations, etc. 2. Movement of fluid or substance within organs or vessels.

mouth pressure A measurement made with a pneumotachograph in pulmonary function studies.

mouth shutter A solenoid-actuated leaf valve which blocks airflow through a pneumotachograph in pulmonary function studies.

moving beam radiotherapy Technique employed to increase the dose incident on the target tissue while reducing the skin dose by rotating the beam in an arc with its center in the target.

MPD Abbreviation for maximum permissible dose.

Mr/m Abbreviation for milliroentgens per minute.

MSB Abbreviation for most significant bit. The highest weighted digit of a binary number.

msec Abbreviation for millisecond.

MSI Abbreviation for medium-scale integration, a technique whereby a complete subsystem or system function is fabricated as a single microcircuit. The subsystem or system is smaller than for LSI (see *LSI*), but whether digital or linear, it is considered to be one that contains 12 or more equivalent gates, or circuitry of similar complexity.

MTBF Abbreviation for mean time before failure.

mu 1. Greek letter μ. 2. Permeability. 3. Amplification factor. 4. Prefix micro (one-millionth or 10^{-6}).

mucosa The membranous tissue which lines tubular structures (i.e., the gastrointestinal tract), and which contains glands that secrete a slimy mucus which lubricates the structural walls and eases passage of material.

mucous membrane A surface which secretes mucus. The lining of the alimentary canal, air passages, and urinogenital organs; merges into true skin at the various orifices of these canals.

mucovisciodosis See *fibrocystic disease.*

multiaddress See *multiple address.*

multichip circuit Microcircuit in which discrete, miniature active electronic elements (transistor and/or diode chips) and thin-film or diffused passive components or component clusters are interconnected by thermocompression bonds, alloying, soldering, welding, chemical deposition, or metallization. See also *monolithic integrated circuit, solid-state circuit,* and *thin-film circuit.*

multichip integrated circuit Hybrid integrated circuit which includes two or more SIC, MSI, or LSI chips. An electronic circuit in which two or more semiconductor wafers that contain single elements or simple circuits are interconnected and encapsulated in a single package to give a more complex circuit.

Multi-Emitter Transistor Logic See *transistor transistor logic.*

multimeter A meter having multiple scales. Usually means a volt-ohm-milliammeter.

multiphasic screening A technique for rapidly testing a comparatively large population to reveal such diverse problems as glaucoma, hypertension, emphysema, diabetes, etc.

multiple Connected in parallel. 1. A circuit which is accessible at several points. 2. The aggregate of such points. 3. The parallel connection between corresponding jacks in several switchboard positions, or corresponding pairs in several cable terminals. 4. To connect in parallel. 5. To make a circuit accessible at several points.

multiple access A computer system with a number of on-line communication channels providing concurrent access to the common system.

multiple address A type of instruction that specifies the addresses of two or more items which may be the addresses of locations of input or output of the calculating unit, or the addresses of locations of instructions for the control unit. The term multiaddress is also used in characterizing computers; e.g. two-, three-, or four-address machines.

multiple address code An instruction code in which an instruction word can specify more than one address to be used during the operation. In a typical instruction of a four-address code, the addresses specify the location of two operands, the location at which the results are to be stored, and the location of the next instructions is dispensed with. The instructions are taken from storage in a preassigned order. In a typical two-address code, the addresses may specify the locations of the operands. The results may be placed at one of the addresses, or the destination of the results may be specified by another instruction.

multiple programming Programming in a digital computer so that two or more arithmetical or logical operations may be executed simultaneously.

multiplex 1. Simultaneous transmission of two or more signals by a common carrier wave by means such as time division, frequency division, or phase division. 2. In a computer operation, the process of transferring data from several storage devices operating at relatively low transfer rates to one storage device operating at a high transfer rate so that the high-speed device is not obligated to wait for the low-speed devices.

multiplexer 1. A device which simultaneously transmits two or more signals over a common transmission medium. 2. A device that will interleave (time-division) or simultaneously transmit (frequency-division) two or more messages on the same communication channel. 3. A device which uses several communication channels at the same time and transmits and receives messages and controls the communication line. This device itself may or may not be a stored-program computer. 4. A device which can combine several low-speed inputs into one high-speed output. Multiplexers can also function in reverse, a process called demultiplexing. 5. Analog or linear—a device that can be used to select one out of a multiple number of inputs and switches its information to the output. The output voltage follows the input voltage with a small error. 6. Digital —a device that can select one of its multiple inputs and pass that input logic level on to the output. Analog data cannot be handled (only digital information). Input channel select information is usually presented to the device in binary weighted form, and decoded internally, selecting the proper input. The device acts as a single-pole multiposition switch that passes digital information in only one direction.

multiplexing The process of combining several measurements for transmission over a single wire or link. There are two widely used methods of multiplexing. The first, time-division multiplexing, utilizes the principle of time sharing among measurement channels. The second method of multiplexing utilizes the principle of frequency sharing among measurement channels and is called frequency-division multiplexing, where the data from each channel are used to modulate sinusoidal signals called subcarriers so that the resultant signal representing each channel contains only frequencies in a restricted, narrow frequency range.

multiplex transmission The simultaneous transmission of two or more signals within a single channel. Multiplex transmission as applied to FM broadcast stations means the transmission of facsimile or other signals in addition to the regular broadcast signals.

multiple x-y recorder Recorder that plots a number of independent charts simultaneously each showing the relation of two variables, neither of which is time.

multiplier A series resistor in a voltmeter to increase the voltage range. Also called *multiplier resistor.*

multiplier phototube Phototube with one or more dynodes between its photocathode and the output electrode. The electron stream from the photocathode is reflected off each dynode in turn, with secondary emission adding electrons to the stream at each reflection. Also called *photomultiplier.*

multiplier resistor See *multiplier.*

multiprocessing The simultaneous or interleaved execution of two or more programs or sequences of instructions by a computer or computer network. Multiprocessing may be accomplished by multiprogramming, parallel processing, or both.

multiprocessor A computer that can

execute one or more computer programs employing two or more processing units under integrated control of programs or devices.

multiprogramming A method by which many programs can be operated on within the same time span. The programs are overlapped or interleaved. This technique is the basis for time-shared operation.

multivibrator A type of oscillator circuit usually using resistors and capacitors for feedback coupling and putting out square waves. There are three main types of multivibrators: *free-running*, which puts out a continuous signal; *monostable*, which puts out a single pulse each time it is triggered (called a "one shot" multivibrator); and *bistable*, which changes states each time it is triggered (called a flip-flop).

murmur A heart sound which is caused by blood flow through an incompetent valve.

Murphy's Law A universal statement in engineering: "Whatever can go wrong— will!" This law summarizes all elements of human fallibility as well as the capricious nature of inanimate objects.

muscle A bundle of individual muscle fibers (or muscle cells), each of which is from 10 to 100 microns in diameter. Skeletal muscle is found in arms, legs, etc., and is completely under voluntary control. Smooth involuntary muscle is found lining the intestine. Skeletal and cardiac muscle are both said to be striated, in that they appear to have striations.

mutual capacitance Capacitance between two conductors with all other conductors connected together and to a grounded shield. (Does not apply to shielded single conductor.)

mutual impedance Between any two pairs of terminals of a network, the ratio of the open-circuit potential difference between either pair of terminals to the current applied at the other pair of terminals, all other terminals being open.

mutual inductance The common inductance of two coupled electrical circuits which determines, for a given rate of change of current in one of the circuits, the electromotive force that will be induced in the other.

mutual interference Man-made interference from two or more electrical or electronic systems which affects these systems on a reciprocal basis.

Mv Abbreviation for megavolt.

mV Abbreviation for millivolt.

mW Abbreviation for milliwatt.

MW Abbreviation for megawatt.

mycosis Disease caused by a fungus.

myelin A fat-like substance which forms a protective, cushioning sheath around certain nerve fibers and bundles.

myelinated axon An axon in which only short, equally spaced sections of its surface (nodes) are excitable, the intervening regions being insulated by the lipoid substance, myelin.

myelogram Radiographic imaging of the spinal cord.

myeloid leukemia A disease in which primitive polymorphonuclear leucocytes appear in large numbers.

myelopathy Any neurological disorder arising from disease of the spinal cord.

mylar A registered trademark of E. I. duPoint deNemours & Co., designating their polyester film.

mylar capacitor A fixed capacitor constructed like a paper capacitor, but using a very thin sheet of my mylar plastic. Smaller than a paper capacitor of the same rating.

myocardial Pertaining to the muscular tissue of the heart.

myocardial electrodes See *implantable pacemaker*.

myocardial infarct An area of necrosis in the myocardium resulting from obstruction of circulation to the area.

myocarditis Inflammation of the heart muscle that is associated with or caused by a number of infectious diseases, toxic chemicals, drugs, and traumatic agents.

myocardium The middle muscular layer of the heart wall.

myoelectric signals Complex pulse potentials of 10 to 1,000 microvolts with durations between one and ten milliseconds. As generated by a muscular effort, recorded from the body surface.

myogenic Originating from muscular tissue.

myograph A recorder of forces of muscular contraction. See also *electromyograph*.

myographic See *myograph*.

myokinesimeter An apparatus for measuring response of a muscle to stimulation by electric current.

myoma A tumorous growth composed of muscular tissue.

myomectomy Removal of a myoma; usually referring to a fibroid from the uterus.

myopathy Any primary disease of muscle.

myophone An instrument for making audible the sound of a muscle contracting.

myositis Inflammation of a muscle. May follow stretching of an injured muscle. Its fibers and hematoma are replaced by cancellous bone. The condition can be prevented by resting the injured muscle.

myriametric waves Signals in the very-low frequency bands (3 kHz to 30 kHz).

myxedema See *hypothyroidism*.

myxoma Tumor of connective tissue containing mucoid material.

N

n Abbreviation for nano (one-billionth or 10^{-9}).

nA Abbreviation for nanoampere.

NAND A function that is true if either A or B is false.

NAND circuit An AND followed by a NOT. Its output is off when all inputs are on. If any one input is off, the output is on.

NAND gate 1. An AND gate followed by an inverter. All its inputs have to be taken to a logic 1 state before the output will fall to a 0. Using the opposite logic polarity, this type of gate becomes a NOR gate. 2. A combination of a NOT and an AND circuit. A binary circuit with two or more inputs and a single output in which the output is logic 0 only if all inputs are logic 1, and is logic 1 if any input is logic 0.

nano A prefix meaning one-thousandth of a millionth, one billionth.

nanoampere (nA) One millimicroampere, 10^{-9} ampere.

nanofarad (nF) One-billionth of a farad, equal to 10^{-9} farad, 0.001 microfarad, and 1,000 picofarads.

nanohenry (nH) One-thousandth of a microhenry, 10^{-9} henry. 1,000 picohenry.

nanosecond (nS) 10^{-9} second, formerly millimicrosecond. A measure of time equivalent to one-billionth of a second.

nanovolt (nV) One-thousandth of a microvolt, 10^{-9} volt.

nanovoltmeter A voltmeter sufficiently sensitive to give readings in thousandth of microvolt.

narcissism Self-love, as opposed to object-love (love of another person). Some degree of narcissism is considered healthy and normal, but an excess interferes with relations with others.

narcosis A state of stupor or unconsciousness produced by a narcotic drug.

narcotic A drug which relieves pain or induces sleep or stupor.

National Electrical Code A set of standards published by the National Fire Protection Association, Boston, Massachusetts, covering the installation of electric conductors and equipment in public and private buildings and on their premises. It has the force of law only when enforced by municipalities or states.

natural frequency The frequency at which a system with a single degree of freedom will oscillate upon momentary displacement from rest position by a transient force in the absence of damping.

natural heart See *heart*.

natural pacemaker A small area of special bioelectrical tissue within the heart, known as the sinoatrial node, that sets the frequency of cardiac contractions. However, in disease, natural pacemakers located in other portions óf the heart can establish slower but regular rhythms.

natural period Period of the free oscillation of a body or system. When the period varies with amplitude, the natural period is the period when the amplitude approaches zero.

natural resonance Resonance in which the period or frequency of the applied agency maintaining oscillation is the same as the natural period of oscillation of a system.

nausea An urge to vomit.

near field In echographic studies, the region of the ultrasonic beam within a distance of a^2/λ of the transducer, where a is the radius of the transducer and λ the wavelength.

near infrared That nonvisible radiant energy with wavelengths nearest the red end of the visible spectrum.

necropsy Examination of a body after death.

necrosis Death of tissue, as individual cells, groups of cells, or in small localized areas.

need Something an individual perceives as being useful or essential.

needle biopsy A procedure used for acquiring a tissue specimen from the liver, kidney, or prostate gland, by means of a hollow needle. The excised tissue is subjected to laboratory study to ascertain the patient's condition, or to determine effects of drugs, or progress in a course of treatment.

needle electrode Small instrument used for subcutaneous electrical recording and stimulating.

negative 1. Said of the pole, a battery or terminal of a circuit which has an **excess** of negative charges called electrons. 2. The opposite of positive. 3. The terminal of a voltage source from which electrons flow, or to which the conventional direct current returns.

negative bias A voltage applied to the control grid of an electron tube to make it negative with respect to the cathode so it operates on the proper portion of its characteristic curve.

negative charge 1. Type of charge in which the object in question has more than its normal number of electrons. 2. Condition in a circuit when the element in question retains more than its normal quantity of electrons.

negative electricity A body is said to possess negative electricity when it contains or possesses an excess of electrons.

negative electrode 1. In a primary cell, the body of conducting material which serves as the anode when the cell is discharging and to which the negative terminal is connected. 2. The cathode of a tube.

negative feedback Feedback from a high-level point to a low-level point of an amplifier, so phased as to reduce the net gain of the amplifier. Also called *degeneration, inverse feedback,* and *reversed feedback.*

negative ion generator A device which bombards air molecules with electrons so that molecules are ionized negatively.

negative logic 1. A form of logic where the more positive voltage logic swing represents a zero and the more negative swing represents a one. 2. A form of logic where zero is assigned to the high level and logic one to the low level.

negative pressure Blood pressure that is less than atmospheric pressure.

negative resistance A property of a circuit in which a reduction in the applied voltage causes an increase in current. The carbon arc, dynatron tube, tunnel diode, and positive feedback amplifier all exhibit negative resistance.

negative-resistance oscillator An oscillator obtained by connecting a resonant circuit to a two-terminal negative-resistance device. Oscillation will be at the resonant frequency.

negative terminal The terminal from which electrons flow, and toward which a conventional current flows in the external circuit.

neon-bulb oscillator A simple relaxation oscillator in which a capacitor repeatedly charges to the ionization potential of the gas within the neon bulb. Ionization rapidly depletes the charge in the capacitor and a new charge cycle commences.

neon glow lamp A glow lamp that contains neon gas, and produces a characteristic red glow. Also called *neon lamp*.

neon lamp See *neon glow lamp*.

neon oscillator Relaxation oscillator in which a neon tube or lamp serves as the switching element.

neoplasm A tumor. An abnormal local multiplication of some type of cell. A neo-plasm may be either benign if it shows no tendency to spread, or malignant if the growing cells infiltrate surrounding tissues and invade other parts of the body.

neper A unit similar to the decibel, but based on Napierian logarithms. The number of nepers is equal to the natural logarithm of the square root of the ratio of the two voltages, or two currents. One neper is equal to 8.686 decibels.

nephritis Inflammatory, acute, or chronic disease of the kidneys, which usually follows some form of infection or toxic chemical poisoning. It impairs renal function, causing headache, dropsy, elevated blood pressure, and appearance of albumin in urine.

nephrocapsulectomy Operation to remove a nephron from the kidney structure.

nephrolithotomy Removal of a stone from the interior of the kidney.

nephron The fundamental filtration unit of the kidney. Each kidney contains about a million nephrons, and each nephron consists of a filtering mechanism (the glomerulus) which removes soluble waste from the blood, and a long tubule which is specialized to reabsorb substances from the urine which pass through it. The modified filtrate of the blood plasma which is thus produced is collected in the renal pelvis and leaves the kidney in the ureter.

nephrosis Disease of the kidney, especially one characterized by degenerative lesions of the renal tubes.

nephrostomy Surgical opening into the kidney, in order to remove fluid from it.

nerve A bundle of nerve fibers which convey the impulses of movement and sensation.

nerve fiber A cylindrical extension of the nerve cell body which conducts the nerve impulse from the cell body to the terminal ending. The length can vary from fractions of an inch to several feet as, for example, in the nerve fiber of the sciatic nerve, extending from the foot to the base of the spinal cord.

nerve impulse An electrochemical process that sweeps along the nerve fiber. As the impulse proceeds down the fiber, electrical currents originating in the fiber spread out into the surrounding tissue fluid which can be recorded with a pair of electrodes and suitable electronic amplifiers as the action potential. In the organism, nerve impulses ceaselessly enter and leave the central nervous system via the peripheral nervous system. Those entering are sensory receptor impulses and those leaving are effector impulses. See *potential, action,* and *action current.*

nervous system Consists of nerve cells and nerve fibers, which are prolongations of cells and have no independent life when the cell dies. The fibers are purely transmitting cables, which lie in bundles, known as nerves, outside the central nervous system. One such cable, the sciatic nerve, reaches from the waist to the sole of the foot. The nerve cells are mostly situated in the central nervous system which consists of the spinal cord, running down inside the backbone, and the brain. The cells establish contact with each other by synapses, junctions at which fibers from two or more cells come in close contact. The whole system is designed to receive impressions from various parts of the body and to act upon them. Thus it has a receiving part (sensors), with relay stations up the cord to the cerebrum and a "handing-out" part running in the reverse direction to all parts of the body and causing movement of muscle groups as and when required. The two parts are connected at various levels. If a message and its reply are connected by conscious effort, the connection between the paths is made in the cerebrum. If the response is given unconsciously, the connection is made at a lower level (spinal cord) and the resulting movement is said to be reflex. One part of the nervous system, the autonomic system, is set apart for dealing with the "housekeeping" affairs which must be attended to, whatever else is missed (breathing, heartbeat, circulation, movements of stomach and bowels, manufacture of urine, and so on).

nesting In a computer including a routine or block of data within a larger routine or block of data.

network 1. A collection of two or more interrelated circuits. 2. An electrical energy distribution system. 3. A group of radio or television transmitter and receiver installations interconnected to disseminate entertainment or information. 4. Organization of stations capable of intercommunications but not necessarily on the same channel. 5. Two or more interrelated circuits. 6. Connected system of nodes and branches geographically or functionally bounded.

network constant 1. One of the established values entering into a functional equation, and corresponding to some characteristic, property, dimension, or degree of freedom. 2. One of the resistance, inductance, mutual inductance, or capacitance values involved in a circuit or network. If these values are constant, the network is said to be linear. See also *parameter.*

network filter A combination of electrical elements. For example, interconnected resistors, coils, and capacitors that present relatively small attenuation to signals of a certain frequency, and great attenuation to all other frequencies.

network synthesis Derivation of the

configuration and element values of a network with given electrical properties.

neural   Relating to nerves.

neuralgia   1. Brief attack of acute and severe shooting pain along the course of one or more peripheral nerves, usually without clear cause. 2. Pain in the distribution of a nerve. The cause may be irritation of the nerve by some bony structure or a tumor but frequently the cause cannot be ascertained. For intractable pain, interruption of the sensory fibers of the nerve is often helpful.

neuritis   Inflammation or degeneration of one or more peripheral nerves, causing pain, tenderness, tingling sensations, numbness, paralysis, muscle weakness, and wasting and disappearance of reflexes in the area involved. The cause may be infectious, toxic, nutritional (vitamin B1 deficiency), or unknown.

neuroelectricity   Electricity generated in a living nervous system.

neurology   Study of diseases of the nervous system.

neuromuscular junction   The synapse formed at a muscle fiber by the juxtaposition of the nerve endings of the motor nerve fiber and the motor end plate of the muscle cell. It is believed that upon arrival of the nerve impulse, a chemical agent, acetylcholine, is released at the endings, diffuses across the interspace, and depolarizes the endplate. The local electrical current flowing outward through adjacent regions of the muscle cell membrane serves as a stimulus and gives rise to a propagated action potential along the muscle fiber.

neuron   1. A nerve cell. Each nerve cell, like all biological cells, begins in the embryonic stage as a spheroidal mass. This mass in time elaborates small filamentous processes, the dendrites, which apparently have some sort of signal-receiving function; the cell body, which is primarily a data processor; and one comparatively long process, the axon or nerve fiber, along which the impulse passes to the effector organ or to other nerves. The axon ends in fine branches or nerve endings. Nerve cells take on different shapes according to their connectivity. Connections between nerves, and between nerves and effector organs are through synapses, the biological analog of capacitors, across which impulses pass without direct connection. 2. A nerve cell with its processes, collaterals, and terminations, considered as a unit of the nervous system.

neuronal   See *neuron*.

neutral conductor   Of a polyphase circuit, or a single-phase, three-wire circuit, that conductor which is intended to have a potential such that the potential differences between it and each of the other conductors are approximately equal in magnitude and are also equally spaced in phase.

neutral ground   A ground connection to the electrically neutral point of a circuit, machine, or system.

neutron   1. A neutral (uncharged) elementary atomic particle that together with protons constitutes the nucleus of the atom. 2. An atomic particle with zero charge and a mass approximately that of a hydrogen atom. Neutrons are highly penetrating and when passing through matter are attenuated exponentially while colliding with nuclei (they do not collide with electrons). In a free state, neutrons decay into a proton and an electron. 3. Fundamental particle of matter discovered by Chadwick in 1932. It has a mass slightly more than that of a proton but no electrical charge. Being neutral it is unaffected by electric forces and so makes a useful projectile for nuclear physics. It occurs in the nuclei of all atoms except that of normal hydrogen.

neutropism Predilection of an infecting organism for nervous tissue.

nF Abbreviation for nanofarad.

nH Abbreviation for nanohenry.

nickel-cadmium cell The most widely used rechargeable sealed cell. Nickel-cadmium gives a flat voltage discharge characteristic, nominal cell voltage of 1.25V, and good low-temperature operation.

nidation Implantation.

nixie tube A gas glow tube having multiple cathodes which are energized in patterns to form any single digit. Also called a *numerical readout tube.*

NMOS 1. N-channel MOS, the opposite of PMOS, where N-type source and drain regions are diffused into a P-type substrate to create an N-channel for conduction. NMOS is from two to three times faster than PMOS. 2. N-channel device. Active carriers are electrons flowing in N-channel between N-type source and drain diffusion beds in a P-type silicon substrate.

nocturia Excessive urination at night.

nodal rhythm A cardiac arrhythmia in which the heart rate is established by the AV nodal tissue, rather than by the normal SA pacemaker. Because the impulses originate in the AV node, they travel upward to the atria and downward to the ventricles. The upward or retrograde conduction in the atria often produces an inverted P wave in the electrocardiogram.

node 1. The terminal of any branch of a network, or a terminal common to two or more branches. 2. That point, line, or surface in a standing wave where some characteristic of the field has essentially a zero amplitude.

noise 1. Random electrical disturbances generated by current flow and thermal electron motion. Electrical noise is a basic limitation of all electrical circuits and becomes particularly severe when dealing with weak signals and high-gain amplifiers. 2. Any unwanted disturbance within a useful frequency band. Noise is often a problem in the evaluation of electroencephalograms and signals recorded from nerve or muscle cells. On occasion, it can cause errors in the interpretation of an electrocardiogram. Sophisticated design is usually required in the development of patient monitoring systems because of noise inherent in signals of biologic origin.

noise figure The ratio of (a) the noise power on the output of a transducer to (b) the portion of that noise which is attributable to thermal noise if the input terminal of the transducer is at the standard noise temperature of 290 degrees Kelvin. It is equal to the input signal-to-noise ratio divided by the output signal-to-noise ratio.

noise filter 1. Combination of electrical components which inhibits extraneous signals from passing through or into an electronic circuit. 2. Combination of one or more choke coils and capacitors inserted between the power cord plug of a radio receiver and a wall outlet to block noise interference that might otherwise reach the receiver through the power line.

noise immunity 1. A measure of the insensitivity of a logic circuit to triggering or reaction to spurious or undesirable electrical signals or noise, largely determined by the signal swing of the logic. Noise can be in either of two directions, positive or negative. 2. The largest noise voltage transient level on any input which a logic circuit is guaranteed to reject; i.e., not respond to. 3. See also *dc noise margin.*

noise margin 1. The voltage amplitude of an extraneous signal which can be algebraically added to the noise-free worst-case input level of a logic circuit before the output voltage deviates from the allowable logic voltage levels. The term input generally refers to logic input terminals, ground reference terminals, or power supply terminals. 2. The difference between the operating voltage of a binary logic circuit and the threshold voltage.

noise pulse A spurious signal of short duration and of a magnitude considerably in excess of the average peak value of the ordinary system noise. Also called *noise spike*.

noise spike See *noise pulse*.

nominal impedance Impedance of a circuit under conditions at which it was designed to operate. Normally specified at center of operating frequency range.

nomogram A graphic display of several interrelated variables. Knowing two variables permits location of a third.

nomography A graphic technique by which the relation between any number of variables may be represented graphically on a plane surface, such as a piece of paper.

noncoherent radiation A form of radiation where there are no definite phase relations between various points in a cross-section of the radiated beam.

nondestructive readout A memory device or system in which the stored data are not lost during the process of being read out. Such memories, therefore, do not require a write operation immediately after each read operation, as do destructive types.

nonerasable storage Storage media in a computer used for containing information which cannot be erased and reused, such as punched paper tapes, and punched cards. See also *fixed storage*.

noninductive capacitor A capacitor in which the inductive effects at high frequencies are reduced to a minimum. Foil layers are offset during winding so that an entire layer of foil projecting at either end is connected together for contact-making purposes. A current then flows laterally rather than spirally around the capacitor and the inductive effect is minimized.

noninvasive techniques Methods of diagnosis and treatment which do not require "invasion" of the patient's body. Echocardiography and phonocardiography are two examples of noninvasive techniques.

noninverting connection The closed-loop connection of an operational amplifier when the forward gain is positive for dc signal (0° phase shift).

noninverting input An input terminal of a differential amplifier that produces an output signal of the same phase.

nonlinear Describing any device in which the output is not related to the input by a simple constant. Not proportional.

nonlinear distortion Distortion which occurs due to the transmission properties of a system, being dependent upon the instantaneous magnitude of the transmitted signal. NOTE: Nonlinear distortion gives rise to amplitude and harmonic distortion, intermodulation and flutter.

nonlinear feedback control system Feedback control system in which the relationships between the pertinent measures of the system input and output signals cannot be adequately described by linear means.

nonlinear impedance An impedance which is not constant, but varies in some manner with the voltage impressed on it or the current through it. A sine-wave voltage impressed on a nonlinear impedance will result in a current containing harmonics of the voltage wave.

nonmagnetic Describing a material which has no effect on a magnetic field, either to attract it or repel it. Such a material has a permeability of 1. See also *diamagnetic, ferromagnetic,* and *paramagnetic.*

nonpolar Describing an insulating material whose dielectric constant does not change with a change in frequency, or a capacitor which can be connected without regard to polarity.

nonvolatile memory 1. A computer memory that does not lose stored information either during readout or as a consequence of power shutdown or failure, e.g., magnetic tapes, drums, or cores. 2. The ability to retain information in the absence of power.

nonvolatile storage A storage medium which retains information in the absence of power and which may be made available upon restoration of power; e.g., magnetic tapes, cores, drums, and discs. Contrasted with volatile storage.

NOR A function that is true if both A and B are false.

NOR device A device which has its output in the logical one state if and only if all of the control signals assume the logical zero state.

NOR element A gate with multiple inputs and one output that is energized only if all inputs are zero. All other logic functions can be formed from combinations of this element.

NOR gate 1. An OR gate followed by an inverter. If any of its inputs are at the logic 1 level, its output will be a logic 0. Using the opposite logic polarity, this type of gate becomes a NAND gate. 2. A combination of a NOT and an OR circuit. A binary circuit with two or more inputs and a single output in which the output is logic 0 if any one of the inputs is logic 1 and is logic 1 only if all the inputs are logic 0.

norm A fixed or ideal standard.

normal electrode Standard electrode used for measuring electrode potentials.

normal impedance See *free impedance.*

normalize In computer programming, to adjust the exponent and fraction of a floating point quantity so that the fraction lies in the prescribed normal standard range. See also *point.*

normally closed contact A contact pair which is closed when the coil of a relay is not energized. Also called *break contact.*

north pole 1. Pole of a magnet at which magnetic lines of force are considered as leaving the magnet. The lines enter the south pole. 2. North–seeking pole of a magnet.

nose clamp In pulmonary function studies, a padded, spring-loaded device which occludes the nostrils so that the subject's inspiration and expiration comes solely through the mouth, a much easier airway to instrument than the nostrils.

NOT A logical operator having the property that if P is a statement, then the not of P is true if P is false, false if P is true.

notch filter An arrangement of electronic components designed to attenuate or reject a specific frequency band with sharp cutoff at either end.

NOT circuit A binary circuit with a single input and a single output in which the output is always the opposite of the input. Also called *inverter circuit.*

NOT device A device which has its output in the logical one state if and only if the control signal assumes the logical zero state. The NOT device is a single input NOR device.

NOT-gate Electronic circuit whose output is not energized if one or more of its inputs are energized.

NOT majority If more than half of the number of inputs in a logic circuit are true, then the output is false; otherwise, the output is true.

NPN semiconductor Double junction formed by sandwiching a thin layer of P-type material between two layers of N-type material of a semiconductor.

ns Abbreviation for nanosecond, 10^{-9} second.

N-type semiconductor A semiconductor material, such as germanium or silicon, which has a small amount of impurity, such as antimony, arsenic, or phosphorous, added to increase the supply of free electrons. Such a material conducts electricity through movement of electrons.

nuclear battery Direct conversion secondary cell or battery using radioisotopes as the energy source.

nuclear device A nuclear explosive used for peaceful purposes, tests, or experiments. The term is used to distinguish these explosives from nuclear weapons, which are packaged units ready for transportation or use by military forces.

nuclear powered pacemaker A cardiac pacemaker which supplies timed impulses to the specialized conduction system of the heart, and is powered by a nuclear energy cell which yields electricity from controlled nuclear bombardment. Overcomes the need for periodic surgical replacement of pacemaker batteries, because the power source is long-lived.

nuclear radiation Neutrons, alpha, beta, and gamma rays from primary or secondary power plants, nuclear weapons, or natural space radiation. Only neutrons and gamma rays penetrate shielding. Neutron energies range to 20 MeV (about 35% at 0.8 MeV). Gamma ray energies range from about 300 keV to at least 8.0 MeV (average about 1.5 MeV).

nuclei Plural of nucleus.

nucleic acid A biochemical substance that is of prime importance in cell duplication. Small changes in the building blocks of the nucleic acids can lead to significant errors in the structure or function of successor cells.

nucleon A constituent of a nucleus, either a proton or a neutron.

nucleus 1. In a biologic sense, the central, dense body occurring in almost all cells that contains the basic genetic code in the deoxoribonucleic acid (DNA) of the cell. 2. In a physical sense, the small centrally placed, positively charged portion of the atom in which is concentrated almost all of its mass. 3. A central mass about which gathering or concentration takes place. 4. A group of nerve cells in the central nervous system.

nuclide Any atomic configuration capable of more than a transient existence. A specific nucleus is characterized by a definite number of neutrons.

null 1. A position of zero or minimum reading on a measuring instrument. 2. A position of minimum tone volume when balancing an audio frequency bridge.

null-balance A condition of balance in a device or system which results in zero output or zero current input.

null detector circuit A circuit which determines when two circuits are in exact electrical balance, having the same voltages or currents.

number systems 1. A systematic method for representing numerical quantities in which any quantity is represented as the sequence of coefficients of the successive powers of a particular base with an appropriate point. Each succeeding coefficient from right to left is associated with and usually multiplies the next higher power of the base. The succeeding coefficient from right to left is associated with and usually multiplies the next higher power of the base. The first coefficient to the left of the point is associated with the zero power of the base. For example, in decimal notation 371.426 represents $(3 \times 10^2) + (7 \times 10^1) + (1 \times 10^0) + (4 \times 10^{-1}) + (2 \times 10^{-2}) + (6 \times 10^{-3})$.

The following are names of the number systems with base 2 through 20: 2, binary; 3, ternary; 4, quaternary; 5, quinary; 6, senary; 7, septenary; 8, octal or octonary; 9, novenary; 10, decimal; 11, undecimal; 12, duodecimal; 13, terdenary; 14, quaterdenary; 15, quindenary; 16, sexadecimal or hexadecimal; 17, septendecimal; 18, octodenary; 19, novemdenary; 20, vicenary; 32, duosexadecimal or duotricinary; and 60, sexagenary. The binary, octal, decimal, and sexadecimal systems are widely used in computers.

numerical control The direct control of a system by conversion of numbers into physical values. The controlled system can be an individual machine or process or a group of machines or processes (or machines and processes).

numerical readout tube See *nixie tube.*

numeric coding A system of abbreviation used in the preparation of information for machine acceptance by reducing all information to numerical quantities, in contrast to alphabetic coding.

nV Abbreviation for nanovolt.

Nyquist interval Maximum separation in time which can be given to regularly spaced instantaneous samples of a wave of a given bandwidth for complete determination of the waveform of the signal.

Nyquist rate The maximum rate at which independent signal values can be transmitted over a specified channel without exceeding a specified amount of mutual interference.

nystagmus Involuntary oscillations of the eyeball; sometimes congenital; sometimes a symptom of brain disease, ocular problem, or lesion in the inner ear.

obese Corpulent; excessively fat.

Obl Abbreviation for oblique.

oblique Slanting, deviating from the perpendicular or horizontal.

oblique muscles 1. A pair of external muscles of the eyeball, arranged above and below it. 2. A pair of large muscles of the abdominal wall, one internal, the other external.

obsession Persistent, unwanted idea or impulse that cannot be eliminated by logic or reasoning.

obturator nerve A spinal nerve which provides controlling impulses to abductor muscles in the region of the thigh, the skin of the hip, and the knee joint.

occipital Relating to the back of the head; also, to a major lobe of the brain.

occiput Relating to the back part of the head or skull.

occlusion 1. The mechanism by which vapors, gases, liquids, or solids become entrapped within the folds of a substance during working or solidification. 2. A press-

ing together of a vessel, thereby closing its lumen. 3. The shutdown of a vessel by accretion of deposits on the vessel wall, until the lumen is filled.

OCR 1. Abbreviation for optical character recognition, a technology which enables a machine to automatically read and convert typewritten, machine-printed, mark-sensed, or hand-printed characters into electrical impulses for processing by a computer. OCR devices are designed to read man-readable data and convert it directly to a computer language. 2. The machine recognition of printed or written characters based on inputs from photoelectric transducers.

octal See *number system.*

octal base A universal base for an electron tube which has eight equally spaced pins, oriented by a central key.

octave A band of frequencies covering a range of 2 to 1. Starting from a given frequency, one octave higher is twice that frequency; one octave lower is half that frequency.

oculomotor nerve The third cranial nerve, arising in the midbrain. It innervates the musculature which moves the eye.

odd-even check In a digital computer, a digit carried along as a check. This digit is 1 if the total number of ones in the machine word is even, and it is 0 if the total number of ones in the machine word is odd.

odd harmonic Any harmonic whose frequency is the fundamental frequency multiplied by an odd whole number. The odd harmonics of 60 Hz are 180 Hz, 300 Hz, 420 Hz, 540 Hz, etc.

odd parity A form of error-detection used with binary-coded data in which one bit is added to the data for each character such that the total number of ones in the data plus the parity bit is always an odd number.

-ode Suffix meaning resemble, like, or form.

odontoid Toothlike.

odontoma Tumor arising from a tooth or a developing tooth.

oersted (Oe) Unit of magnetic field strength (magnetic intensity, magnetizing force) H in centimeter-gram-second electromagnetic system. At any point in a vacuum, the value of the magnetic intensity in oersteds is equal to the force in dynes exerted on a unit magnetic pole placed at that point.

off-delay A circuit that retains an output signal some definite time after the input signal is removed.

off-ground The voltage above or below ground at which a device is operated.

off-line 1. Pertaining to equipment or programs not under the direct control of the central processor. 2. Pertaining to a computer that is not actively monitoring or controlling a process or operation, or pertaining to a computer operation performed while the computer is not monitoring or controlling a process or operation. 3. Pertaining to the use or review of recorded, processed data, in peripheral equipment, so that the central processor need not be occupied thereby.

off-line equipment The peripheral equipment or devices not in direct communications with the central processing unit of a computer. Synonymous with *auxiliary equipment*.

off-line operation 1. Computer operation independent of the time base of the actual inputs. 2. Operation of peripheral equipment independently of the central processor of a computer system.

offset The change in input voltage required to produce a zero output voltage in a linear amplifier circuit. In digital circuits, it is the dc voltage on which a signal is impressed.

offset current A dc error current appearing at either input terminal of a dc amplifier.

off-the-shelf Available for fast delivery from stock, as opposed to requiring custom design or manufacturing to order.

ohm (Ω) The unit of electrical resistance. One volt will force one ampere of current through one ohm.

ohm-centimeter The unit of resistivity; the reciprocal of conductivity. Numerically, it is the resistance in ohms between opposite faces of a cube of the material which is one centimeter on each side.

ohmic contact An electrical contact whose resistance is directly proportional to the current flowing through it.

ohmmeter A measuring instrument which

indicates directly in ohms the resistance of a circuit to which it is connected. Incorporates a source of measuring potential.

Ohm's law Current in an electric circuit is directly proportional to the electromotive force in the circuit. NOTE: The fundamental law of electrical circuits and is true of all metallic circuits and most circuits containing resistance. The most common form of the law is $E = IR$, where E is the electromotive force or voltage across the circuit, I is the current flowing in the circuit, and R is the resistance of the circuit.

ohms per square The resistance of any square area of thin-film resistive material as measured between two parallel sides.

ohms per volt Sensitivity rating for measuring instruments, obtained by dividing the resistance of the instrument in ohms at a particular range by the full-scale voltage value at that range.

-oid Suffix meaning resemble, like or form.

-ol Suffix meaning oil.

olfactory Relating to the sense of smell.

olfactory nerve The first cranial nerve, arising in the forebrain. It innervates the nasal mucosa and conveys the sensory impulses associated with smell.

oligo Prefix meaning deficiency.

oliguria Diminished amount of urine.

olivary body An oval mass of gray matter behind the anterior pyramid of the medulla oblongata.

-oma Suffix meaning swelling or tumor.

omentum A fold of the peritoneum, consisting of fatty tissue, principally in the anterior part of the abdominal cavity, suspended from the stomach. The greater omentum is suspended from the greater curvature of the stomach and hangs in front of the gut. The lesser omentum passes from the lesser curvature of the stomach to the transverse fissure of the liver.

on-delay A circuit that produces an output signal some definite time after an input signal is applied.

one address 1. A single address. 2. A system of machine instruction such that each complete instruction explicitly describes one operation and involves one storage location. Synonymous with *single address* and related to one address instruction.

one-address code In computers, a code using one-address instructions.

one-address instruction In computers, an instruction consisting of an operation part and only one address.

one-level code See *absolute code* and *specific code*.

one-plus-one instruction In computers, a two-address instruction in which one of the addresses always specifies the location of the next instruction to be performed.

ones complement A radix-minus-one complement with the radix equal to two.

one-shot 1. A circuit that produces an output signal of fixed duration when an input signal of any duration is applied. See also *monostable* and *multivibrator*.

one-shot multivibrator See *monostable multivibrator*.

on-line computation A method of study in which a computer is fed raw data through a biophysical amplifier which continuously process vital signs inputs from an organism. The computational results are therefore continually available during the course of the experiment, and conditions can be modified so as to obtain maximum information about the system. This procedure is often used in the cardiac catheterization laboratory.

on-line computer system A computer system in which the input data enter the computer directly from their point of origin, and output data are transmitted directly to where they are used. The intermediate stages such as punching data into cards or paper tape, writing magnetic data into cards or paper tape, writing magnetic tape, or off-line printing are largely avoided.

on-line data processing In computers, processing of data as fast as the information becomes available for processing.

on-line data reduction The processing of information as rapidly as the information is received by the computing system or as rapidly as it is generated by the source.

on-line equipment Descriptive of a system and of the peripheral equipment or devices in a system in which the operation of such equipment is under control of the central processing unit and in which information reflecting current activity is introduced into the data processing system as soon as it occurs. Thus, directly in-line with the main flow of transaction processing. Synonymous with in-line processing and on-line processing.

on-line operation 1. Operation in which input data are fed directly from measuring devices into the computer. Results are obtained in real time, i.e., computations are based on current values of operating data, and answers are obtained in time to permit effective control action to be taken. Can also mean the operation of peripheral equipment in conjunction with the central processor of a computer system. 2. In control terminology, operation in which data are fed directly from the controlled process into the computer and the computer exer-

cises direct control based upon these data. Since the computer's reaction is moment-to-moment without appreciable time lag, results are said to be in real time. 3. A type of system application in which the input data to the system are fed directly from the measuring devices and the computer results obtained during the progress of the event; e.g., a computer receives data from catheterization measurements during a procedure and the computations of dependent variables are performed during the procedure, enabling a change in the conditions so as to produce particularly desirable results.

on-line processing See *on-line equipment*.

on-line system 1. In teleprocessing, a system in which the input enters the computer directly from the point of origin and/or in which output data are transmitted directly to where it is used. 2. In the telegraph sense, a system of transmitting directly into the system.

on-line unit In a computer, the input/output device or auxiliary equipment under direct control of the computer.

op amp Short form for operational amplifier.

open 1. Condition in which conductors are separated so that current cannot pass. 2. Break or discontinuity in a circuit which can normally pass a current.

open core A magnetic core which is within a coil or winding, with the portion of the magnetic path external to the coil or winding being through air. Such a path has a high reluctance, and the magnetic flux is correspondingly low.

open heart surgery Surgical procedures that involve section through the wall of at least one atrium or one ventricle.

opening snap Heart sound which often

precedes the mid-diastolic murmur of mitral stenosis.

open loop   1. Pertaining to a control system in an interactive system in which there is no self-correcting action for error of the desired operational condition, as there is in a closed-loop system. 2. Implies operational amplifier without feedback.

open-loop bandwidth   Without feedback, the frequency limits at which the voltage gain of the device drops off 3 dB below the gain at some lower reference frequency.

open-loop differential voltage gain   See *differential voltage gain*.

open-loop gain   1. The gain of an amplifier having no external feedback. 2. The ratio of the (loaded) output of an amplifier without any feedback to its net input at any frequency. Usually implies voltage gain.

open-loop output impedance   The complex impedance seen looking into the output terminals of an operational amplifier with no external feedback and in the linear amplification region. In closed-loop operation, the output impedance is equal to the open-loop impedance divided by the loop gain. If the open-loop impedance is not more than a few hundred ohms and the loop gain is high enough for good gain accuracy and stability, the closed-loop impedance will be on the order of an ohm or less which can be neglected in most applications.

open-loop output resistance   The resistance looking into the output terminal, of an operational amplifier, operating without feedback and in the linear amplification region.

open-loop system   A control system that has no means of comparing input with output for control purposes.

open-loop voltage gain   Ratio of the change in voltage between differential input terminals of an operational amplifier to the corresponding change in output voltage, with no feedback.

open routine   A routine in a computer which can be inserted directly into a larger routine without a linkage or calling sequence.

operand   Any one of the quantities entering into or arising from an operation. In computer programming, an operand may be an argument, a result, a parameter, or an indication of the location of the next instruction.

operating point   Point on a family of characteristic curves of a vacuum tube or transistor where the coordinates of the point represent the instantaneous values of the electrode voltages and currents for the operating conditions under study or consideration. Also called *quiescent point*.

operating system   A computer program which acts as a supervisor for applications programs and as an aid in using editors, compilers, assemblers, etc. Usually the operating system has a variety of programs available for all users, especially for handling input/output or communications with mass storage devices. An operating system is essential in a time-sharing system but it is not essential in a system that is dedicated to a single application.

operating time   The time period between turn-on and turn-off of a system, subsystem, component, or part during which time operation is as specified. Total operating time is the summation of all operating time periods.

operation   1. A specific action which a computer will automatically perform whenever the instruction calls for it. 2. A surgical procedure. 3. In mathematics, determining a resulting quantity according to some rule when given two or more other quantities. Examples: addition and multiplication in conventional arithmetic, and the Boolean AND and OR operations in Boolean algebra.

operational amplifier An important amplifier in biomedical measurement that performs various mathematical operations. It is designed to be used with external circuit elements to perform a specified computing operation, such as adding, subtracting, integrating, or to provide other specified transfer functions.

operational differential amplifier An operational amplifier which utilizes a differential amplifier in the first stage. Most biomedical amplifiers (e.g., ECG, EEG, EMG) use this amplifier type because it offers high gain with excellent common mode rejection for artifact elimination.

ophthalmia Inflammation of the eye; especially, severe inflammations of the conjunctiva. There is an acute infectious form which occurs in epidemics, especially in schools and military camps.

ophthalmic Pertaining to the eye.

ophthalmoscope An instrument, fitted with a lens, used to examine the interior of the eye.

ophthalmotonometer Instrument to measure the intraocular tension of the eye. Used to detect glaucoma.

optic Relating to the sight.

optical character reader (OCR) A photoelectric device which scans printed or typed copy and produces electrical signals that can be fed directly to a tape puncher or to a telegraph line facility. Obviates the need for manual tape punching, but requires copy using special machine readable

characters, such as those on the lower edge of bank checks.

optical character recognition See *OCR*.

optical coupling 1. Coupling between two circuits by a light beam or light pipe, having transducers at opposite ends, to isolate the circuits electrically. The technique is being used to isolate patient circuits in biomedical instruments, so as to minimize accidental shock hazard. 2. A fiber coupling method in a multichannel recorder, the cathode-ray tube, and light-sensitive recording paper. The fiber optic coupling affords transmission characteristics superior to the best lens, and allows use of minimal writing spot size for optimum resolution.

optical fiber bundles See *fiber optics*.

optical fibers See *fiber optics*.

optically pumped laser See *laser*.

optical pumping The use of visible light to raise the energy level of electrons in a laser material. Typically, short-duration, high-intensity output from an electronic flash lamp is used. This energy is mostly absorbed by the impurity atoms, in an otherwise transparent host solid, e.g., ruby.

optical spectrometer Instrument with an entrance slit, a dispersing device, and with one or more exit slits, with which measurements are made at selected wavelengths within the spectral range, or by scanning over the range. The quantity detected is a function of radiant power.

optical twinning See *twinning*.

optic foramen The opening where the optic nerve enters the skull.

optic nerve The second cranial nerve, arising in the forebrain. It innervates the retinas and conveys the sensory impulses associated with sight.

optimization A method by which a

process is continually adjusted to the best obtainable set of operating conditions.

optimum code A computer code which is particularly efficient with regard to a particular aspect, e.g., minimum time of execution, minimum or efficient use of storage space, and minimum coding time. Related to minimum access code.

optoelectronic integrated circuit An integrated component that utilizes electroluminescence in combination with photoconductivity to perform all or at least a major portion of its intended function.

optoelectronics 1. Circuitry involving solid-state light emitters and detectors. 2. Technology dealing with the coupling of functional electronic blocks by light beams.

optoelectronic transistor A transistor that has an electroluminescent emitter, a transparent base, and a photoelectric collector.

optophone A photoelectric instrument that converts printing and letters to sounds, used as a reading device by the blind.

OR A logical operator which has the property such that if P or Q are two statements, then the statement P OR Q is true or false varies according to the following table of possible combinations:

P	Q	P or Q
False	True	True
True	False	True
True	True	True
False	False	False

oral antidiabetes agents Insulin replacement for some types of diabetes (e.g., isophane insulin and tolbutamide).

orchidectomy Removal of one or both testicles. Castration.

orchis Testicles.

orchitis Inflammation of the testicles.

OR device A device which has its output in the logical zero state if and only if all of the control signals assume the logical zero state.

organ A structure consisting of cells and tissues and performing some specific function.

organic 1. Of, relating to, or containing living organisms. 2. Relating to the organs; thus, organic disease of the heart means that the structure itself is affected.

organic chemistry Chemistry relating to the carbon compounds.

organomercurial diuretic An agent used to remove fluid from the body in certain diseases of the heart, liver, and kidneys.

OR gate (nonexclusive) A multiple-control device that evidences the appearance of any combination of positive-pressure control signals with a single positive output signal. With no positive control signal, the positive OR output signal ceases. An OR gate with a single control connection acts as a straight-through connection.

ortho Prefix meaning straight or correct.

orthogonal At right angles to.

orthogonality See *orthogonal*.

orthopnea The inability to breathe except when in a sitting position.

orthostatic Pertaining to or caused by standing upright.

-ory Suffix meaning place or thing where.

os 1. Bone. 2. Mouth or opening.

osciducer A transducer in which information pertaining to the stimulus is provided in the form of deviation from the center frequency of an oscillator.

oscillate 1. To vary around a mean value. 2. To swing back and forth. 3. To vibrate in a uniform manner. 4. Of an electronic circuit, to generate a varying electrical output. 5. To repeat a cycle or state with strict periodicity.

oscillating transducer A transducer in which information pertaining to the stimulus is provided in the form of deviation from the center frequency of an oscillator.

oscillation 1. A periodic variation between maximum and minimum values, as an electric current. 2. A single swing of an oscillating object between the two extremes of its arc.

oscillator 1. Electronic device which generates alternating current power at a frequency determined by the values of certain constants in its circuits. An oscillator may be considered an amplifier with positive feedback with circuit parameters that restrict the oscillations of the device to a single frequency. 2. Nonrotating device which is capable of setting up and maintaining oscillations at a frequency determined by the physical constants of the system, such as an amplifier, sparks, or arc generator. 3. A self-excited circuit having a loop gain greater than unity, capable of converting direct current into alternating current, of a frequency determined by the inductive and capacitive constants of the circuit. 4. Device used to generate and repetitiously oscillate at radio frequencies. 5. A mechanical device which changes position in a regular, repeating manner, as a pendulum.

oscillatory discharge Alternating current of gradually decreasing amplitude which, under certain conditions, flows through a circuit containing inductance, capacitance, and resistance when a voltage is applied. See also *damped wave*.

oscillogram A photograph of the luminous trace or image produced by an oscilloscope.

oscillograph An instrument for producing a permanent record (called an oscillogram) of the instantaneous values of rapidly varying electrical quantities as a function of time, or of another electrical or a mechanical quantity. All light-beam recorders are oscillographs. Some employ reflecting mirror galvanometers, while others use the direct light output of a cathode-ray tube. Frequency response of a medical oscillograph may be from 0 (dc) to 5,000 Hz, or greater.

oscillograph tube Cathode-ray tube used to produce a visible pattern, which is the graphical representation of electric signals, by variations of the position of the focused spot or spots according to these signals.

oscillometer An instrument for measuring oscillations (periodic variations) of any kind.

oscilloscope 1. Measuring instrument which uses a cathode-ray tube and associated circuitry to provide a visual display of an electronic waveform or signal. 2. An instrument in which the horizontal and vertical deflection of the electron beam of a cathode-ray tube are, respectively, proportional to a pair of applied voltages. In the most usual application of the instrument, the vertical deflection is a signal voltage and the horizontal deflection is a linear time base. 3. An instrument consisting of a cathode-ray tube on the fluorescent screen of which is produced a visible pattern or waveform of some fluctuating electrical quantity such as voltage. It is employed to reveal the detailed variations in rapidly changing electric currents, potentials, or pulses. 4. A device which displays the shape of a voltage or current wave on the screen of a cathode-ray tube.

-ose Suffix meaning full of.

OSHA Abbreviation for Occupational

Safety and Health Administration. An agency of the U.S. Government created to ensure that workers perform tasks in an environment which is free of work-related hazards.

-osia Suffix meaning condition or state of.

-osis Suffix meaning condition or state of.

osmol Unit which expresses the ability of disolved substances to cause osmosis and osmotic pressure. One osmol is equal to one gram molecule of nondiffusible and nonionizable dissolved substance.

osmosis The phenomenon of diffusion transfer across a semipermeable membrane separating two solutions of different concentration. For example, if two saline solutions of different strength are separated by such a membrane, water will diffuse from one to the other, until the two solutions are equivalent in strength.

osmotic pressure The force existing across a semipermeable membrane separating two solutions of different strengths.

osseous Like bone, bony.

ossicle 1. Small bone. 2. The three tiny bones of the middle ear, which conduct sonic vibrations from the tympanic membrane to the cochlea.

osteitis Inflammation of bone.

osteoarthritis Disorder due to excessive wear and tear to joint surfaces. Affecting chiefly weight-bearing joints, late in life, and resulting in pain, especially at night, deficient movement, and, in some cases, deformity.

osteoarthrotomy Excision of joint and neighboring bone.

osteoarthropathy Damage or disease affecting the bones and joints.

osteolytic That which is destructive to bone.

osteoma A tumor of the bone.

osteomalacia Softening of bones in adults.

osteomyelitis Inflammation of the bone marrow.

osteopath One who practices osteopathy.

osteopathy A system of diagnosis which attributes diseases to structural derangement of skeletal parts and seeks to relieve symptoms by manipulative and mechanical means of therapy.

osteophony Conduction of sound by bone.

osteoplastic Pertaining to the repair of bones.

osteotomy Operation of cutting through a bone.

otologist Specialist in hearing and the ear.

otology Study of diseases of the ear.

otosclerosis A thickening of the structures of the internal ear which leads to a fixation of the innermost ossicles of the ear (the *stapes*), thus causing conduction deafness. The disease is chronic and progressive.

-ous Suffix meaning to be full of.

out-of-phase Having waveforms that are of the same shape but which do not pass through corresponding values at the same instants.

output 1. Current, voltage, power, or driving force delivered by a circuit or device. 2. Terminals or other places where current, voltage, power, or driving force may be delivered by a circuit or device. 3. In computers, information transferred from internal storage to external storage or to any device outside of the computer. 4. Data that have been processed. 5. State or sequence of states occurring on a specified output channel. 6. Device or collective set of devices used for taking data out of a device. 7. Channel for expressing a state of a device or logic element.

output block 1. In a computer, a segment of the internal storage reserved for data to be transferred out of the machine. 2. Block of computer words considered as a unit and intended to be transferred from an internal storage medium to an external destination. 3. Block used as an output buffer.

output impedance 1. The impedance presented by a device to the load. 2. The impedance looking into the output terminals of a power source, a circuit, or device.

output offset voltage In a differential amplifier, the difference between dc voltages present at the two outputs of a double-ended amplifier (or between one output and ground, for amplifiers having single-ended output), when the two input terminals are grounded.

output resistance The small signal ac resistance of an operational amplifier seen looking into the output with no feedback applied and the output dc voltage near zero.

output saturation voltage Lowest voltage level to which the collector of the output transistor can be reduced without degrading circuit performances.

output unit In computers, a unit which delivers information from the computer to an external device or from internal storage to external storage.

output voltage The maximum output voltage which an amplifier will develop in the linear operating region; i.e., before the onset of saturation.

ovaries Two small oval bodies situated on either side of the uterus in which ova are formed. The ovaries are also endocrine glands and are the source of the female hormone.

overbreathing See *hyperventilation*.

overcompensation A conscious or unconscious process in which a real or fancied psychological deficit inspires exaggerated correction.

overcoupling Coupling between two resonant circuits which is greater than critical coupling. This produces two peaks in the response curve, and provides a wider bandwidth.

overflow The result of an arithmetic operation that exceeds the capacity of the number representation in a digital computer.

overflow storage Additional storage provided in a store and forward switching center of a computer to prevent the loss of messages (or parts of messages) offered to a completely filled line store.

overlay A system whereby segments of a large program are stored on mass storage (usually discs or flexible discs) and called into memory as they are needed, thereby erasing the program segment that previously occupied those locations in memory.

overload 1. In electronics, that quantity of power from an amplifier or other component or from a whole transmission system which is sufficient to produce unwanted waveform distortion. 2. Load greater than the rated load of an electric device. 3. Electrically operated counter which registers the number of times all trunks are busy between the various office units. 4. In an analog computer, a condition existing within or at the output of a computer element that causes a substantial computing error because of saturation of one or more of the parts of the computing element.

overload level Level above which operation ceases to be satisfactory as a result of signal distortion, overheating, damage, etc.

overload margin In an amplifier, the safety margin prior to the onset of overload to avoid clipping on transients. This also enhances the reproduction, giving it a smoother quality in many cases.

overload protection Effect of a device operative on excessive current but not necessarily on short circuit, to cause and maintain the interruption of current flow to the device governed. Also called *overcurrent protection*.

overpotential A voltage above the normal operating voltage of a device or circuit.

overt homosexuality Homosexuality that is consciously recognized or practiced.

ovulation The process of development and discharge of an ovum from an ovary.

ovum The egg cell produced in an ovary of the female.

Owen bridge Four-arm ac bridge circuit used to measure self-inductance in terms of capacitance and resistance. Bridge balance is independent of frequency.

oxidation 1. The process of combining with oxygen; more generally, the process by which atoms lose valence electrons or begin to share them with more electronegative atoms. 2. The reaction of oxygen on a compound, usually detected by a change in the appearance or feel of the surface or by a change in physical properties, or both.

oxide isolation The method of producing electrical isolation of a circuit element by forming a layer of silicon oxide between it and the substrate.

oximeter An instrument for measuring the oxygenation of the blood, usually by measuring light transmission through the ear lobe.

oximetry The measurement of oxygen saturation of the blood.

oxygen capacity The amount of oxygen which blood could carry if its hemoglobin were fully saturated. Each gram of normal hemoglobin has an oxygen capacity of 1.34 ml.

oxygen saturation The amount of O_2 actually combined with hemoglobin as compared to the oxygen capacity, expressed as a percentage. This is the percent of total hemoglobin which exists as oxyhemoglobin. Normally, blood leaving the lungs via the pulmonary vein is about 97 percent saturated.

oxyhemoglobin Hemoglobin which has been oxidized through circulation of blood through the lungs. The amount of oxygen taken up by the hemoglobin depends upon the oxygen tension of the medium surrounding it. At tensions of 100 mmHg and higher, hemoglobin is fully saturated with oxygen.

ozone (O_3) Extremely reactive form of oxygen, normally occurring around electrical discharges and present in the atmosphere in small but active quantities. It is faintly blue and has the odor of weak chlorine. In sufficient concentrations, it can break down certain rubber insulations under tension (such as a bent cable).

ozone-producing radiation Ultraviolet energy shorter than about 220 nanometers which decomposes oxygen (O_2), thereby producing ozone (O_3). Some ultraviolet sources generate energy at 184.9 nanometers which is particularly effective in producing ozone.

P

p Abbreviation for the prefix pico (10^{-12} A).

pA Abbreviation for picoampere.

PA Abbreviation for posteroanterior or pulmonary artery.

pacemaker 1. The heart's natural pacemaker, a cluster of cells (the sinoatrial node) which periodically depolarize, exciting the atrioventricular node. 2. An instrument used for starting and/or maintaining the heartbeat. This instrument is essentially a pulse generator and its output is applied either externally to the chest or internally to the heart muscle. In cases of cardiac standstill, the use of the pacemaker is temporary—just long enough to start a normal heart rhythm. In cases requiring long-term pacing, the pacemaker is surgically implanted in the body, and its electrodes are in direct contact with the heart. Also called *pacer.*

pacer See *pacemaker.*

pacing threshold The minimum voltage level needed to pace the heart.

pack 1. In computers, to include several short items of information into one machine item or word by using different sets of digits to specify each brief item. 2. To compress data in a storage medium by taking advantage of known characteristics of the data in such a way that the original data can be recovered, e.g., to compress data in a storage medium by making use of bit or byte locations that would otherwise go unused. 3. A surgical kit, including disposable sterile drapes, usually intended for a specific type of surgical procedure.

packaging The process of physically locating, connecting, and protecting devices or components.

packaging density The number of devices or equivalent devices per unit volume in a working system or subsystem.

PaCO$_2$ Carbon dioxide content of the blood as measured in the pulmonary artery.

pad 1. Nonadjustable passive network which reduces the power level of a signal without introducing appreciable distortion. 2. Assembly of resistors which presents the proper input and output impedances to the circuits with which it is connected and which provides a fixed value of energy loss. 3. Device which introduces transmission loss into a circuit. It may be inserted to introduce loss or to match impedances. See also *switching pad*. 4. An absorptive, loose, sterile fabric, used in surgery and feminine hygiene. 5. A resilient cushion used on surgical tables and chairs. 6. A protective covering for diathermy electrodes and internal defibrillator paddles. See *land*.

paddle button The switch (or switches) on the handles of defibrillator electrodes, by which the defibrillating energy dosage is discharged into the patient.

paging 1. The separation of a computer program and data into fixed blocks, often 1,000 words, so that transfers between disc and core can take place in page units rather than as entire programs. 2. The transmission of a modulated radio-frequency carrier, bearing tone combinations which selectively actuate one of many battery-operated, pocket-carried receivers. The user whose receiver is activated then calls a central station to receive his message.

pain threshold The level of stimulation at which pain is first experienced.

pain tolerance An individual's ability to withstand pain.

pair Two wires of a single circuit associated together by twisting, binding, or by an overall braid or cover.

palate The roof of the mouth.

palliative A medicine which relieves symptoms only and does not affect the underlying disease.

pallidectomy Operation used in Parkinson's disease to decrease the activity of part of the lentiform nucleus in the base of the brain.

pallidotomy A surgical procedure used in Parkinson's disease to produce lesions in the globus pallidus, with the objective of reducing activity in that area. This reduces the involuntary movements associated with that disease.

pallor Paleness.

palpation Examination by touch or feel, as in palpating the pulse, breasts, liver, or spleen.

palpitation Unusually rapid heartbeat, sufficiently fast and strong to make the patient aware of its action.

pancreas A large gland situated behind the stomach. It secretes four principal digestive enzymes: renin, trypsin, amylopsin, and steapsin. These pass into the duodenum through the pancreatic duct. Its most important secretion is insulin, which helps to regulate the metabolism of carbohydrates.

pancreatitis Inflammation of the pancreas, either mild or acute and fulminating. The chronic form is characterized by recurrent attacks of diverse severity. Symptoms are sudden abdominal pain, tenderness and distention, vomiting and, in severe cases, shock and circulatory collapse.

panophthalmia Generalized inflammation of the eyeball.

panotitis Inflammation of the middle and internal ear.

paper capacitor A fixed capacitor consisting of two strips of metal foil separated by oiled or waxed paper, and rolled together in compact tubular form.

paper electrophoresis Analytical instrument for technique in which ion migrates along a strip of porous filter paper, saturated with an electrolyte, when a potential gradient is applied across the length of the strip. It is used to identify ion types in analysis of serums, proteins, biochemicals, inorganic ions, rare earths, etc.

paper tape Perforated or printed tape to store data. Various widths (5, 6, 7, or 8 information bits or characters) are accommodated. Standard density is 10 characters per lined inch. Playback speed is up to 1,000 characters per second. Some paper tapes are capable of being read by the input device of a computer which senses the hole patterns corresponding to coded information. Entire programs can be stored on such tapes and read into the computer at will.

paper tape punch Device which places binary characters on a paper tape by punching holes in appropriate channels on the tape. A binary one is placed on the tape by punching a hole. A zero is indicated by the absence of a punched hole.

paper tape reader An input which one accepts paper tape which has been punched with ASCII code, and converts it to a parallel (or serial) digital electrical output. Available to handle eight-track one-inch tape, or five-track $1\frac{1}{16}$, $\frac{7}{8}$, or one-inch tape at speeds of 150, 300, 600, or 1200 bauds.

papilla A small nipple-shaped elevation or projection.

pap smears (Papanicolau smears) Method of staining smears of various body secretions—especially vaginal but also respiratory, digestive, or genitourinary—to detect cancer by examining the normally shed cells in the smear. The procedure is named for its developer.

papule A small solid pimple.

para Prefix meaning beside.

paracentesis The surgical puncture of a body cavity for the purpose of aspirating a fluid. The instrument used for this purpose is usually a trocar.

paracorporeal See *heart.*

paracusis Hearing disorder.

parallel 1. Said of circuit elements which are connected to the same pair of terminals. 2. Transmitted simultaneously. 3. The technique for handling a binary data word which has more than one bit. All bits are acted upon simultaneously. 4. Said of lines or planes which are the same distance apart at every point.

parallel access In a computer, the process of obtaining information from, or placing information into, storage where the time required for such access is dependent on the simultaneous transfer of all elements of a word from a given storage location. Also called *simultaneous access.*

parallel adder A conventional technique for adding where two multibit numbers are presented and added simultaneously parallel.

parallel addition In a computer, a form of addition that operates on each set of corresponding digits simultaneously in the addition of two numbers.

parallel buffer Electronic device (magnetic cores or flip-flops) used to temporarily store digital data in parallel, as opposed to serial storage.

parallel circuit A circuit whose elements are all connected across the circuit or to the same pair of terminals, so that the same voltage is across all the elements and the current divides between the elements in inverse proportion to their impedance.

parallel computer A computer in which the digits or data lines are handled concurrently by separate units of the computer. The units may be interconnected in different ways as determined by the computation to operate in parallel or serially. Mixed serial and parallel machines are frequently called serial or parallel according to the way arithmetic processes are performed. An example of a parallel computer is one which handles decimal digits in parallel although it might handle the bits which comprise a digit either serially or in parallel. Contrasted with serial computer.

parallel connection Elements in an electrical circuit are said to be connected in parallel when the same voltage appears across each element, and the current divides among the elements in inverse proportion to their resistance or impedance.

parallel data Simultaneous handling of data of two or more bits, digits, or channels.

parallel load See *shift*.

parallel operation 1. Type of information transfer whereby all digits of a word are handled simultaneously, each bit being transmitted on separate lines in order to speed operation; as opposed to serial operation, in which the bits are transmitted one at a time along a single line. 2. Pertaining to the manipulation of information within computer circuitry, in which the digits of a word are transmitted simultaneously on separate lines. Faster than serial operation, but requires more equipment. 3. Performance of several actions, usually of a similar nature, simultaneously through provision of individual similar or identical devices for each such action. Parallel operation, particularly for flow or processing of information, but usually requires more equipment. Contrasted with serial operation. 4. The connecting of two or more power supplies so that their outputs are tied together permitting the accumulated flow of current from all units to a common load. In regulated power supplies, interconnections other than the output terminals themselves may be required. For example, the amplifiers of all units but one may be made inoperable and this single amplifier would control all regulating elements. Also called *master/slave operation*.

parallel output Simultaneous availability of two or more bits, channels, or digits.

parallel resonance Condition which exists in a parallel circuit of inductance and capacitance when the inductive reactance is equal in magnitude (but opposite in sign) to the capacitive reactance. The circuit thus appears to be a pure resistance at its resonant frequency. Under these conditions, a voltage at the resonant frequency of the circuit will be at a maximum, while voltages at frequencies higher and lower than resonance will be at a minimum.

parallel search storage A storage device in a computer in which one or more parts of all storage locations are queried simultaneously. See also *associative storage*.

parallel-T oscillator An oscillator tuned by resistive and capacitive elements, which generates a sine wave by providing phase inversion at only one frequency. The oscillator is so connected that positive feedback results only when this phase inversion occurs.

parallel transmission Method of information transfer in which all bits of a character are sent simultaneously.

paralysis Loss of ability to contract a muscle due to injury or disease of the muscle or its nerve supply.

paralytic ileus Intestinal obstruction due to paralysis of the muscles which give rise to persistalsis.

paramagnetic A substance which is attracted by a magnetic field, but which has a permeability only slightly greater than one. See also *ferromagnetic* and *diamagnetic*.

paramagnetism Magnetism that involves a permeability somewhat greater than unity.

parameter 1. Literally, a measuring device that sits beside something—para (beside) + meter (a device for measuring), as in a coefficient which sits beside the variable it multiplies. A parameter is either a dimensionless constant or a function of time, in either case derived from the mathematical relationship between a group of properties whose values are fixed. It then takes the place of these properties and represents them as one family in subsequent physical relationships. 2. A variable (or a constant) that plays a role in some type of data processing. Physiologic parameters (variables) constitute the inputs to patient-monitoring systems, while program parameters (constant or variable) are used in computers. 3. A derived or measured value which expresses performance for use in calculation. When considered together, all the parameters of a device describe its operational and physical characteristics. 4. A quantity which varies with the circumstances of its application, such as input voltage, frequency, or maximum allowable current. 5. Quantity in a computer subroutine whose value specifies or partly specifies the process to be performed. It may be given different values when the subroutine is used in different main routines or in different parts of one main routine, but it usually remains unchanged throughout any one such use. See also *preset parameter* and *program parameter*. 6. Quantity used in a generator to specify machine configuration, designate subroutines to be included, or otherwise to describe the routine to be generated.

parametric testing Testing based on reasonably precise measurements of voltage or currents.

paranoia A personality disorder marked by extreme suspicion of the motives of others, often taking the form of belief that a plot is being carried out against the person.

paranoid Relating to paranoia, or describing the behavior of an individual suffering from paranoia.

paranoid state Characterized by delusions of persecution. A paranoid state may be of short duration or chronic.

paraplegia Paralysis of the legs and the lower part of the body, affecting motion and sensation. Usually, the result of spinal cord transection. Paraplegics are rarely able to adequately void and empty their urinary bladders. Some degree of obstructive kidney disease often with accompanying infection is then likely.

parasite 1. An organism which lives in or on an organism of another species, from which it derives nutriment and shelter. 2. Current in a circuit, due to some unintentional cause. The cause may be a high-frequency oscillation in a low-frequency amplifier, brought about by poor lead layout or lack of shielding.

parasitic 1. Pertaining to a parasite. 2. A signal in an electronic circuit caused by the inherent characteristics of the components, or of the circuit design or construction. Usually undesirable.

parasitic component In a monolithic integrated circuit, the capacitors and diodes which are formed between the planned circuit elements and the substrate during processing. Circuit design must allow for the functional effects of these parasitic components.

parathyroid Two pairs of small endocrine glands which control calcium and phosphorus metabolism. They are situated in the vicinity of the thyroid gland, in the neck.

parasympathetic system A subdivision of the autonomic system that generally functions to conserve the resources of the body. It acts antagonistically to the sympathetic system.

paravertebral To either side of the spinal column.

parenchyma The functional part of an organ.

parenteral Not by way of the alimentary tract; outside the intestine.

parietal Relating to or forming the upper posterior wall of the head.

parietal lobe The part of the brain's cerebral cortex lying immediately behind the central sulcus. It contains areas involved in somesthesis and somesthetic memory.

parietes The walls of any cavity of the body.

parity 1. Method by which binary numbers can be checked for accuracy. An extra bit, called a parity bit, is added to numbers in systems using parity. If even parity is used, the sum of all ones in a number and its corresponding parity bit is always even. If odd parity is used, the sum of ones in a number and its corresponding parity bit is always odd. 2. Method of verifying the accuracy of recorded data.

parity check Addition of noninformation bits to data, making the number of ones in a grouping of bits either always even or always odd. This permits detection of bit groupings that contain single errors. It may be applied to characters, blocks, or any convenient bit grouping.

Parkinsonism (paralysis agitans) A usually chronic condition, marked by muscular rigidity, immobile face, excessive salivation, and tremor. These symptoms characterize Parkinson's disease; however, they are also observed in the course of treatment with psychopharmaceutical drugs or following encephalitis or trauma.

parotid Near the ear; applied to a salivary gland under the ear.

parotitis Inflammation of the parotid gland. 1. Mumps. 2. Spread of infection from a septic mouth.

paroxysm A sudden temporary attack or seizure.

paroxysmal nocturnal dyspnea Attacks of breathlessness occurring at night due to accumulation of fluid in the lungs. Heart failure is the typical cause.

partial carry In parallel addition, a technique in which some or all of the carriers are stored temporarily instead of being allowed to propagate immediately.

parturient In the condition of giving or being just about to give birth to a child.

Paschen's law Sparking potential between two given terminals in a given gas is proportional to the product of pressure and spark length. For a given voltage, this means that spark length is inversely proportional to pressure.

passband The frequency range in which an electrical filter network is intended to pass electrical signals.

pass element An automatic variable resistance device, either a vacuum tube or power transistor, in series with the source of dc power. The pass element is driven by an amplifier error signal to increase its resistance when the output voltage needs to be lowered, or to decrease its resistance when the output must be raised. See *series regulator*.

passivation Growth of an oxide layer on

the surface of a semiconductor to provide electrical stability by isolating the transistor surface from electrical and chemical conditions in the environment. This reduces reverse current leakage, increases breakdown voltages, and raises power dissipation rating.

passive Describing a device which does not contribute energy to the signal it passes.

passive device A component that does not provide rectification, amplification, or switching, but reacts to voltage and current; e.g., resistor, capacitor.

passive film circuit A thin-film or thick-film circuit network consisting entirely of passive circuit elements and interconnections.

passive network 1. A network whose net influx (or efflux) of available energy is stored or dissipated within the network. There may be no sources of energy other than those explicitly bookkept as influxes. 2. A network with no source of energy and without gain elements.

passive substrate A substrate for an integrated component which may serve as physical support and thermal sink to a thick-film or thin-film integrated circuit, but which exhibits no transistance. Examples of passive substrates are glass, ceramic, and similar materials.

passive transducer 1. A transducer whose output waves derive their power solely from the input waves. 2. A transducer that does not require any local source of energy other than the received energy.

patch In a computer, to correct or change the coding at a particular location by inserting transfer instructions at that location and by adding elsewhere the new instructions and the replaced instructions. Usually used during checkout.

patch panel In computers, a panel containing means of changing circuit configurations, usually receptacles into which jumpers can be inserted.

patella The kneecap. A sesamoid bone in front of the knee joint.

patent 1. Open; usually said of a vessel or the lumen of a catheter. 2. A governmental grant of monopoly rights to an invention, in return for its publication. The grant lasts for a stipulated time (17 years in the U.S.). Also called *letters patent*.

patho Prefix meaning disease.

pathogen A disease-producing microorganism.

pathognomonic Characteristic of, or peculiar to, a particular disease.

pathological Relating to pathology. Morbid, abnormal.

pathology The branch of medical science which is concerned with the structural and functional changes in organs and tissues, resulting from disease.

patient monitor An oscilloscopic display device which presents a continuously updated trace of a patient's electrocardiogram. Often combined with a hot-stylus recorder for hard-copy demand writeout, as well as tachometer (heart rate meter) and alarm functions.

pattern recognition In a computer, the process of examining records for certain combinations of code elements.

patulous Open wide.

pC Abbreviation for picocoulomb.

pc Abbreviation for picocure.

PCM Abbreviation for pulse-code modulation, modulation of a pulse train in accordance with a code.

PCW Abbreviation for pulmonary capillary wedge (pressure). A blood pressure measured in the pulmonary artery by means of a flow-directed catheter of the Swann-Ganz type.

PDM Abbreviation for pulse-duration modulation. Pulse-width modulation or pulse-length modulation. A form of pulse modulation in which the durations of pulses are varied.

PE Abbreviation for photographic effect.

peak The maximum instantaneous value of some quantity, such as a voltage or current.

peak amplitude The maximum amplitude attained by a wave during a period.

peaking Adjusting a component so as to increase the reponse of a circuit at a desired frequency or band of frequencies.

peak level The maximum instantaneous level that occurs during a specific time interval.

peak limiter Device which automatically limits the magnitude of its output signal to approximate a preset maximum value by reducing its amplification when the instantaneous signal magnitude exceeds a preset value.

peak load Maximum electrical power consumed or produced in a stated period of time. It may be the maximum instantaneous load or the maximum average load over a designated interval of time.

peak signal level Expression of the maximum instantaneous signal power of voltage as measured at any point in a transmission system. This includes auxiliary signals.

peak spectral emission The wavelength at which a lamp radiates its highest intensity.

peak-to-peak Describing the measurement of an alternating current or voltage,

from positive maximum to negative maximum value.

peak value Of any quantity, such as current or voltage, which varies with time, the maximum value of which occurs during a period of time. If the quantity is a sine wave, the peak value is 1.4142 times the effective value.

peak voltage Of a time-varying voltage, the maximum value which occurs during a period of time. If the voltage is a sine wave, the peak value is 1.4142 times the effective value. See also *period*.

pectoral Relating to the chest.

pectoral muscles The muscles on the anterior surface of the chest.

pectus excavatum See *funnel chest*.

pedicle The stalk of a collection of tissue which contains a supply of vessels and nerves.

-pellent Suffix meaning to drive.

pellicle A thin skin or membrane.

Peltier effect Current flowing through the junction of two different metals produces or absorbs heat at the junction. Reversing the current flow changes a production of heat to an absorption, and vice versa.

Peltier electromotive forces The boundary EMFs produced across the junctions of two different metals, associated with the heating and cooling effects of the two junctions.

pentode An electron tube having five elements; an anode, a cathode, a control grid, a screen grid, and a suppressor grid.

pentode field-effect transistor A five-lead transistor with three gates. It can be connected like a pentode if each of the gates is supplied from an independent bias source.

peptic Pertaining to digestion.

perceptron A particular class of machines that are designed to recognize and discriminate certain categories of visual patterns. The machines begin with randomly connected elements and proceed to higher levels of organization as a result of recognized patterns, thus offering positive reinforcement to the particular circuitry at that time. Eventually, efficient pattern recognition is achieved.

percussion Examination by tapping upon the body.

percutaneous Through unbroken skin, as in absorption by inunction.

perfuse 1. To pour a fluid. 2. To pass a fluid through spaces. 3. To introduce a fluid into tissue by injecting it into an artery.

perfusion A liquid flowing through something, usually the blood vessels of an organ.

peri Prefix meaning outside, around, or near.

pericardial Pertaining to the pericardium.

pericardial effusion The accumulation of fluid in the pericardial sac. In an extreme case, normal heart action is affected by the strangulating force of the surrounding fluid.

pericarditis Acute or chronic inflammation of the pericardium (fibrous sac surrounding the heart), caused by infection, trauma, myocardial infarction, cancer, or complication from other diseases.

pericardium A sac of membranous tissue which surrounds and encloses the heart, leaving a potential space and thus enabling changes in shape and volume of the heart to take place during contraction.

perineum The area between the anus and the posterior part of the genitals; the entire anogenital area.

periodic wave A wave which repeats itself at regular intervals of time.

peripheral 1. A device through which the computer communicates to the outside world. The term may also include auxiliary memories, such as tape, disc, and drum. 2. Those portions of the body away from the center; hence, the arms and legs. 3. Outward from or toward of the surface.

peripheral control unit An intermediary control device which links a peripheral unit to the central processor; or, in the case of off-line operation, to another peripheral unit.

peripheral data processing equipment Automatic data processing equipment associated with, but separate from, the main frame and interconnected equipment. For example, a punched-card-to-magnetic-tape converter.

peripheral nervous system The assembly of nerve cells having nerve fibers extending from the central nervous system to regions throughout the organism. It is the function of some fibers to conduct sensor impulses and others to conduct effector impulses, the peripheral system is also referred to as the *input-output channels* of the nervous system.

peripheral processor A general term for a lesser computer associated with a large machine. Among the functions may be multiplexing, data formating, concentrating, pollings, and the handling of simple routines to increase the capacity of a communications channel or to relieve the main (often called host) computer.

peripheral resistance The steady-state fluid impedance of the peripheral vascular bed to the flow of blood through its vessels.

Peripheral resistance is largely controlled by sympathetic autonomic nerves that regulate the diameter of vessel lumina. Arteries and arterioles with small diameters have high vascular resistances.

peripheral transfer The process of transmitting data between two peripheral units of a computer system.

peripheral units Equipment devices that work in conjunction with a data terminal or computer but are not a part of that unit, such as a card or paper tape reader or punch or keyboard.

peristalsis Wavelike contractions and movements by which the alimentary tract propels its contents along.

peritoneal cavity The potential space between the layers of the peritoneum.

peritonitis Acute or chronic inflammation of the serous membrane lining abdominal walls and covering the contained viscera. Its symptoms are abdominal pain and tenderness, nausea, vomiting, moderate fever, and constipation. It is usually caused by infectious agents or foreign matter entering the abdominal cavity from the intestinal tract (perforation), female genital tract, blood dissemination, or the outside (wounds, surgery).

permanent memory In computers, a storage device for keeping data intact when the computer has been shut down, e.g., storage on a magnetic drum.

permanent storage See *fixed storage*.

permeability 1. A measurement of the ability of a material to conduct the lines of force of a magnetic field. The permeability of air is considered to be unity and the permeability of other materials is measured in relation to air. 2. The quality of being permeable to the passage of ions, gases, or liquids.

permeable Porous so as to permit the passage of ions, gases, or liquids.

permeance A measure of the ease with which a magnetic field can be set up in a magnetic circuit. It is the reciprocal of reluctance.

permeate To pass through; as by mechanical passage through pores or interstices, or through ionic or molecular transfer occurring between two solutions separated by a membrane.

permissible dose The amount of radiation which may be received by an individual or material within a specified period with expectation of no harmful result to person or the material.

permittivity The property of a material which determines how much electrostatic energy can be stored per unit volume per unit voltage.

permutation Any of the combinations or changes that are possible within a group.

pernicious 1. Destructive; harmful. 2. Denotes a disease of a severe character which is usually fatal without specific treatment.

peroneal Pertaining to the fibula or to the outer side of the leg.

persistence The time it takes a cathode-ray tube or TV tube phosphor to decline to 10 percent of its peak intensity after it has been struck by a beam of electrons.

perspiration Sweat.

petechia A pinpoint spot of blood in the skin or mucous membrane.

petit mal Minor epilepsy.

petrous Stony; a term given to a hard part of the temporal bone.

-pexy Suffix meaning fixation.

PFX Abbreviation for photofluorographic X-ray.

pH 1. Symbol used to express the concentration of hydrogen ions in a solution. The pH is the negative logarithm (to the base 10) of the hydrogen-ion concentration. 2. The pH of a solution is an index of its acidity—literally the concentration of hydronium ions (H+). The scale ranges from 0 to 14. Numerically, indices of pH may vary enormously, a strong acid having a pH of 1, the strongest base having a pH value of 14. pH is expressed on an inverse logarithmic scale, e.g., a neutral concentration of 10^{-7} has a pH of 7.

pH is a primary quantity in practically all body solutions. For blood, normal pH is a neutral 7; for clinical laboratory measurements on other solutions, it depends greatly upon determination of initial values, particularly in buffer solution capacities where acid-base variations bear heavily on final analytical results.

Nonelectronic, nonquantitative indicators use solutions whose colors react by changing from one hue to another upon the addition of acid or alkali. Sets of these colorimetric indicators are commercially available and consist of sealed ampules containing buffer solutions to which have been added appropriate indicator chemicals. Complicated colorimetric indicators have largely been superseded by electronic methods. Electronically, the determination of pH is based on the fact that the difference of electric potential between two suitable electrodes in a solution containing hydronium ions depends upon the concentration or activity of the ions.

phagocyte An ameboid cell (leukocyte) of the blood which acts to engulf and destroy invading pathogens and foreign material, providing a line of defense against infection.

phalanges The small bones of the fingers and toes.

pharmacogenetics Study of genetically determined variations in response to, and metabolism of, drugs.

pharmacokinetics The study of drug effects on living organisms. Also called *pharmacodynamics.*

phase 1. The time displacement between two currents or two voltages or between a current and a voltage measured in electrical degrees, where an electrical degree is 1/360 part of a complete cycle. 2. The number of separate voltage waves in a commercial alternating current, designated as single-phase, three-phase, etc. Abbreviated as the Greek letter Phi (Φ). 3. One of the stages in which a thing appears during its course of change or development. 4. The state which a substance assumes, depending upon conditions. For example: water has solid, liquid, and gaseous phases which are temperature–dependent.

phase angle 1. The angle between two vectors representing two simple periodic quantities which vary sinusoidally and which have the same frequency. 2. A notation for phase position when the period is designated by 360 degrees.

phase-controlled rectifier A rectifier circuit using a thyratron as a rectifying element and a variable-phase sine wave for grid bias.

phase detector An electronic circuit that provides an output signal indicating the relative phase, or timing difference, between input signals. It is often used to operate a voltage-controlled oscillator, thus keeping it synchronized with a reference signal. Also called *phase discriminator.*

phase discriminator See *phase detector.*

phase distortion See *delay distortion* and *distortion.*

phase-locked Said of two alternating currents whose frequencies are locked together so that their phase is at all times identical.

phase-locked loop 1. In communications, a circuit technique by which a local oscillator is synchronized in phase and frequency with a signal being received. 2. A closed-loop electronic servomechanism whose output will lock onto the track of a reference signal. Phase lock is accomplished by comparing the phase of the output signal (or multiple thereof) with the phase of the reference signal. Any phase difference between these two signals is converted to a correction voltage that changes the phase of the output signal to make it track the reference.

phase modulation Modulation in which the phase angle of the carrier wave is caused to vary by an amount proportional to the instantaneous value of the modulating wave.

phase multiplexing The process of encoding two (or more) information channels on a single tone.

phase sensitive detector A system which produces a dc output signal in response to an ac input signal of a defined frequency equal to the frequency of an ac reference signal. The dc output is proportional to both the amplitude of the ac input signal and the cosine of its phase angle relative to that of the reference signal. Used as synchronous rectifiers in chopper dc amplifiers and for the accurate measurement of small ac signals obscured by noise.

phase shift The changing of phase of a signal as it passes through a filter. A delay in time of the signal is referred to as *phase lag* and in normal networks, phase lag increases with frequency, producing a positive envelope delay (see *envelope delay*). It is possible for an output signal to experience a time shift ahead of the input signal and this is called *phase lead*. The phase shift is always dependent on frequency.

phase-shift oscillator An oscillator consisting of an amplifier which is caused to oscillate by positive feedback. The positive feedback is provided by connecting the amplifier output to its input through a resistance-capacitance network which shifts the phase of the signal by 180 degrees.

phelbotomy An incision into a vein made in order to remove blood.

pH electrode Transducer sensitive to hydrogen-ion (H+) concentration, comprising thin-walled glass membrane (glass electrode) or spongy platinum exposed to gaseous hydrogen (hydrogen electrode) or platinum exposed to quinhydrone (quinhydrone electrode), all of which develop an electric force proportional to the hydrogen-ion concentration of a solution when immersed in the solution.

phenol Carbolic acid (C_6H_5OH). An antiseptic and disinfectant, occurring in its natural state as a colorless crystalline compound. Generally used in a 10 percent water solution where a powerful germicide is needed. It is highly poisonous and has a distinctive odor.

phenylophrine A drug that is used to shrink swollen nasal mucous membranes and to clear nasal passages.

phlebectomy Excision of a vein.

phlebitis Condition caused by inflammation of a vein wall, usually resulting in the formation of a blood clot inside its cavity. Phlebitis produces pain, swelling, and stiffness of the affected part, generally a limb.

phlebothrombosis Thrombosis in a vein, particularly the veins of the legs, due to prolonged stagnation of the blood due to poor circulation.

pH Meter Instrument for measuring hydrogen-ion concentration (relative acidity) of a solution.

phobia An obsessive, unrealistic fear of an external object or situation. Some of the common phobias are acrophobia, fear of heights; agoraphobia, fear of open places; claustrophobia, fear of closed spaces; mysophobia, fear of dirt and germs; xenophobia, fear of strangers.

-phobia Suffix meaning fear.

phon A subjective unit defining the loudness of a sound. It is that level of a complex sound which is judged to be equal to 0.0002 microbar of a 1000 Hz sine wave.

phonation The act of forming and uttering vocal sounds.

phonocardiogram The written graphic record of heart sounds resulting from the use of phonocardiograph.

phonocardiograph An instrument used for producing a strip-chart record of cardiac sounds—consists essentially of a microphone, an audio amplifier, and a graphic recorder. Often, the instrument contains highly selective audio frequencies from the recorded trace, so that specific phenomena may be analyzed free from the masking effects of other frequencies in the heart's vibratory spectrum, which extends nominally from 0.1 to 2000 cps.

phonocardiography The process of recording and interpreting the sounds of the heart. Typically, the instrument consists of a microphone, an amplifier, and either a cathode-ray tube or strip-chart recorder. A loudspeaker or headset may also be included to permit listening to the sounds being displayed or recorded.

phonocatheter Catheter-microphone combination that is inserted through vessel into heart. It picks up inner cardiac sounds.

phonoelectrocardioscope A dual-beam oscilloscope which displays both ECG signals and heart sound signals.

phonoelectroscope Stethoscopic device which suppresses low frequencies (characteristic of normal heart function) to permit detection of higher frequency sounds.

phonons Packets of sound energy vibrating in a solid at ultrahigh frequencies—so high that the energy is commonly thought of as heat.

phosphor Layer of luminescent material, applied to the inner face of a cathode-ray tube, which fluoresces during bombardment by electrons and phosphoresces after bombardment.

phosphorescence Emission of radiation by a substance as a result of previous absorption of radiation of shorter wavelength. Differs from fluorescence in that emission may continue for a considerable time after cessation of the exciting irradiation.

phosphors Chemical substances that exhibit fluorescence when excited by ultraviolet radiation, X-rays or an electron beam. The amount of visible light is proportional to the amount of excitation energy. If fluorescence decays slowly after the exciting source is removed, the substance is said to be phosphorescent.

phot (pt) The unit of illumination when the centimeter is taken as the unit of length; it is equal to one lumen per square centimeter.

photobiology The study of the effects on living matter (or substances derived therefrom) of electromagnetic radiation extending from the ultraviolet through the visible light spectrum into the infrared. The conversion of electromagnetic energy into chemical energy, photosynthesis is an important branch of photobiological investigation.

photocell Either (1) a photoconductive cell, (2) a photovoltaic cell, or (3) a phototransistor. See *phototransistor*.

photochemical radiation Energy in the ultraviolet, visible, and infrared regions to produce chemical changes in materials.

photocoagulator Device used in surgery that uses electrical current or light to stop bleeding.

photoconductive cell A semiconductor device whose resistance varies with the amount of light falling on it.

photoconductivity 1. Electrical conductivity which varies with illumination. 2. The ability of a material to hold a charge of electricity in the absence of light (insulator), yet act as a conductor of electricity when exposed to light. 3. That property of a material which causes its resistivity to change when it is bombarded by photons.

photoconductor 1. A light-sensitive resistor whose resistance decreases with increase in intensity of illumination. Consists of a thick, single-crystal, or polycrystalline films of compounded semiconductor substances. 2. A material whose resistance decreases as light strikes it, such as cadmium sulphide (cds).

photodetector A device which senses incident radiation.

photodiode A PN semiconductor diode designed so that light falling on it greatly increases the reverse leakage current, so that the device can switch and regulate electric current in response to varying intensity of light.

photoelectric Pertaining to the effect in which photons (light) impinging on certain metals will release bound electrons and thus cause a flow of current. Photoelectric effect does not include photoconductivity or photovoltaic effect.

photoelectric cathode A cathode whose primary function is photoelectric emission.

photoelectric cell A device which indicates intensity of light incident upon it by some electrical characteristic, usually a voltage generated or resistance change. Cell whose electrical properties are affected by light.

photoelectric control Control of a circuit or piece of equipment in response to a change in incident light impinging on a photosensitive device.

photoelectric effect The emission of an electron from the surface of a material that has been exposed to light (or photons). Photoelectric devices are used clinically in the evaluation of cardiac output, in measurement and analysis of blood parameters, and in pulse rate counting (via a finger plethysmograph).

photoelectric multiplier A phototube in which the primary photoemission current, prior to being extracted at the anode, is multiplied many times.

photoelectric reader A device that reads information stored in the form of holes punched in paper tape or cards, by sensing light passed through the holes.

photoelectric relay A relay combined with a phototube (and amplifier if necessary) so arranged that changes in incident light on the phototube cause the relay contacts open to close. Also called *light relay*.

photoemissive detector Electronic tube device whose anode current varies with incident light intensity on the cathode.

photofluorometer Instrument using fluorescence of an irradiated substance for analysis of the substance.

photoglow tube   Gas-filled phototube used as a relay by making the operating voltage sufficiently high so that ionization and a flow discharge occur, with considerable current flow when a certain illumination is reached.

photographic writing speed   A figure of merit which describes the ability of a particular camera, film, oscilloscope, and phosphor to record a fast-moving trace. This figure expresses the maximum single-event spot velocity (usually in centimeters per microsecond) which may be recorded on film as a trace just discernible to the eye.

photojunction battery   Nuclear type battery in which the radioactive material, promethium 147, irradiates a phosphor which converts nuclear energy into light. The light is then converted to electrical energy by a small silicon junction.

photomacrograph   A moderately magnified or unmagnified picture of a small object. No compound microscope is used and the magnification is usually less than 20 times (a photograph of a coin twice life size is a photomacrograph).

photomagnetelectric effect   The production within a semiconductor of an electromotive force perpendicular both to an applied magnetic field and to a photon flux of proper wavelength. Hole-electron pairs generated by the photons diffuse in a direction perpendicular to the surface because of the concentration gradient.

photometric brightness (luminance)   The luminous flux per unit projected area, per unit solid angle, either leaving a surface at a given point in a given direction, or arriving at a given point from a given direction; or, the luminous intensity of a surface in a given direction, per unit of projected area of the surface, as viewed from that direction.

photomultiplier   See *multiplier phototube*.

photomultiplier tube   A vacuum tube that converts light signals into electrical pulses. The tube consists of a light-sensitive surface that gives off electrons when light is incident on it. The electrons then pass through many successive stages, with amplification at each stage. The result is a considerably stronger signal than would be obtainable from a single-stage photoelectric tube.

photon   A quantum of electromagnetic radiation having no rest mass and traveling at the speed of light. Unit of radiation associated with an energy quantum. The energy of a photon in ergs is the product of its frequency in cycles per second and flanck's constant ($E = h\nu$). Optical photons have energies corresponding to wavelengths between 1800 and 120 nanometers.

photon coupled isolator   Circuit coupling device, consisting of an infrared emitter diode coupled to a photon detector over a short shielded light path, which provides extremely high circuit isolation.

photophobia   An abnormality characterized by an intolerance of light.

photoprotection   The protection of some cells by exposure to near ultraviolet light prior to exposure to light in the far ultraviolet.

photoptarmosis   The stimulation of sneezing by the influence of light.

photoradiometer   Instrument for measuring intensity and penetrating power of radiation.

photoresist   A solution which after exposure to ultraviolet light becomes extremely hard and resistant to etching solutions that normally can dissolve materials such as silicon dioxide.

photoresistive or photoconductive transduction Conversion of the measurand into a change in the resistance of a semiconductor material (by changing the illumination incident on the material).

photosensitive field-effect transistor A special unipolar field-effect transistor (FET) structure that is positioned to receive illumination transmitted through a lens in the top of its header can. It combines the circuit and device characteristics of a photodiode and a high-impedance low-noise amplifier.

photosensitization Tendency of tissues to react abnormally to light, usually as the result of the presence in the tissues of certain chemicals which magnify the damaging effect of the incident radiation.

photoswitch Solid-state device which acts as a high-speed power switch; activated by incident radiation.

photosynthesis The production of carbohydrates by the combination of carbon dioxide (CO_2) and water (H_2O) in the presence of light energy.

phototransistor A junction transistor device in which light collected by a tiny lens impinges on the base. The transistor collector current increases with increased light intensity.

phototube An electron tube whose current output depends upon the number of photons which strike its photocathode.

photovoltaic Pertaining to the electron flow which can be caused by exposing the junction between a metal and a semiconductor to radiation. Selenium is a common photovoltaic substance.

photovoltaic transduction Conversion of the measurand into a change in the voltage generated when a junction of dissimilar material is illuminated.

photran A triode PNPN-type switch of the reverse blocking PNPN-type switch class. The photran provides optical triggering in addition to standard gate terminal triggering.

photronic cell Type of photovoltaic cell in which a voltage is generated in a layer of selenium during exposure to visible or other radiation.

-phrenia Suffix meaning the diaphragm, or the mind.

phrenic Relating to the mind, or to the diaphragm.

phrenic nerve The nerve that controls movement of the diaphragm. Electrical stimulation of the phrenic nerve has been attempted in cases of neurological damage involving those portions of the brain that originate respiratory command signals.

PHT Abbreviation for primary of the high-tension circuit.

physical configuration The arrangement of materials, components, and wires on a structural assembly for physical strength, conservation of size, and minimal electrical interaction.

physiological monitoring The continuous or periodic measurement of quantitative physiological data in a form suitable for evaluating, recording, or storage.

physiological patient monitor A device for automatically measuring and/or recording one or several physiological variables and responses of a patient, including heart potentials, blood pressure, pulse rate, respiration rate, temperature. May include automatic alarms to signal for help in case any monitored variable goes out of limits. Most monitors include a scope (nonfade or conventional) to allow evaluation of clinically significant vital sign traces.

physiological phenomenon A response,

observable activity, or unusual occurrence, in relation to the function of the body. (The ECG is an observable phenomenon.)

physiological simulator 1. Device for producing selected sound and/or wave signals such as normal and abnormal ECG, used in teaching and demonstration. See also *electrocardiographic simulator*. 2. A device used with a high-speed recorder to study nerve and muscle characteristics. The stimulator provides single-pulse, paired-pulse, and pulse-train outputs, dimensioned in voltage and current characteristics. As a stimulus is applied to the tissue, a triggered sweep output is applied to the recorder, so that response characteristics of the instrumented tissue sample can be recorded with a common base.

physiology The study of the structure and functions of the body, living things, and the normal vital processes of animal and vegetable organisms.

physiotherapy Treatment of diseases—especially those involving body motion—by the use of exercise, heat, water, air, and light.

-physis Suffix meaning nurture or to grow.

pia mater The soft, fine membrane surrounding the brain and spinal cord. It is the innermost of the three membranes of the brain and spinal cord.

pica Morbid appetite. Craving for unnatural articles of food (e.g., plaster, paint flakes).

pico (p) Prefix meaning 10^{-12} (formerly micromicro).

picoammeter A sensitive ammeter that indicates current values in picoamperes.

picoampere (pA) One-millionth of one microampere; 10^{-12} ampere.

picofarad (pF) A very small unit of capacitance, equal to one-millionth of a microfarad. Previously called a micromicrofarad; 10^{-12} farad.

picosecond (ps) A micromicrosecond; 10^{-12} second.

Pierce oscillator Oscillator in which a piezoelectric crystal unit is connected between the grid and the plate of an electron tube, in what is basically a Colpitts oscillator, with voltage division provided by the interelectrode capacitances of the circuit.

piezoelectric 1. Capable of developing an electrical voltage when a mechanical stress is applied. 2. Capable of reacting mechanically when a voltage is applied.

piezoelectric accelerometer A crystalline material, which, when force is applied, generates a charge. Through the incorporation of a mass in direct contact with the crystal, an acceleration transducer is produced.

piezoelectric crystal A substance such as quartz or barium titanate that has the ability to become electrically polarized and has strong piezoelectric properties. It is so cut as to emphasize the coupling to some distinct mechanical mode of the crystal. Used as an electromechanical transducer and as a precision frequency-regulating medium in oscillator circuits.

piezoelectric effect The effect of producing a voltage by placing a stress either by compression, expansion, or twisting on a crystal and conversely, producing a stress in a crystal by applying a voltage to it.

piezoelectricity Electricity or electric polarity due to pressure especially in a crystalline substance.

piezoelectric transducer A transducer in which the transduction is accomplished by means of the piezoelectric properties of certain crystals or salts.

piezoresistance Resistance that changes with pressure.

pigtail 1. A short, very flexible braided or stranded wire used to carry current from a movable member, such as a generator commutator. 2. The splice made by twisting together the bared ends of two conductors laid side by side.

pile 1. An enlarged vein about the anus; a hemmorhoid. 2. A source of energy, as in a voltaic pile (a chemical generator of current), thermoelectric pile (a temperature-responsive generator of current), or an atomic pile (a nuclear source).

pilot light A lamp which illuminates to call attention to any of several indicators or conditions in a particular bay or panel of equipment.

pin 1. An electrical terminal in a connector or on an electron tube base. Pushes into a socket to make a connection. 2. An insulator pin. 3. A surgical implant designed to secure bony parts.

pinch-off voltage The gate voltage of a field-effect transistor that blocks current flow for all source-drain voltages below the junction breakdown value. Pinch-off occurs when the depletion zone completely fills the area of the device.

pin diode Diode consisting of a silicon wafer containing nearly equal P-type and N-type impurities, with additional P-type impurities diffused from one side and additional N-type impurities from the other side, leaving a lightly doped intrinsic layer in the middle, to act as a dielectric barrier between the N-type and P-type regions.

pinguecula Small yellow patch of connective tissue on conjunctiva occurring in old age.

pink noise Broadband noise having constant energy per octave.

pipet See *pipette*.

pipette A graduated glass tube for taking up small quantities of liquid or gas. Also called *pipet*.

placebo Medicine given to please the patient, composed of inert material and flavoring. Also, used as a control in certain drug experiments in which emotions or psychological influences must be offset.

planar transistor A diffused transistor in which emitter, base, and collector regions come to the same plane surface. Their junctions are protected by a material such as silicon oxide. The manufacturing process consists of an oxide-masking technique where the silicon oxide is formed by adding oxygen or water vapor to the atmosphere of a diffusion furnace. The thickness of the oxide layer is a function of time, temperature, and the amount of oxidizing agent.

planchet A small metal container or sample holder that is usually used to hold radioactive materials that are being checked for the degree of radioactivity in a proportional counter or scintillation detector.

plantar Pertaining to the sole of the foot.

plantar response Reflex movement of toes when the sole of the foot is stroked.

-plasia Suffix meaning formation.

plasma 1. A wholly or partially ionized gas for which the densities of positive ions and negative electrons are roughly equal. 2. A gas in which an important fraction of the molecules are dissociated into ions and electrons, the gas as a whole remaining electrically neutral. 3. A gas at an extremely high (20,000 degrees Kelvin) temperature and completely ionized, which is therefore conductive and affected by magnetic fields. 4. The liquid in which the corpuscles of the blood are suspended.

plasma frequency A natural frequency for coherent motion of the electron in a plasma.

plasma oscillation Electrostatic or space charge oscillations in the plasma which are closely related to the plasma frequency. There is usually enough damping of the oscillations due to electron collisions so they are not self-generated. They can be excited, however, by such techniques as shooting a modulated beam of electrons through the plasma.

plasma proteins Fibrinogen, albumin, and globulins.

plastic deformation The change in the dimensions of an object under load that is not recovered when the load is removed.

plastid Any specialized organ of the cell other than the nucleus.

-plasty Suffix meaning mold, shape, or form.

plate 1. One of the electrodes in a storage cell. 2. One of the metal surfaces in a capacitor. 3. The anode in an electron tube. 4. A type of surface ECG electrode. 5. To deposit, through electrolysis, a coating of one metal on another.

-plegia Suffix meaning stroke or paralysis.

pleomorphism Having several different forms, as in the case of an organism. Also, the chemical property of crystallizing in two distinct forms.

plethora Fullness; usually, of the vascular system, characterized by swelling of the vessels, engorgement with blood, and a very strong pulse.

plethysmogram A graph of the changes in size or volume of an organ or of the amount of blood present or passing through a vessel. The transducer for a plethysmogram can be a pressure transducer, a pair of surface electrodes, or a photoelectric device, and others.

plethysmograph Device that measures change in volume of tissue, organ, toe, finger, or body resulting from changing quantity of blood. Types include body capacitance, electrical, impedance, hydraulic, photoelectric, and pneumatic.

plethysmography Detecting changes of blood volume in the tissues as the volume alternately increases and decreases during the cardiac cycle. See *electrical-impedance cephalography* and *finger plethysmograph*.

pleura A thin serous membrane which covers the exterior of the lung and lines the inner surface of the thoracic cavity. The two pleura are distinct from one another. Each encloses a space within which the lung can expand and contract during respiration.

pleural cavity The potential space between the pleura.

pleural rub A squeaking sound produced by friction between the pleura and surrounding tissue, as detected by auscultation.

pleurisy Acute or chronic inflammation of the pleura (serous membrane lining the thoracic cavity and lungs). It often accompanies inflammatory lung diseases and may be caused by infection (tuberculous, viral, or other), cancer, or cardiac infarct.

plexor A small, hammer-like instrument used in performing percussion, and in the testing of reflexes.

plexus A network of vessels or nerves.

-plexy Suffix meaning stroke or paralysis.

pliers A small pair of pincers: an instrument having two short handles extended into pivoted jaws suitable for grasping or cutting.

plotter A visual display x-y recorder in which a dependent variable is graphed by an automatically controlled pen or pencil as a function of one or more variables.

plug A contact member on the end of an electrical cord which terminates the cord conductors. It can be inserted into a fixed jack, connector, or receptacle to make temporary connections with the conductors they terminate. It may make one or many contacts. The familiar switchboard plug has three contacts, called tip, ring, and sleeve. Once used extensively in biomedical equipment, this plug has given way to other types more consistent with the needs for patient input isolation.

PMOS 1. P-channel MOS where P-type source and drain regions are diffused into an N-type substrate to create a P-channel for conduction. 2. MOS devices made on an N-type silicon substrate where the active carriers are holes (P) flowing between P-type source and drain controls. 3. An MOS (unipolar) transistor in which the working current consists of positive (P) electrical charges.

-pnea Suffix meaning to breathe.

pneumatic Relating to, or using air or other gas either moved or worked by air pressure or adapted for holding or inflated with compressed air.

pneumatocele A sac, tumor, or swelling which contains gas.

pneumocardiography The recording of cardiac data from sensors that monitor respiratory parameters, which in turn reflect cardiac variations. This is possible because the electrodes placed for measurement of respiration-induced changes in transthoracic impedance are excited from a

high-frequency source, easily filtered out of the ECG signal which these same electrodes are picking up from the subject's heart.

pneumogram The tracing or graphic record of respiratory movements, usually through measurement of transthoracic impedance change. Pneumographic equipment detecting either respiratory rate or depth is utilized in patient-monitoring systems.

pneumograph An instrument which records changes in transthoracic impedance, corresponding to thoracic volume changes with inspiration and expiration.

pneumonectomy Surgical removal of a lung.

pneumonia An acute inflammation of the lungs, usually caused by bacterial or viral infection. Chills, sharp chest pain, shortness of breath, cough, rusty sputum, fever, and headache are primary symptoms.

pneumoradiography X-ray examination of a part after it has been injected with air or oxygen.

pneumotachograph A respiratory flow measurement device comprising two funnel-shaped chambers, joined at their large ends, with a platinum wire mesh separating the two. A differential pressure transducer communicates with the chambers on either side of the mesh. As the subject draws in or expires air, the mesh impedes the flow slightly, creating a differential pressure which the transducer converts to a signal representing flow.

pneumotachography The measurement of breathing mechanics and pulmonary function by transduction of flows and volumes.

pneumothorax Accumulation of air or gas in the pleural cavity (between the chest wall and the lung), resulting in lung collapse. It may result from a penetrating chest wound or some disease, or may be deliberately induced for treatment of lung ailments (tuberculosis).

PN junction In a semiconductor, the boundary surface between P-type and N-type materials. It has the properties of a rectifying diode.

PNPN (four-layer) diode A semiconductor device which may be regarded as a two-transistor structure with two separate emitters feeding a common collector. This combination constitutes a feedback loop which is unstable for loop gains greater than unity. This instability results in a current which increases until ohmic circuit resistances limit the maximum value. This gives use to a negative resistance region which may be utilized for switching, or for waveform generation.

PNPN-type switch A bistable semiconductor device comprising three or more junctions at least one of which can switch between reverse and forward voltage polarity within a single quadrant of the anode-to-cathode voltage-current characteristic.

PNP transistor A junction transistor in which a thin layer of N-type semiconductor (the base) is sandwiched between two pieces of P-type material. Conduction is by the movement of holes; therefore, the emitter is positive to the base, and the collector is negative to the base.

point source Radiation source whose maximum dimension is less than one-tenth the distance between source and receiver.

point-to-point 1. Describing communication between two fixed stations. 2. A method of wiring electronic circuits, where component layout is subordinate to short lead lengths.

polar 1. Polarized. 2. Having a dielectric constant which decreases with increasing frequency.

polarity 1. Condition in an electrical circuit by which the direction of the flow of current can be determined. Usually applied to batteries and other direct voltage sources. 2. Two opposite charges, one positive and one negative. 3. Quality of having two opposite magnetic poles, one north and the other south. 4. Polarity of a video image is the sense of the potential of a portion of the signal representing a dark area of a scene relative to the potential of a portion of the signal representing a light area. Polarity is stated as *black negative* or *black positive*.

polarization 1. The formation of chemical products near the electrodes of an electric cell (such as a dry cell) which inhibits further flow of current through the cell. 2. The restriction of the vibration of an electromagnetic wave to a single plane. 3. The formation of chemical products near the electrodes applied to a patient for ECG monitoring. An appreciable potential (0.5 volt or more) thus constantly appears between any two electrodes. As long as this voltage does not change, the ECG amplifier rejects it as common mode voltage. However, movement by the patient can modulate this voltage, creating wandering traces and annoying artifact.

polarize To design the two mating halves so that only a particular combination of halves can be assembled, thus preventing accidental incorrect assembly.

polarized 1. Flowing in one direction. 2. Sensitive to the direction of current flow. 3. Vibrating in a particular direction or manner. 4. In a dry cell, restricted in current flow by the formation of electrolysis products around the electrodes.

polarized plug Plug which is so constructed that it can be inserted into a jack or receptacle in only one position.

polarized receptacle   Receptacle which is constructed so that only a polarized plug can be inserted.

polar relay   A relay whose operation depends upon the direction of the current through its operating winding.

polar signal   A signal whose information is contained in current reversals in the circuit.

pole face   The end of a relay core next to the armature.

pole (magnetic)   In a bar magnet, the attractive forces are concentrated at points near the ends and these are called the poles of the magnet. The pole of the magnet which is always attracted to the north is known as the north-seeking pole of the magnet. Similarly the other pole is the south pole. The magnetic North and South Poles of the Earth are those points on the Earth's surface to which magnetic compass needles arrange themselves and where the angle of magnetic dip is 90 degrees.

pole-piece   Either extremity of a magnet or of the core of a relay, at which the magnetic flux is concentrated.

polioencephalitis   A disease afflicting the gray matter of the brain, which becomes inflamed as a result.

poly   Prefix meaning many.

polycythemia   An abnormal increase in the number of circulating red blood cells.

polydipsia   Excessive thirst.

polygraph   Recorder of several signals simultaneously, such as blood pressure, respiratory motion, galvanic skin resistance; commonly used for study of emotional reactions involving deception (lie detection).

polymer   1. A material formed by the joining together of many (poly) individual units (mer) or a monomer; synonymous with elastomer. 2. A high-molecular-weight organic compound, natural or synthetic, whose structure can be represented by a repeated small unit, the mer; e.g., polyethylene, rubber, cellulose. Synthetic polymers are formed by addition or condensation polymerization of monomers. If two or more monomers are involved, a copolymer is obtained. Some polymers are elastomers, some plastic.

polymerization   A chemical reaction in which the molecules of a monomer are linked together to form large molecules whose molecular weight is a multiple of that of the original substance. When two or more monomers are involved, the process is called copolymerization or heteropolymerization.

polyopia   Seeing multiple images of the same object.

polyp   A protruding excrescence or growth from a mucous membrane, usually of the nasal passage but also of the uterine cervix, alimentary tract, or vocal cords.

polyphase   Describing an electrical circuit or electrical equipment which uses two or more phases. Polyphase circuits having two, three, and six phases are common.

polypnea   An abnormal increase in the respiratory rate.

polypus   A small, finger-like projecting tumor usually occurring in the nose, uterus, rectum, or ear. It has several forms, most of which are benign.

polystyrene   A water-clear thermoplastic produced by the polymerization of styrene (vinyl benzene). The electrical insulating properties of polystyrene are outstandingly good and the material is relatively unaffected by moisture. In particular, the power loss factor is extremely low over the frequency range 10^3 to 10^8 Hz.

polyuria The production and elimination of abnormally large volumes of urine.

polyvinyl chloride (PVC) A thermoplastic material composed of polymers of vinyl chloride. A tough, nonflammable, water resistant insulator much used for wire insulation. It has higher dielectric losses than polyethelene. Cautionary notices issued by the FDA have warned, however, that the material is carcinogenic.

popcorn Slang for undesired noise.

popliteal space Posterior surface of the knee.

portal hypertension Hypertension in the hepatic portal system, usually resulting from cirrhosis of the liver.

portal vein A vein carrying blood between capillary networks, usually the hepatic portal vein.

positive 1. Any point to which electrons are attracted. 2. Opposite of negative. 3. More than zero; symbolized by a plus sign (+).

position feedback A feedback signal which is proportional to the position or deflection of some object. It is an essential part of a closed-loop servo system in which the mechanical position of a recording pen is monitored by a position sensor and compared with the input signal. The discrepancy or error in position is amplified and used to drive the pen precisely to the position dictated by the accuracy of the position-sensing element and the drive-mechanism torque.

positive feedback 1. Process by which a part of the power in the output circuit of an amplifying device reacts upon the input circuit in a way to reinforce the initial power, thereby increasing the amplification. Also called regeneration; regenerative feedback. 2. Recycling of a signal that is in phase with the input to increase amplification. Used in digital circuits to standardize the waveforms in spite of anomalies in the input.

positive logic 1. A form of logic in which the more positive voltage logic swing represents a one and the more negative swing represents a zero. 2. A form of logic where one is assigned to the high level and logic zero to the low level.

positive rays Streams of positive ions which are started in motion in an evacuated tube in a direction from the anode to the cathode. Also called *canal rays*.

positron Fundamental particle equal in mass and energy to an electron, but having a positive charge. It has a very short life, being usually lost in the formation of a photon by combination with an electron. It occurs in the radiations from a few radioactive isotopes.

postaccelerating electrode See *intensifier electrode*.

posterior The rearmost parts of the body.

potential The difference in voltage between two points of a circuit. When a point is assumed to be ground, or reference, it has a zero potential.

potential difference or gradient A term used in electrical engineering. The electrical potential of a body is a measure of its condition and determines whether electricity will flow to or from it to earth when a connection is made. If a conductor which carries a positive charge, i.e., has a deficiency of electrons, and another which is charged negatively, i.e., has a surplus of electrons, are connected together, an electric current will flow and heat energy will be released. The current flows because of the potential difference between the two conductors.

potential drop Difference in potential between the two ends of a resistance with a current flowing through it.

potentiometer 1. Instrument for measuring an unknown electromotive force or potential difference by balancing it, wholly or in part, by a known potential difference produced by the flow of known currents in a network of circuits of known electric constants. 2. Three-terminal rheostat, or a resistor with one or more sliding contacts, which functions as an adjustable voltage divider. 3. A resistor provided with a tap that can be moved along it in such a way as to put the tap effectively at the junction of two resistors whose sum is the total resistance, the ratio of the two effective resistors being a function of the position of the tap. 4. A measuring instrument in which a potentiometer is used as a voltage divider in order to provide a known voltage that can be balanced against an unknown voltage. 5. A variable voltage divider, a resistor which has a variable contract arm so that a portion of the potential applied between its ends may be selected. 6. A variable voltage divider used for measuring an unknown electromotive force or potential difference by balancing it, in whole or in part, by a known potential difference. 7. An instrument used to measure or compare voltages.

potentiometer recorder A null-balance-type recorder using a servo-operated voltage-balancing device; the sliding contact of a precision measuring potentiometer is adjusted automatically by a servomechanism so that the difference in voltage of the circuit becomes zero. Main feature is high sensitivity.

potentiometric transducer A transducer in which the displacement of the force-summing member is transmitted to the slider in a potentiometer, thus changing the ratio of output resistance to total resistance. Transduction is accomplished in this manner by means of the changing ratios of a voltage divider.

potentiometric transduction Conversion of the measurand into a change in the position of a contact on a resistance element across which excitation is applied, the output usually being given as a voltage ratio.

poultices Hot moist material applied as an external counterirritant.

power The time rate at which work is done. Power in watts equals the amount of work in joules divided by the time during which the work was done in seconds. See *power (alternating current)* and *power (direct current)*.

power (alternating current) (ac) Power in an alternating current circuit is obtained by multiplying three factors: (a) the effective voltage, (b) the effective current, and (c) the cosine of the phase angle between the current and the voltage: $P = EI \cos \theta$.

power amplifier An amplifier designed to produce a gain in signal power, as distinguished from a voltage amplifier.

power density Power in watts per hertz, or the total power in a band of frequencies divided by the bandwidth in hertz.

power (direct current) (dc) The power in a direct current circuit, assuming P = watts, E = volts, I = amperes, and R = ohms, can be obtained in any of these ways:

$$\text{(a)} \quad P = EI$$

$$\text{(b)} \quad P = I^2 R$$

$$\text{(c)} \quad P = E^2/R$$

power dissipation Dispersal of the heat

generated in a component or circuit when current flows through it.

power factor   The ratio of the true power in watts of an alternating current circuit as measured by a wattmeter, to the apparent power obtained by multiplying the current and the potential, in amperes and volts. The power factor is usually less than unity because the current and potential are seldom in phase. If the current and voltage waveforms are sinusoidal, the power factor is equal to the cosine of the phase angle between current and potential.

power ground   The ground between units which carries the main source of power to, or from, these units. It is usually at a negative potential.

power supply   1. The commercial ac power provided by a power company. 2. A rectifier or other device which provides power for equipment. 3. A source of constant dc voltage for a circuit, usually obtained by conversion of ac input. May also use batteries, converters, generators, and the like. 4. A regulated source of direct current, containing circuitry which acts in a compensatory sense to oppose a change in output voltage because demands of the load circuit change.

ppm   Abbreviation for parts per million.

pps   Abbreviation for pulses per second.

preamplifier   A low-noise amplifier (usually differential input), designed to take a very-low-level signal, such as that from ECG or EEG electrodes and boost it to a level which can be accepted by a monitor or recording device.

precipitation   The process of separating solids from the liquids which hold them in solution, by the application of heat or cold, or the addition of chemicals.

precision   1. The degree of exactness with which a quantity is stated. 2. The degree of discrimination or amount of detail; e.g., a three-decimal digit quantity discriminates among 1,000 possible quantities. A result may have more precision than it has accuracy; e.g., the true value of pi to six significant digits is 3.14159; the value 3.14162 is precise to six figures, given to six figures, but is accurate only to about five.

precordium   The area of the chest over the heart.

prefix multiplier   Prefix which designates a greater or smaller unit than the original. Some common prefixes and the factors they represent are:

Prefix	Symbol	Factor
tera	T	10^{12}
giga	G	10^{9}
mega	M	10^{6}
kilo	k	10^{3}
hecto	h	10^{2}
deka	da	10
deci	d	10^{-1}
centi	c	10^{-2}
milli	m	10^{-3}
micro	μ	10^{-6}
nano	n	10^{-9}
pico	p	10^{-12}
femto	f	10^{-15}
atto	a	10^{-18}

premature atrial contraction (PAC)   An atrial ectopic beat produced by irritable tissue in the atrial wall.

premature ventricular contraction (PVC)   An ectopic beat produced by irritable tissue in the ventricular wall.

premonitory   Giving a warning beforehand. Some arrhythmias are considered predictive of acute coronary episodes.

presbyopia   Long-sightedness due to inability to make the lens of the eye convex by contraction of the ciliary muscles.

pressure areas Parts of the body where the bone is near the skin surface and where a bedsore may occur if there is prolonged pressure owing to diminished blood supply to these parts.

pressure pad A felt pad which applies a gentle but firm force to keep a moving tape in good contact with the heads of a tape recorder.

pressure points Points on the body at which pressure may be applied to check hemorrhage from wounds further downstream of the intentional occlusion.

pressure transducer A manometer employing a diaphragm and strain-gauge resistance elements to convert arterial, venous, or intracardiac blood pressures into representative voltage analogs. A hollow catheter is often used to connect one side of the diaphragm into the patient's cardiovascular system. The catheter may be filled with saline solution to act as a pressure-transmissive medium and to minimize patient's blood loss.

presumptive address See *reference address.*

presystole Period in the cardiac cycle before systole.

preventive maintenance A form of maintenance of any system which aims to prevent failures rather than correct malfunctions.

prf Abbreviation for pulse repetition frequency.

primary First. Fundamental. Principal. 1. A primary winding. 2. A primary power circuit. 3. A primary cell.

primary cell Cell designed to produce electric current through an electrochemical reaction which is not efficient reversible. Hence, the cell, when discharged, cannot be efficiently recharged by an electric current.

primary failure A failure which occurs under normal environment stresses and has no significant relationship to a prior failure, but whose occurrence imposes abnormal stress on some other part(s) which may cause secondary failure.

primary hemorrhage See *hemorrhage.*

primary hypothyroidism See *hypothyroidism.*

primary storage The main internal storage in a computer.

primary voltage 1. Voltage applied to the terminals of the primary winding of a transformer. 2. Voltage generated by a primary cell.

primary winding A transformer winding which receives energy from a supply source, and uses it to create a magnetic flux in the transformer core.

primordial Pertaining to the beginning.

printed circuit A copper foil circuit formed on one or both faces of an insulating board, to which circuit components are soldered. The copper foil pattern serves as connections between components, and is produced either by etching or plating.

printed contact That portion of a printed circuit used to connect the circuit to a plug-in receptacle and to perform the function of a pin in a male plug.

printed element An element in printed form, such as a printed inductor resistor, capacitor, or transmission line.

printed wiring A conductive network pattern created by plating or etching on the surface of an insulating baseboard, for the purpose of interconnecting components.

printer A teletypewriter or similar character-producing device used as the output of a computer. The device converts coded input signals into alphanumeric printed characters, usually on paper.

P-R interval That period in the surface ECG between atrial and ventricular contraction, during which impulses are traveling through the specialized conduction system of the heart, to the ventricular musculature.

print-through The transfer of magnetic signals between adjacent layers of magnetic tape when stored on the reel for long periods of time.

probe 1. A test prod, sometimes with circuit components like an RF detector in its handle. 2. A wire loop inserted into a resonant cavity or waveguide, used to couple an external circuit.

process 1. A prolongation or eminence of a part. 2. The act of converting and manipulating data to achieve high-speed answers.

process control Descriptive of systems in which computers, most frequently analog computers, are used for the automatic regulation of operations or processes. Typical are operations in the production of chemicals wherein the operation control is applied continuously and adjustments to regulate the operation are directed by the computer to keep the value of a controlled variable constant. Contrasted with numerical control.

processor 1. In hardware, a data processor. 2. In software, a computer program that includes the compiling, assembling, translating and related functions for a specific programming language. Includes the functions of logic, memory, arithmetic, and control (e.g., COBOL processor, FORTRAN processor). See *data processor* and *multiprocessor*.

processor (CPU) Circuits of a computing system that control the interpretation and execution of instruction (does not include interface, core memory, or peripherals).

proctoscope An instrument for viewing the interior of the rectum.

proctoscopy Examination of the rectum and the anus by means of a lighted instrument.

prodromal period The period that elapses in an infectious disease between the appearance of the first symptoms and the development of the rash, e.g., in smallpox, three days.

prognosis 1. The considered opinion as to the course of a disease. 2. A prediction of the probable outcome of an illness.

program 1. To plan a computation or process from the original statement of the problem to the delivery of the results, including the integration of the operation of the resulting program into an existing system. 2. A sequence of instructions that embodies the logic to be carried out by a computer or other automaton, usually for the solution of a problem. 3. In a calculator, a sequence of detailed instructions for the operations necessary to solve a problem. Programmable electronic calculators can "learn" the steps of a problem so that, after the first sequence of entries, only the variable numbers need be entered on the keyboard without manual activation of control keys. Some programmable machines store programs on cards or tapes. 4. A set of coded instructions which direct a computer or system to perform some specific function or yield the solution to some specific problem. 5. A series of actions proposed in order to achieve a certain result. 6. To prepare such a set of coded instructions. 7. To design, write, and test a program.

program-address counter A register in which the address of the current instruction is recorded. Also called *instruction counter*.

program control A system in which a computer is used to direct an operation or process and automatically to hold or to make changes in the operation or process on the basis of a prescribed sequence of events.

program counter See *program register*.

program line See *code line*.

programmability Ability of a device to accept external control signals.

programmable calculator 1. A calculator whose operation is controlled by programs stored in its memory. 2. Electronic calculator capable of performing preset sequences of computations. 3. One that can "learn" a repetitive series of operations or be programmed by various means to handle a series of steps so that only variable information need be entered into a calculator.

programmable communications processor A digital computer that has been specifically programmed to perform one or more control and/or processing functions in a data communications network. As a self-contained system, it may or may not include communications line multiplexers, line adapters, a computer system interface, and on-line peripherals. It always includes a specific set of user-modifiable software for the communications function.

programmable counter A device that can be logically programmed to count to any number from zero to its maximum possible modulus. Also called *module N counter*.

programmable ROM See *PROM*.

programmed check A means of testing for the correctness of a computer program and machine functioning, either by running a similarly programmed sample problem with a known answer (including mathematical or logical checks) or by building a checking system into the actual program being run.

programmed machines A computer which recognizes patterns by comparing them on a bit-by-bit basis with patterns stored in memories or by sequential scrutinizing to determine whether the object has certain properties which are stored by the machine's memory.

programmed marginal check Computer program that varies its own voltage to check a given circuit or other piece of electronic computer equipment during a preventive maintenance check.

programmed operators Computer instructions which enable subroutines to be accessed with a single programmed instruction.

programmer A person who prepares the planned sequence of events that the computer must follow to solve a problem, but who need not necessarily convert them into detailed instructions (coding).

programming 1. Definition of a computer problem resulting in a flow diagram. 2. Preparing a list of instructions for the computer to use in the solution of a problem. 3. To select various circuit patterns by interconnecting or "jumping" appropriate contacts on one side of a panel connector plug.

programming device A device by which a series of mechanical or electrical operations or events may be present to be performed automatically in a predetermined sequence and at specified time intervals.

programming language 1. The original form in which a program is prepared prior to processing by the computer. 2. A language which is used by programmers to write computer routines.

program register The computer control unit register into which the program instruction is being executed, is stored. Hence, controlling the computer operation during the cycles required to execute that instruction. Also called *program counter,* or *control register.*

program storage In a computer portion of the internal storage reserved for the storage of programs, routines, and sub-routines. In many systems, protection devices are used to prevent inadvertent alteration of the contents of the program storage. Contrasted with temporary storage.

projective tests Psychological tests used as a diagnostic tool. Among the most common projective tests is the Rorschach (inkblot) test.

prolapse The collapse or falling of an organ, such as the uterus.

proliferation The reproduction or duplication of similar forms, especially of cells.

PROM 1. Abbreviation for programmable ROM. A ROM that can be programmed by the user but only once. After a PROM is programmed, effectively it becomes a ROM. 2. Similar to the conventional ROM (read-only memory). A write-once memory. When an instruction is written via a memory write cycle into the programmable ROM, certain kinds of fusing take place and the data are written permanently into the memory.

promontory A projecting part. An eminence.

propagation 1. In electrical practice, the travel of waves through or along a medium. 2. Traveling of a wave along a transmission path. 3. Travel of electromagnetic waves or sound waves, through a medium. Propaga-

tion does not refer to the flow of current in the ordinary sense. Also called *wave propagation.* 4. Perpetuation of a species.

propagation delay 1. A measure of the time required for a change in logic level to propagate through a chain of circuit elements. 2. A measure of the time required for a logic signal to travel through a device or a series of devices forming a logic string. It occurs as the result of four types of circuit delays; storage, rise, fall, and turn-on delay, and is the time between when the input signal crosses the threshold voltage point and when the responding voltage at the output crosses the same voltage point.

propagation time The time necessary for a unit of binary information (high voltage or low) to be transmitted or passed from one physical point in a system or subsystem to another. For example, from input of a device to output.

prophylactic Tending to prevent disease.

proportional control A control system in which corrective action is always proportionate to any variation of the controlled process from its desired value. For example, instead of snapping directly open-closed in the manner of two-position control, a proportionally controlled valve will be always positioned at some point between open and closed, depending on a system's flow requirement at any given moment.

proportional temperature control A method of stabilizing (an oscillator) by providing heater power that is directly proportional to the difference between the desired operating temperature and the ambient temperature.

propranolol A drug for heart-rhythm disorders.

proprioceptor Sensory end organ which detects changes in position, of muscles or joints or fluid in the balancing apparatus of the inner ear.

prostate A gland associated with the male reproductive system. Its size and secretion are under the influence of androgens. Its function is to supply supportive substances to the spermatozoa in the seminal vesicles.

prostatectomy Operation of removing the prostate gland. The operation may be suprapubic when the bladder is first incised, or retropubic.

prostatitis Inflammation of the prostate gland.

prosthesis Aritificial eye, tooth, limb, or other body part. See also *heart valve*.

prosthetics The field devoted to development of replacements parts and organs for the human body.

prostration Extreme exhaustion.

proteinuria The presence of protein in the urine.

proton A positively-charged elementary particle which is part of the nucleus of the atom. Its mass is 1,837 times greater than that of an electron.

protoplasm 1. The complex of protein, other organic and inorganic substances, and water that constitutes the living nucleus, cytoplasm, plastids, and mitochondria of the cell and is regarded as the only form of matter in which the living phenomena are manifested. 2. The living substance of a cell, thus excluding material which has been ingested or excreted.

protoplasmic See *protoplasm*.

prototype Original design or first operating model.

protozoa A class of unicellular organisms, forming the lowest division of the animal kingdom, e.g., an amoeba, or paramecium.

proud flesh Excessive granulation tissue in a wound.

proximal 1. Closest to the point of reference or to the point of attachment of two points. 2. Nearest to the median line; toward the center of the body.

pruritus Intense itching.

ps Abbreviation for picosecond (10^{-12} second).

pseudo Prefix meaning false.

pseudocode Computer instructions written in mnemonic or symbolic language by the programmer. These instructions do not necessarily represent operations built into the computer. They must be translated into machine language and have absolute addresses assigned before they can become a finished and assembled program which the computer can use to process data. See also *symbolic code*.

pseudoprogram A program written in a pseudocode that may include short coded logical routines.

pseudorandom Having the property of being produced by a definite calculation process, but at the same time satisfying one or more of the standard tests for statistical randomness.

psoas An important muscle attached above to the lumbar vertebrae and below to the femur. It flexes the femur on the trunk.

psoriasis Chronic, occasionally acute, recurrent skin disease of unkown cause, characterized by thickened red skin patches that are covered with whitish shiny scales. Psoriasis usually affects the scalp, elbows, knees, back, and buttocks.

psychiatry The study and treatment of mental disorders.

psychoacoustical research Research on

the sensory effects of sound waves on the ear.

psychoanalysis A theory of human development and behavior, a method of research, and a system of psychotherapy, originally described by Sigmund Freud (1856–1939). Through analysis of free associations and interpretation of dreams, emotions and behavior are traced to the influence of repressed instinctual drives in the unconscious. Psychoanalytic treatment seeks to eliminate or diminish the undesirable effects of unconscious conflicts by making the patient aware of their existence, origin, and inappropriate expression.

psychogalvanic See *psychogalvanometer*.

psychogalvanometer A galvanometer for recording the electric agitations produced by emotional stresses.

psychogenic Originating in the mind; psychic.

psychologist One who studies psychology but is not a doctor of medicine.

psychology The study of behavior patterns.

psychoneurosis One of the two major categories of emotional illness, the other being the psychoses. It is usually less severe than a psychosis, with minimal loss of contact with reality.

psychopath A mentally deranged person.

psychosis A major mental disorder of organic and/or emotional origin in which there is a departure from normal patterns of thinking, feeling, and acting. Commonly characterized by loss of contact with reality, distortion of perception, regressive behavior and attitudes, diminished control of elementary impulses and desires, and delusions and hallucinations. Chronic and generalized personality deterioration may

occur. A majority of patients in public mental hospitals are psychotic.

psychosomatic 1. Adjective to denote the constant and inseparable interdependence of the psyche (mind) and the soma (body). Most commonly used to refer to illnesses in which the manifestations are primarily physical with at least a partial emotional cause. 2. Denotes physical conditions which have psychological causes.

psychotherapy The term for any type of mental treatment that is based primarily upon verbal or nonverbal communication with the patient, in distinction to the use of drugs, surgery, or physical measures such as electric or insulin shock.

ptyalin An enzyme found in salvia which can digest starch (amylase).

P-type semiconductor A semiconductor material which has been doped so that it has a net deficiency of free electrons. It therefore conducts electricity through movement of holes. See also *N-type semiconductor*.

puberty The age of sexual maturity.

pull 1. To cause an oscillator to depart from its designed frequency of operation. 2. To depart from the designed frequency of operation, as an oscillator.

pull-down resistor A resistor connected to a negative voltage or ground.

pull-up resistor A resistor connected to the positive supply voltage. For example, from V_{cc} to the output collector of a transistor.

pulmonary 1. Relating to, functioning like, or associated with the lungs. 2. Relating to the respiratory function of the lung.

pulmonary edema Usually an acute condition in which there is a waterlogging of the lung tissue, including its alveolar cavities. Respiration is impaired. If inadequately treated, it may lead to rapid death; it is often a complication of chronic heart disease.

pulmonary emphysema Disease characterized by reduced elasticity, ruptured alveolar septa, increase in air sac size and decrease in pulmonary capillary bed.

pulmonary function measurements See *functional residual capacity, inspiratory capacity, expiratory reserve volume, forced expiratory flow, forced expiratory volume, forced midexpiratory flow, forced vital capacity, inspiratory reserve volume, total lung capacity, maximal breathing capacity, minute volume,* and *tidal volume.*

pulmonary hypertension Increase of the pressure in the pulmonary circulation.

pulmonary stenosis Narrowing of the pulmonary valve of the heart.

pulmonary valve The valve at the exit of the right ventricle into the pulmonary artery. See also *heart valve.*

pulsating current Current which varies in amplitude but does not change polarity.

pulse 1. Brief excursion of a quantity from normal. 2. Signal characterized by the rise and decay in time of a quantity, the value of which is normally constant. 3. In radio, surge of electrical energy of short duration. 4. Voltage level of short duration used in computers to represent a bit. 5. Single disturbance characterized by the rise and decay in time or space of a quantity whose value is normally constant. 6. Single impulse of a telephone dial or similar signal. 7. In relay operation, sudden change of brief duration produced in the current or voltage of a circuit to actuate or control a switch or relay. 8. Electrical signal of short duration used for the transmission of control information. 9. The rhythmic change in volume of an artery, which is

sensible to the finger or detectable by a plethysmograph, resulting from each contraction of the heart.

pulse amplifier A wideband amplifier capable of amplifying pulses without distortion.

pulsed ruby laser A laser that uses ruby as the active material. The extremely high pumping power required is obtained by discharging a bank of energy storage capacitors through a special high-intensity flash tube.

pulse duration 1. Time interval between the first and last instants at which the instantaneous amplitude reaches a stated fraction of the peak pulse amplitude. 2. Time duration of transmission of a pulse of energy, measured in time or equivalent distance. Also called *pulse length, pulse width.*

pulse equalizer A circuit that produces output pulses of uniform size and shape in response to input pulses which may vary in size and shape.

pulse generator A device that produces a single pulse or a train of repetitive pulses.

pulse-height analyzer An instrument capable of indicating the number or rate of occurrence of pulses falling within each of one or more specified amplitude ranges. Also called *kick-sorter.*

pulse length See *pulse duration.*

pulse meter Device that counts pulse rate; may use a finger or toe cuff, ear lobe sensor, etc.

pulse modulation 1. Use of a series of pulses, modulated or characterized to convey information. Types of pulse modulation include amplitude (PAM), position (ppm), and duration (PDM) systems. 2. Modulation of a carrier by a pulse train. NOTE: In this case, the term describes the process of generating carrier frequency pulses. 3. Modulation of one or more characteristics of a pulse carrier. NOTE: In this sense, the term describes methods of transmitting information on a pulse carrier.

pulse pressure The difference between systolic and diastolic blood pressure (typically 40 mmHg).

pulse radiation A rapid increase and decrease in radiation intensity over a very short period of time.

pulse resolution The minimum time separation betwen input pulses that will permit a circuit or component to respond properly. Usually specified in microseconds or nanoseconds.

pulse sampling A method whereby samples are taken from successive pulses in order to reduce a very high-frequency wave.

pulse scaler A device that produces an output signal whenever a prescribed number of input pulses has been received. It frequently includes indicating devices for interpolation.

pulse shaping Taking a pulse through a filter to remove some of the frequency components, thus making it easier to send over a transmission facility.

pulse stretcher A circuit in a computer which generates a long pulse when triggered by a short pulse. The width of the output pulse is determined by the value of the coupling capacitor. The maximum width of the output pulse cannot exceed 50 percent of the clock rate.

pulse-time modulation Modulation in which the values of instantaneous samples of the modulating wave are caused to modulate the time of occurrence of some characteristic of a pulse carrier.

pulse train A succession of pulses which follow each other, usually at equal intervals.

pulse transformer A transformer capable of passing a wide band of frequencies, and thus of passing a pulse without unacceptable distortion.

pulse width See *pulse duration.*

pulse-width discriminator Device that measures the pulse length of signals and passes only those whose time duration falls into some predetermined design tolerance.

punch card A heavy stiff paper of constant size and shape, suitable for punching in a pattern that has meaning, and for being handled mechanically. The punched holes are sensed electrically by wire brushes, mechanically by metal fingers, or photoelectrically by photocells. Related to eighty-column card and ninety-column card.

punched paper tape A tape on which information is recorded by punched holes.

punched-tape recorder A recorder that records data in the form of holes punched in tape strip.

pupil 1. The opening in the center of the iris of the eye. 2. The contractile aperture in the iris of the eye.

Purkinje fibers 1. The lowest branches of the heart's specialized conduction system which distribute pacing impulses to the myocardium. Named for J. E. von Purkinje, 18th century anatomist and physiologist. 2. Beaded, muscular fibers forming a network in the subendocardial tissue of the ventricles of the heart, which carry signals originating in the natural pacemaker from the atria to the ventricles and thereby permit regular ventricular contraction.

purpura Purple-colored spots due to hemorrhage into the tissues. There are many causes of purpura which may be roughly categorized as due to: (a) deficiency of clotting mechanism, e.g., essential thrombocytopaenia; (b) capillary damage; (c) both.

purulent Containing pus.

pus The thick liquid product of inflammation, composed of leukocytes, liquid, tissue debris, and microorganisms.

pushbutton A normally open (or normally closed), nonlocking two-pole switch which can be operated by the pressure of a finger on a button.

pushdown list A list of items in which the last item entered is the first item of the list, and the relative position of the other items is pushed back one.

push-pull amplifier A balanced amplifier using two identical output stages operating 180 degrees out of phase to feed a center-tapped output transformer. Distortion produced by one is canceled by the other stage.

push-push configuration A harmonic oscillator design in which the signals from each output transistor or tube operating at a given frequency f_0 are combined to produce an output signal at $2 f_0$. The main advantage of this configuration is the extension of a transistor's operating frequency limits without the use of an extra frequency doubler circuit.

pushup list A list that is constructed and maintained so that the next item to be retrieved and removed is the oldest item still in the list, i.e., first in, first out.

pustulation The formation of pustules.

pustule A pimple containing pus.

PVC Abbreviation for polyvinyl chloride.

pyarthrosis Suppuration in a joint.

pyelitis Inflammation of the pelvis of the kidney.

pyemia A general septicemia in which there is pus in the blood.

pyoderma Any septic skin lesion.

pyogenic Pus-producing, forming pus.

pyorrhea A condition in which pus oozes from the gums around the roots of the teeth. This is also known as *Rigg's disease*.

pyrexia Fever. Elevation of the body temperature above the normal.

pyroelectric material One which produces an electrical output when subjected to a change in temperature.

pyrogen A fever-producing subtsance.

pyrometer An instrument used to measure elevated temperatures (beyond the range of mercury thermometers) by electric means. These include immersion, optical radiation, resistance, and thermoelectric pyrometers.

pyuria Pus in the urine.

Q

Q 1. The ratio of energy stored to energy dissipated. 2. In a capacitor, the ratio of susceptance to effective shunt conductance at a specified frequency. 3. In an inductor, the ratio of reactance to effective series resistance at a given frequency. 4. A measure of frequency selectivity or the sharpness of resonance.

Q-factor A measure of the excellence, or lack of resistance losses, of a coil. It is equal to the coil reactance divided by the coil resistance, both in ohms.

Q output The reference output of a flip-flop. When this output is one, the flip-flop is said to be in the one state. When it is zero, the output is said to be in the zero state.

$\overline{Q}$ (Q-bar) output The second output of a flip-flop. It is always opposite in logic level to the Q output.

QRS complex That portion of the waveform in an electrocardiogram extending from point Q to point S; it includes the maximum amplitude shown in an ECG trace, encompassing the full complex of ventricular muscle depolarization.

Q-T interval That portion of the ECG wave which describes the electrical forces attending one complete ventricular contraction, including the phases of depolarization and depolarization of both ventricles.

quad latch A group of four flip-flops, each of which can store a true or false level, and that are all normally enabled by a single control line. When the flip-flops are all enabled, new data may be stored in each of them.

quadrature The state or condition of two related periodic functions (i.e., voltage and current) occurring at the same frequency, but being separated in phase by a quarter of a cycle, or 90 electrical degrees.

quadri Prefix meaning fourfold.

quadriceps Four separate muscles, covering the front of the thigh, which extend from the tubercle of the tibia to the knee.

quantitizer A device which partitions a continuum of analog values into discrete ranges to be represented by a digital code. An analog-to-digital converter.

quantity of electricity 1. The quantity of

electricity is measured in coulombs. A coulomb is the quantity of electricity which passes a point in a circuit during one second when a current of one ampere is flowing. See *electrical charge*.

quantization Process in which the continuous range of values of an input signal is divided into subranges, and to each subrange a discrete value of the output is uniquely assigned. Whenever the signal value falls within a given subrange, the output has the corresponding discrete value.

quantize To convert a continuous variable, such as a waveform, into a series of levels or steps. There are no in-between values in such a quantized waveform. All values of signal are represented by the nearest standard value or code position.

quantizing Expressing an analog value as the nearest one of discrete set of prechosen values.

quantizing error The error inherent in digital data, in the worst case equal to one-half quantum.

quarantine A period of separation of infected persons or contacts from others, to prevent the spread of disease.

quarter-wave Having an electrical length of one-quarter wavelength, expressed in meters or fractions thereof.

quartz Crystal of pure silicon dioxide. The original piezoelectric material, widely used to control the frequency of oscillators and in timing applications.

quasi-analog signal A digital signal which is suitable for transmission over an analog (voice) channel. Implies that the frequency, bandwidth, distortion, and noise requirements can be met by a voice channel.

quibinary code A binary coded decimal code for representing decimal numbers in which each decimal digit is represented by seven binary digits which are coefficients of 8, 6, 4, 2, 0, 1, 0, respectively.

quick-acting relay A relay which operates immediately after its operating circuit is closed and restores immediately after its operating circuit is opened, without any significant time lag.

quick-connect Describing a terminal or terminal block on which insulated conductors can be terminated quickly.

quick-disconnect A type of connector which permits rapid locking and unlocking of two connector halves or contacts.

quiescence The operating condition that exists in an amplifier circuit when no input signal is applied to the circuit.

quiescent 1. Inactive. 2. Without an input signal.

quiescent current See *idling current*.

quiescent input voltage The dc voltage present at the input of an amplifier having one input terminal when the input terminal is not connected to any source.

quiescent operating point The dc voltages, currents, and dissipation associated with a circuit while it is not being subjected to an input signal.

quiescent output voltage The dc voltage present at the output terminals of an amplifier when the input is ac grounded through a resistance representing the signal source resistance.

quiescent point See *operating point*.

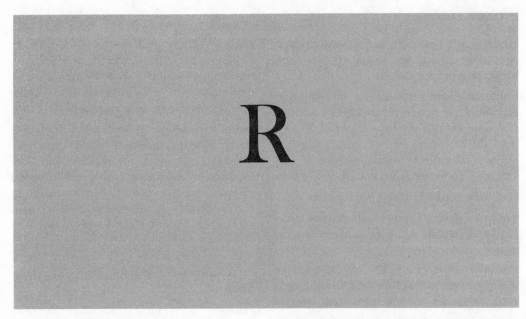

R

R Abbreviation for roentgen.

raceway A channel used for holding and routing wires or cables in a structure.

rad Abbreviation for radiation absorbed dose, the quantity of radiation that delivers 100 ergs of energy to 1 gram of substance (almost equivalent to a roentgen when referred to body tissue). Approximately 1.2 roentgens.

radial lead A lead extending from the side of a component. Opposed to axial lead.

radian The angle at the center of a circle formed between two radii which cut off an arc of the circle whose length is equal to the radius. One radian is equal to 57.2958 degrees.

radiance The radiant intensity per unit solid angle per unit of projected area of an extended source.

radiant energy 1. Any energy which radiates in the form of radio waves, infrared (heat) waves, light waves, X-rays, etc. 2. Electromagnetic radiation.

radiant flux Time rate of flow of radiant energy. Expressed in watts or in ergs per second.

radiant intensity Power radiated from a source per unit solid angle in the direction considered. Also called *radiation intensity*.

radiate 1. The spreading out of radiant energy, as from a center. 2. To transmit energy in waves which are sharply directional.

radiation Emanation of energy from a source. The energy may be in a number of different forms which, according to their frequency of emission, are known as radio waves, light, X-rays, gamma rays, etc. Subatomic particles (such as electrons, neutrons, protons, and positrons), mesons, and many subnuclear particles may also be radiated. Helium nuclei (two protons and two neutrons) have a short track length but are sometimes used to irradiate biological material. They are a natural product of certain radioactive elements and are known as alpha rays. High-energy neutrons may be produced by atomic explosions and certain special equipment. Electrons are also a natural product of radioactivity and are known as beta rays. Radiation can damage living tissues, resulting in illness from overexposure to radiation.

radiation counter A device for counting radiation particles (alpha, beta, gamma, neutrons, etc.) or photons of energy (X-ray, etc.) usually using either scintillation or ionization resulting from presence of particle or photon to be measured.

radiation counter tube See *counter tube*.

radiation dose The total amount of ionizing radiation absorbed by material or tissue.

radiation dose rate The radiation dose absorbed per unit of time. A radiation dose rate can be set at some particular unit of time (e.g., H + 1 hour would be called H + 1 radiation dose rate).

radiation hazard Health hazard arising from exposure to ionizing radiation.

radiation intensity See *radiant intensity*.

radiation monitor A device for determining amount of exposure to radioactivity. May be periodic or continuous, may monitor an area, or an individual, his breath, clothing, etc.

radiation resistance The ability of a material to retain useful properties during or after exposure to nuclear radiation.

radiation scattering The diversion of radiation (thermal, electromagnetic, or nuclear) from its original path as a result of interactions or collisions with atoms, molecules, or larger particles in the atmosphere, or other media between the source of radiation (for example, a nuclear explosion) and a point at some distance away. As a result of scattering, radiation (especially gamma rays and neutrons) will be received at such a point from many directions instead of only from the direction of the source.

radiator An emitter of radiant energy.

radical A group of atoms, replaceable by a single atom, that is capable of remaining unchanged during a series of reactions, or that may show a definite transitory existence in the course of a reaction.

radioactive bolus A large pill, usually soft in consistence, made of radioactive materials. The pill is swallowed, with the radioactive substance lining the mucosal surface, and studies can then be made on the continuity of the gastrointestinal tract using radiographic (X-ray) equipment. The secondary absorption of certain materials with radioactive tags is also an objective of radioactive bolus studies using other techniques.

radioactive decay The spontaneous transformation of one nuclide into a different nuclide, or into a different energy state of the same nuclide. The process results in a decrease, with time, of the number of the original radioactive atoms in a sample. It involves the emission from the nucleus of alpha particles, beta particles, or electrons, or gamma rays, or the nuclear capture or ejection of orbital electrons, or fission. Also called *radioactive disintegration*.

radioactive disintegration See *radioactive decay*.

radioactive fallout Radioactive isotopes distributed in the atmosphere as the result of atomic explosions. It constitutes a biological hazard since isotopes may be ingested and passed along in the food chain.

radioactive isotopy An isotopic form of an element characterized by the property of spontaneous decomposition through emission of alpha, beta, or gamma radiation. Mildly radioactive isotopes, administered by injection or ingestion, are often used as tracers in studies of metabolism.

radioactive tag A radioisotope that has

been incorporated within a biological chemical by metabolic or other processes.

radioactivity A process whereby certain nuclei undergo a spontaneous disintegration in which energies are liberated and the emission of corpuscular (particle) or electromagnetic radiation takes place, generally resulting in the formation of new nuclides.

radiobiology The study of the effects on living matter (or substances derived therefrom) of high energy radiation extending from X-rays to gamma rays, including high energy beams of neutrons and charged particles, e.g., alpha particles, electrons, protons, deuterons.

radiocardiogram X-ray recording of concentration of radioisotopes injected into heart chambers.

radiocardiography A diagnostic method whereby the patient wears, strapped onto his body, a box containing a transistorized radio transmitter whose telemetry signals are keyed by ECG electrodes attached to the skin. The device transmits the ECG of the heart. The transmitted ECG data is picked up by a receiver and displayed on a cathode-ray tube or recorded on chart paper. A patient so equipped can move about freely while transmitting a continuous record of his heart's activity. See *ECG telemetry*.

radioelectrocardiograph An electrocardiograph employing a radio link (telemetry) so that the patient is free to move about while the electrocardiogram is being recorded.

radioelectroencephalograph An electroencephalograph employing a radio link so that the patient is free to move about while the electroencephalogram is being recorded.

radio frequency (RF) That portion of the electromagnetic spectrum lying between about fifty thousand and a few billion hertz. Radio-frequency waves are used in many physiologic and medical applications, such as ECG telemetry, urinary bladder stimulation, and recharging nickel-cadmium batteries implanted within the body.

radio-frequency choke An inductor which presents a considerable impedance to radio-frequency currents, but very little to direct currents. Generally uses an air core or a powdered-iron core.

radio-frequency induction The transfer of electrical energy through the intact skin by electromagnetic coupling from an extracorporeal coil to an implanted one. Radio-frequency pacemakers utilize this principle. Recently, urinary bladder stimulators have also been designed to receive electromagnetically coupled radio-frequency energy.

radio-frequency interference Any electrical signal capable of being propagated into, and interfering with, the proper operation of electrical or electronic equipment. The frequency range of such interference may be taken to include the entire electromagnetic spectrum.

radiography Science of examination by means of X-ray radiation.

radioisotope Radioactive isotope produced artificially from an element or from a stable isotope of the element by the action of neutrons, protons, deuterons of alpha particles (in chain-reacting atomic piles, cyclotrons, etc.). Radioisotopes are used in radiation treatment of cancer and as tracers. 2. An isotope exhibiting radioactivity. Radioisotopes have many different medical applications. Radioactive tracers are used to follow biologic processes, since a tagged substance, even when changed chemically, will retain its radioactivity for a time and with a degree proportional to its half life. When combined with a stable element, the stable element so treated is said to be labeled or tagged. Radioisotope scanning is useful in visualizing certain tissues when a substance is selectively absorbed, for example, by a tumor. Radioisotopes are also used therapeutically, for example, in treating tumors and hyperthyroidism. Also called *radioactive isotope*.

radiologist A physician who has made a special study of radiology and radiographic techniques.

radiology The study of diagnosis by means of radiography. Study of the properties of X-rays and their applications. It also includes related subjects, such as the use of nuclear materials in medicine. Radiology is now sharply divided into diagnostic radiology and nuclear medicine, and radiologists tend to specialize in one or the other of these branches. The diagnostic radiologist is concerned with the use of X-rays to detect disease and to discover its nature. This is done principally by two methods; radioscopy, in which a fluorescent screen reveals the passage of X-rays through the patient's body, and radiography, in which the rays, after passage through the body are brought into contact with an exposed film, so as to make, after developing, a permanent record to be interpreted later. At first, radiology was concerned with bones and foreign bodies, but the use of swallowed or injected substances opaque to X-rays have made possible investigation of the stomach, bowels, bladder, kidneys, lungs, womb, and the heart cavities. The radiotherapist uses X-rays and the rays emitted by nuclear materials (such as radium) to treat disease. The bulk of his work is concerned with treatment of cancer, particularly in cases where the surgeon's knife cannot remove the growth.

This science now covers a wide spectrum of diagnostic and therapeutic applications, such as diagnosis using roentgen rays, diagnosis using radioisotope scanning techniques, genetics radioisotopes in research, roentgen cinematography, angiography, fluoroscopy, pneumoencephalography, spectroscopy, and treatment of tumors using roentgen rays, implanted radioactive pellets and cyclotrons.

radiometer Device for measuring radiant flux density. Generally employs a blackened thermocouple or bolometer. The simplest is the rotating-vane-type radiometer.

radiometric Relating to the measurement of radiation.

radiometry 1. The science of radiation measurement concerned with the detection and measurement of radiant energy either at separate wavelengths or integrated over a broad wavelength band, and the interaction of radiation with matter such as absorption, reflectance, and emission. 2. The measurement of radiation in the infrared, visible, and ultraviolet portion of the spectrum.

radiopaque A substance or material which appears as a solid, reflective object to radiant energy, such as X-rays. Indocyanine green dye is such a material. It is used in cineangiographic studies, to outline and make visible the coronary arteries of the heart, which would otherwise be transparent to X-rays. Perfusing the heart's arterial system with this dye reveals atherosclerotic deposits which occlude the blood flow to the myocardium.

radioscopy Direct vision of X-ray picture by means of a fluorescent screen. This has been improved by the introduction of the image intensifier, which provides a brighter image, with a substantial reduction in the irradiation energy required.

radiosensitive Term applied to a structure, especially a tumor, responsive to radiotherapy.

radiotelemetering Telemetering by means of radio waves.

radiotherapy The treatment of disease with radiations, especially ultraviolet, infrared, X-rays, and gamma rays.

radiotherapy apparatus See *X-ray therapy apparatus*.

radix The base of a numbering system. It is the total number of distinct marks or symbols used in the numbering system. For example, the decimal numbering system uses 10 symbols (0, 1, 2, 3, 4, 5, 6, 7, 8, 9). The radix here is 10. We obtain decimal numbers by using various powers of 10. In the binary numbering system, there are only 2 marks or symbols (0, 1). The radix is 2, and we obtain binary numbers by using various powers of 2.

radix point The dot that delineates the integer digits from the fractional digits of a number; specifically, the dot that delineates the digital position involving the zero exponent of the radix from the digital position involving the minus-one exponent of the radix. The radix point is often identified by the name of the system; e.g., binary, octal, or decimal. In the writing of any number in any system, if no dot is included, the radix point is assumed to follow the rightmost digit.

rads The units of absorbed dose of any material. One rad liberates 100 ergs per gram of absorbing material.

rads/sec The units of absorbed dose rate of any material.

rales Abnormal discontinuous breath sounds, detectable by auscultation, caused by air passing through secretions.

RAM Abbreviation for random-access memory.

ramus A branch; thus, a branching or ramification of vessels.

random access 1. Process of obtaining information from computer storage where access time is independent of information location. 2. Technique of storing information so that its recovery time will not be influenced by how recently it was stored. A random access device permits information recovery at will, with no appreciable time penalty for the order in which it was last handled. Contrasted with sequential address.

random access device A device in which the access time is effectively independent of the location of the data. Also called *direct access device*.

random access memory (RAM) 1. A computer storage technique in which the time required to obtain information is independent of the location of the information most recently obtained. This strict definition must be qualified by the observation that we usually mean relatively random. Thus, magnetic drums are relatively non-random access when compared to magnetic cores for main storage, but are relatively random access when compared to magnetic tapes for file storage. 2. A memory system for which the access time is constant regardless of the location being addressed. A magnetic core memory is a random access memory. 3. A device that permits individual interrogation of any memory cell in a completely random sequence. Any point in the total memory system can be accessed without looking at any other bit. 4. A memory that has the stored information immediately available when addressed, regardless of the previous memory address location. As the memory words can be selected in any order, there is equal access time to all. 5. A memory which may be written into, or read from, any address location, in any order. May refer specifically to the integrated circuit method of implementation. 6. A read/write memory that stores information in such a way that each bit of information may be retrieved within the same amount of time as any other bit. As opposed to serial or sequential address memory. 7. Any memory in which a word (sometimes a byte) can be read at random. For a mass storage device, this usually means a small block of words (e.g., 256). Core and certain semiconductor memories are examples of RAMs as are magnetic discs and flexible discs. Shift register memories and magnetic tapes are not random access devices.

random access programming Programming a problem for a computer without regard to the access time to the information in the registers called for in the program.

random access storage A form of storage in which information can be recovered immediately, regardless of when it was stored. For example, magnetic core memory devices will usually yield any bit of information with almost no time penalty, regardless of where it is located, while magnetic tape must be run until the required information can be found.

random error The inherent imprecision of a given process of measurement; the unpredictable component of repeated independent measurements on the same object under sensibly uniform conditions, usually of an approximately normal (Gaussian) frequency distribution; other than systematic or erratic errors and mistakes, sometimes called *short-period error*.

random experiment An experiment that can be repeated a large number of times, under similar circumstances, but which may yield different results at each trial.

random failure Early breakdown caused by inherent weaknesses in material and/or by damage incurred during the manufacturing process. The random failure rate pattern is characterized by a high initial failure rate followed by a leveling-off period. 2. Any change failure, the occurrence of which is unpredictable. Also called *chance failure*.

randomness An equal chance for any of the possible outcomes.

random noise 1. Noise generated in an electronic circuit by random movement of electrons, caused by thermal agitation. In tape recording, it can be caused by uneven distribution of magnetized particles and is reproduced as a background hiss. 2. Noise in which the frequency and phase of the components vary at random. It is characterized by a peak-to-average-noise-level ratio in the order of 4:3 to 5:4. This is a broadband type noise. Thermal and shot noise are typical. Also called fluctuation noise. See also *broadband electrical noise* and *grass*.

random number 1. A set of digits constructed of such a sequence that each successive digit is equally likely to be any of n digits to the base n of the number. 2. A number formed by one or a set of digits selected from an orderless sequence of digits.

random processing The treatment of data without respect to its location in external storage, and in an arbitrary sequence governed by the input against which it is to be processed.

random sample A sample in which each item in the lot has an equal chance of being selected in the sample.

random sampling A sampling process involving significant time uncertainty between the signal being sampled and the sample-taking operation.

random-sampling oscilloscope An oscilloscope employing the random-sampling process, together with means for constructing a coherent display from the samples taken at random.

random sequential memory A memory in which one reference can be found immediately; the other is found in a fixed sequence.

RAO Abbreviation for right anterior oblique.

raphe Fibrous junction between muscles.

rapid storage In computers, storage with a very short, read-in access time. Rapid access is generally gained by limiting storage capacity. Also called *rapid memory, fast-access storage* and *high-speed storage*.

rapid memory See *rapid storage.*

RAPO Abbreviation for right anteroposterior oblique.

raster 1. The pattern of lines followed by the scanning beam in a television camera tube. 2. The area which is scanned. 3. The line pattern produced on the screen of a cathode-ray tube through deflection of its electron beam by horizontal and vertical control circuits, whether or not the beam is sufficiently intensity modulated to create a light dot on the phosphor screen. In effect, the raster forms a point-addressable matrix on the phosphor screen.

ratemeter 1. A tachometer, providing a reading of beats or pulses per minute. 2. A type of radiation detector whose output is proportional to instantaneous radiation intensity (rate of radioactivity emission).

rate of rise timer A percentage time as applied in a temperature control system for controlling rate of temperature rise. Usually a standard percentage timer, with dial marked in various scales of degrees–per-hour temperature rise.

rating 1. The number of watts that a circuit component can dissipate without becoming overheated. 2. The rated output. 3. The voltage which may be safely applied to a circuit or device.

Rauwolfia and Veratrum alkaloids Principles of plant origin used in treating hypertension or as a tranquilizer.

raw data Data which have not been processed or reduced.

raw tape Virgin tape, i.e., factory-fresh magnetic tape that has never been used.

Rayleigh distribution A mathematical statement of the frequency distribution of random variables, for the case where the variables have the same variance and are not correlated.

Raynaud's disease A syndrome of vascular spasms in the arteries nourishing the tissues of the fingers. Typically, the condition is aggravated by cold or emotion, and symptoms of pain, cyanosis, and loss of color are usual. First noted and described by Maurice Raynaud in the early 1860s.

RBE Abbreviation for relative biological effectiveness.

RC constant The time constant of a resistor-capacitor circuit, equal to the resistance in ohms times the capacitance in microfarads, expressed in microseconds. It is the time required for current in an RC circuit to rise to 63 percent of its final steady-state value, or to fall to 37 percent of its initial steady-state value.

RC coupling Coupling between two or more circuits, usually amplifier stages, by means of a combination of resistive and capacitive elements. See also *coupling*.

RC differentiator A passive network of resistance and capacitance elements, capable of producing a first derivative (rate of change with respect to time) analog of any applied voltage function. Generally used in pressure measurements, where dp/dt function is an important parameter of cardiac performance. See *resistance-capacitance differentiator*.

RC filter See *resistance-capacitance filter*.

RC network Circuit containing resistances and capacitances arranged in a particular manner to perform a specific function.

RC oscillator See *resistance-capacitance oscillator*.

reactance When an alternating current circuit has inductive or capacitative components, back EMFs are set up which tend to limit the flow of current in a similar manner to resistance. The effect is treated as a resistance and measured in apparent ohms. The ratio of voltage to current in an ac circuit is known as reactance and given the symbol X. Inductive reactance is directly proportional to the frequency of the applied alternating current, while capacitive reactance is inversely proportional to frequency. Thus, the reactance of a circuit is usually expressed by the equation:

$$X = 2\pi fL - \frac{1}{2\pi fC}$$

where L = inductance, f = frequency of ac, and C = capacitance.

reactance drop The voltage drop in quadrature with the current.

reactionary hemorrhage See *hemorrhage*.

reactive Possessing reactance, either capacitive or inductive.

reactive component That part of an im-pedance which is due solely to reactance, either capacitive or inductive.

reactive power The power in an alternating current circuit obtained by multiplying voltage, current, and the sine of the phase angle between voltage and current. Reactive power is delivered to the circuit during part of the cycle, but returned to the source during another part of the cycle.

reactive volt-amperes Component of the apparent power in an alternating current circuit which is delivered to the circuit during part of a cycle, but is returned to the source during another part of the cycle. The practical unit of reactive power is the var, equal to one reactive volt-ampere. Also called *wattless power*.

read 1. To obtain information from a storage device. 2. To sense the characters involved in a program which is prepared on a suitable input media (punched tape, magnetic tape, punched cards). 3. To transmit information from an input device to a computer. 4. The process of acquiring information from a storage device.

reader A device in a computer which converts information in one form of storage to information in another form of storage.

read head In a computer, a device that converts digital information stored on a magnetic tape, or on a drum, into electrical signals usable by the computer arithmetic unit.

read-in To sense information contained in some source and transmit this information to an internal storage element.

read-in program Computer program that can be put into a computer in simple binary form and allows other programs to be read into the computer in more complex forms.

read-only memory (ROM) 1. Devices

(especially programmed chips) which look to the computer like memory, but whose contents cannot be changed by the computer. They are used for storing programs which never change (such as loader programs to load other programs). 2. A memory that cannot be altered in normal use of a computer. Usually a small memory that contains often-used instructions such as microprograms or system software, as firmware. Peripheral equipment uses ROM for character generation, code translation, and for designing peripheral processors. 3. A memory in which information is stored permanently, e.g., a math function or a microprogram. A ROM is programmed according to the user's requirements during memory fabrication and cannot be reprogrammed. A ROM is analogous to the dictionary, where a certain address results in predetermined information output.

read-only storage   See *fixed storage.*

readout   1. The manner in which a computer displays the processed information. May be digital visual display, punched tape, punched cards, automatic typewriter, etc. 2. The physical presentation of the output of a computer, such as punched cards, lights, printed words, or numerals.

read pulse   A pulse which causes information to be read out of a memory cell.

read/write memory   A memory whose contents can be continuously changed, quickly and easily during system operation. It differs from a read-only memory (ROM), whose contents are fixed and not subject to change, and a reprogrammable ROM, whose contents can be changed, but only periodically.

real time   1. Pertaining to the actual time during which a physical process transpires. 2. Pertaining to the performance of a computation during the actual time that

the related physical process transpires, in order that results of the computation can be used in guiding the physical process. 3. In random-sampling oscilloscope technique, the time scale associated with the signal events, themselves. 4. Time in which reporting on events or recording of events is simultaneous with the events.

real-time computer   A computer supplying information and computational capacity as needed during the actual operation of a process.

real-time data   Data presented in usable form at essentially the same time the event occurs. The delay in presenting the data must be small enough to allow a corrective action to be taken if required.

real-time input   Input data inserted into a system at the time of generation by another system.

real-time operation   1. Concurrent operations for data processing and physical processing so that the results of the computing operations are available whenever needed by the physical processing operations and vice versa. 2. Operations performed on a computer, in time with a physical process, so that the answers obtained are useful in controlling that process.

real-time output   Output data removed from a system at time of need by another system.

real-time spectrum analyzer   A spectrum analyzer that performs a continuous analysis of the incoming signal with the time sequence of events preserved between input and output.

real-time system   A computer system that processes data so that the results are available in time to influence the process being controlled or monitored.

receptacle The fixed or stationary half or a two-piece multiple-contact connector.

receptaculum chyli The lower expanded portion of the thoracic duct.

recessive Tending to disappear. In genetics, a gene which tends not to be expressed unless it is present in a homozygous, dominant state.

rechargeable Capable of being recharged. Usually refers to secondary cells or batteries.

recombination The process whereby a free electron fills a hole. The action eliminates two charge carriers, the electron is no longer free and the hole no longer exists.

recombination coefficient In an ionized gas, the value obtained by dividing the time rate of recombination of ions by the product of the positive–ion density and the negative–ion density.

record 1. The reviewable result of an analog or digital data recording process. It may be the strip chart containing physiologic waveforms, an X-ray film, or cine-angiographic motion picture film, or a magnetic tape or disc-stored collection of procedural data, which can be replayed for study. 2. One or more characters that are grouped together in the flow of data in a system. A record, for example, can represent one line of type or the contents of a punched card. Records may be fixed length, as they are with punched cards, or variable length as they are with the line of type. 3. A group of related facts or fields of information treated as a unit; thus, a listing of information, usually in printed, printable, or graphic form. Also, in computer terminology, the process of putting data into a storage device.

record density See *bit density*.

recovery time 1. Resumption of a normal operating condition. 2. Interval of time, following a sudden decrease in input signal to a system or component, to attain a specified percentage of the ultimate change in amplification or attenuation due to this decrease. 3. Of ultrasound or echocardiographic component, the time required, after the end of the transmitted pulse, for recovery to a specified relation between receiving sensitivity or received signal and the normal value.

rectification The process of converting alternating current into a current which flows in only one direction. Unidirectional current is not necessarily a constant-voltage noise-free direct current.

rectification factor Quotient of the change in average current of an electrode by the change in amplitude of the alternating sinusoidal voltage applied to the same electrode, the direct voltages of this and other electrodes being maintained constant.

rectified value of an alternating quantity Average of all the positive (or negative) values of the quantity during an integral number of periods.

rectifier 1. An electrical device for converting alternating current to direct current. Implies an output which is not pulsating, and is reasonably free from noise components. The term rectifier does not apply to a rotary machine, such as a rotary converter. 2. Device having an asymmetrical conduction characteristic employed in a way to convert alternating current into unidirectional current. In amplitude modulation detection, recovery of original signals is frequently accomplished by a rectifier. 3. Device that converts alternating current into unidirectional current by permitting appreciable current flow in one direction only. 4. A two-element, or solid-state device, which is used to convert alternating current to direct current. Usually rated above one-half ampere.

rectifier stack   The series of metal discs having oxide coatings which form the rectifying element of a dry disc rectifier. The discs have a hole in the center and are compressed on a threated rod.

rectifier transformer   A transformer whose primary operates at the fundamental frequency of the ac system and whose secondary has one or more windings conductively connected to the main electrodes of the rectifier.

rectify   To convert an alternating current into a direct current.

rectoscope   See *proctoscope*.

rectovesical   Of the rectum and bladder.

rectus   Straight; applied to certain muscles, for example, the two external abdominal muscles, one each side of the midline, running from pubic bone to ensiform cartilage and the fifth, sixth, and seventh ribs. It is enclosed in a strong sheath. Also, four short muscles of the eye: external, internal, superior, and inferior rectus.

recumbent   Lying down.

recycling circuit breaker   The property of some power circuit breakers which, when tripped by a fault, will reset one or more times automatically before tripping out completely.

red blood cell (corpuscle)   These are the blood cells which contain hemoglobin (erythocytes). There are about 5,000,000 in each ml of blood and they carry nearly all the oxygen required by the body cells. Red blood cells are formed in the bone marrow, normally at a rate of about a million a second and are notable for the absence of a nucleus in the mature state. They do not divide and have a lifetime of

about 120 days when they are destroyed in the spleen.

redundancy   1. Any deliberate duplication or partial duplication of circuitry or information designed to decrease the probability of a system or communication failure. Redundancy is also used to refer to inadvertent communication of more information or detail than is required with a consequent waste of available communication capacity. 2. The existence of more than one functional means for accomplishing a given task, where all means must fail before there is an overall failure of the system. 3. Any excess of signal elements over those required to carry the message. Speech is a highly redundant form of message coding. Certain data signals have no redundancy in themselves, and must use an extra redundant data bit for error checking. 4. The fraction of the total information content of a message which can be eliminated without loss of meaning.

redundancy check   An automatic or programmed check based on the systematic insertion of extra components or characters used expressly for checking purpose. Related to parity check and to forbidden combination check.

redundant   1. Exceeding what is necessary or normal. 2. Containing more information than is needed for intelligibility. About 75 percent of the information content of normal speech is redundant. 3. Said of the elements of equipment which exists in duplicate so that if one fails the second continues operation without interruption.

redundant character   A character especially added to a group of characters to insure conformity with certain rules which can be used to detect computer malfunction.

redundant code   A code using more signal

elements than necessary to represent the intrinsic information. Used for checking purposes.

redundant system See *duplexed system*.

reed frequency meter See *vibrating-reed meter*.

reed relay 1. A device that uses two strips of magnetizable metal, enclosed in glass, as the contacts. The control member is a coil or magnet surrounding the glass capsule. 2. One or more reed switches operated by a coil or magnet.

reed switch Two overlapping, flat, canti-levered reeds of ferromagnetic material sur-rounded by a dry inert gas, and sealed in a glass envelope. A small airgap separates the free overlapping ends of the reeds—either by permanent magnets or an operating coil surrounding the glass envelope—the mag-netic induction in a gap causes the reeds to attract each other.

reference acoustic pressure Magnitude of any complex sound that will produce a sound level meter reading equal to that produced by a sound pressure of 0.0002 dyne per square centimeter at 1000 hertz. Also called *reference sound level*.

reference address A number that appears as an address in a computer instruction but which serves as the base, index, initial, or starting point for subsequent addresses to be modified. Also called *presumptive ad-dress*.

reference electrode An electrode, usually filled with hydrogen, used to provide a reference potential in pH determinations. See also *Glass electrode*.

reference sound level See *reference acous-tic pressure*.

reflected impedance 1. Impedance value that appears to exist across the primary of a transformer due to current flowing in the

secondary. 2. Impedance which appears at the input terminals as a result of the char-acteristics of the impedance at the output terminals.

reflective code See *gray code*.

reflex A predictable, involuntary response to a stimulus.

reflux Flowing back, e.g., oesophageal re-flux is the flow of stomach acid into the esophagus.

refraction The bending of a sound, radio, or light wave as it passes obliquely from a medium of one density to a medium of another density in which its speed is different.

refractory period Interval of time re-quired for recovery of the nerve fiber be-fore it is capable of producing a nerve impulse again.

refresh display A CRT device that re-quires the refreshing of its screen presenta-tion at a high rate in order that the image will not fade or flicker. The refresh rate is proportional to the decay rate of the phosphor.

regeneration 1. Renewal of damaged tissue such as regenerating nerve fibers. 2. A method of gaining increased power by coupling a high-level point to a low-level point in an amplifier so that part of the output is fed back to the amplifier input, thus reinforcing the original input signal. See also *positive feedback*.

regenerative feedback See *positive feed-back*.

regimen A regulated pattern of activity.

register 1. In an electric computer, a de-vice capable of retaining information, often that contained in a small subset—for example, one word of the aggregate in-formation in a digital computer. 2. Accurate matching of two or more patterns, such as the three images in color television. 3. Part of an automatic switching telephone system which receives and stores the dialing pulses which control the further operations necessary in establishing a telephone connection. 4. An interconnection of computer circuitry, made up of a number of storage devices (usually flip-flops) to store a certain number of digits, usually one computer word. For example, a four-bit register requires four flip-flops.

register length In a computer, the number of characters which a register can store.

regression Reverting to a more primitive stage. In psychology, reverting to childlike behavior as is often seen in illness.

regulated power supply Power supply device containing means for maintaining constant voltage or constant current under changing load conditions.

regulator A device for holding an electrical voltage or current at a preset, steady value despite variations in the input voltage or the load. The most common types of regulators are magnetic transformers, gas tubes, and Zener diodes.

regurgitation The return of food from the stomach into the mouth. The amount may be small, or it may be large, as in vomiting. Also, the backflow of blood through inadequately closed heart valves.

rehabilitation The restoration of the ill or injured to function at their full capacity.

reignition voltage That voltage which is just sufficient to reestablish conduction of a gas tube if applied during the deionization period. This voltage varies inversely with time during the deionization period. Also called *restriking voltage.*

reject In an echograph, a control function which increases the baseline level, thus eliminating low-amplitude echoes, making it easier to distinguish larger echoes.

rel The unit of magnetic reluctance, equal to one ampere-turn per magnetic line of force.

relative biological effectiveness (rbe) A number expressing how much greater absorbed dose of X-ray or gamma radiation is needed to produce the same biological damage as the radiation in question.

relative humidity The ratio of the quantity of water vapor present in the air to the quantity which would saturate it at a given temperature.

relaxation oscillator Device which generates a nonsinusoidal wave by gradually charging and quickly discharging a capacitor or an inductor through a resistor.

relaxing factor system A particulate fraction obtained from muscle that causes the glycerol extracted muscle preparation to become extensible in the presence of ATP.

relay 1. An electromagnetic device in which current through a coil moves an armature to operate spring contacts which open and close circuits. 2. An intermediate station on a multilink radio system.

relay driver A circuit that has the high-voltage and high-current switching capability needed to actuate electromechanical relays.

reliability Quality or property built into, or inherent in, a device that indicates that the device will probably perform its specified function without failure under given conditions for a specified period of time.

reliability data 1. Data related to the frequency of failure of an item, equipment or system. These data may be expressed in terms of failure rate, mean time between failures (MTBF), or probability of success. 2. Data contained in comprehensive documents that provide a detailed history of the reliability evaluation of component parts, components assemblies, etc., or the entire program during the design, development, production, and major product improvement phases of an equipment, in which engineering studies have been performed to select the most reliable item for the intended application.

reliability test Tests and analyses in addition to other type tests which are designed to evaluate the level of reliability in a product, parts, or systems as well as the dependability or stability of this level with time, and use under various environmental conditions.

relocate In computer programming, to move a routine from one portion of storage to another and to adjust the necessary address references so that the routine in the new location can be executed.

reluctance The property of a magnetic circuit which determines the total magnetic flux (lines of force) which will result when a given magnetomotive force is applied. Reluctance is to a magnetic circuit what resistance is to a direct current circuit.

rem Abbreviation for roentgen equivalent man.

remanence The magnetic induction which remains in a magnetic circuit after the removal of an applied magnetomotive force. If there is an air gap in the circuit, the remanence will be less than the residual induction; if there is no air gap, the remanence will equal the residual induction. A substance with remanence is known as *ferromagnetic*.

remission The time during which a disease subsides and shows no symptoms.

remote station Data terminal equipment for communicating with a data processing system from a location that is time, space, or electrically distant.

renal Relating to the kidney.

renal calculus Stone situated in the kidney.

renal threshold A concentration of a substance within the blood, which, when reached, causes the substance to appear in the urine.

rep Abbreviation for roentgen equivalent physical.

repetitive firing 1. Recurrent generation of action potentials by a neuron. 2. Condition in a silicon controlled rectifier switching circuit where multiple triggering inputs are applied to the gate circuit.

repolarization The process by which a biological cell regains its normal resting potential after it has "fired" or depolarized. See *depolarization*.

representative calculating time 1. In computers, a time used in evaluating the performance speed of a computer, i.e., the time required to perform a specified operation or series of operations. 2. One-tenth of the time required to perform nine complete additions and one complete multiplication, including in each instance, the time required to procure two operands from storage, perform the operation, and store the result, and the time required to select and execute the required instructions.

repression A defense mechanism, operating unconsciously, that banishes unacceptable ideas, emotions, or impulses from consciousness or that keeps out of consciousness what has never been conscious.

reproduce head An electromagnetic transducer which converts the remnant flux pattern in a magnetic tape into electrical signals during the reproduce (playback) process.

repulsion The mechanical force which tends to push apart adjacent conductors carrying currents in opposite directions, or tends to separate bodies having like electric charges or magnetic poles of like polarity.

rerun 1. In a computer, a system that will restart the running program after a system failure. Snapshots of data and programs are stored at periodic intervals and the system rolls back to restart at the last recorded snapshot. 2. To replay a procedure, as in the recorded results of a cardiac catheterization study, through the computer, for purposes of review, data manipulation, or editing. 3. A repeat of a machine run, usually because of a correction, an interrupt, or a false start. 4. To repeat a complete transmission.

rerun point One of a set of preselected points located in a computer program such that if an error is detected between two such points, the problem may be rerun by returning to the last such point instead of returning to the start of the problem.

rerun routine A computer routine designed to be used in the wake of a malfunction or a mistake, to reconstruct a routine from the previous rerun point.

resection A complete removal by surgery.

resectoscope Instrument to view and remove pieces of tissue in transurethral prostatectomy.

reset 1. An input to a binary counter or register which forces all binary elements to the zero logic state or the minimum binary state. 2. A control which "unlatches" a triggered circuit. 3. The act of returning to normal state, following the termination of abnormal conditions and the resumption of normal conditions.

reset terminal The flip-flop input terminal that triggers the circuit back from its second state to its original state. Also called *clear terminal* or *zero-input terminal.*

residual volume The quantity of air remaining in the lungs after a deep exhalation (approximately 1.2 liters).

resist An enamel or lacquer impervious to process chemicals or plating that is selectively applied by printing, stenciling, or screening to mask surfaces selectively for plating or etching. It is removable by solvent bath. The term also is applied to metal plating over copper that will permit selective copper etching in the manufacture of printed circuits.

resistance 1. The nonreactive opposition which a device or material offers to the flow of direct or alternating current. The opposition results in production of heat in the material carrying the current. Resistance is measured in ohms, and is usually designated by the symbol Ω. Conductors usually have low resistance; insulators have high resistance. 2. Circuit element designed to offer a predetermined opposition to current flow. 3. That property of a substance which impedes the flow of current and results in the dissipation of the power in the form of heat.

resistance-capacitance differentiator A passive resistor/capacitor circuit which produces an output voltage whose amplitude is proportional to the first derivative (rate of change of the input voltage with respect to time). Thus, a square-wave input voltage produces sharp output voltage spikes. Abbreviated RC differentiator.

resistance-capacitance oscillator Abbreviated RC oscillator. An oscillator consisting of an amplifier with controlled positive feedback, in which the frequency of oscillation is determined by the phase of the current fed back as obtained from a variable resistance-capacitance network.

resistance coupling 1. Method of interconnection between stages in an amplifier which connects the output of one stage to the input of the following stage by means of a resistor. 2. Method of transferring energy from one circuit to another by means of resistance common to both circuits.

resistance drop Voltage drop occurring between two points on a conductor due to the flow of current through the conductor resistance existing between those points.

resistance grounded Grounded through a resistance which will limit the current flow to ground in case of a ground fault.

resistance thermometer Thermometer using variation of resistance with temperature of some material, usually nickel or platinum, considered the standard for temperature measurement over its range.

resistance thermometery A temperature-measuring technique that utilizes the temperature coefficient of a wirewound resistor. Known as a resistance thermometer, this resistor consists of a spiral of nickel or platinum wire. Since the ohmic value of the wire varies with temperature, the resistance of the spiral is an indication of temperature.

resistive component That part of an impedance which is due solely to resistance.

resistivity A measure of a material's resistance to the flow of electric current. Expressed in ohm-cm. 2. The resistance in ohms of a block of material one square cm in area and one cm thick. The smaller the value of resistivity, the better the material is as an electrical conductor. 3. The ability of a material to resist the flow of current. The reciprocal of conductivity.

resistor A component made of a material (such as carbon or nichrome) that has a specified resistance, or opposition to the flow of electrical current. Resistors are used to limit the amount of current flowing in a circuit or to provide a voltage drop.

resistor color code Code adopted by the Electronic Industries Association to mark the values of resistance on resistors in a readily recognizable manner. The first color represents the first significant figure of the resistor value, the second color the second significant figure, and the third color represents the number of zeros following the first two figures. A fourth color is sometimes added to indicate the tolerance of the resistor.

resolution 1. The limit to the amount of detail of which a system is capable. In a photograph the resolution is limited by the size of the particles in the image. In a television picture, it is limited by the number of lines per inch, and in an information system by the maximum frequency it can handle. 2. Of a measuring instrument, the minimum scale value which an be read. Not the lowest scale value, but to how many significant figures. See also *accuracy*.

resonance 1. Condition in an electrical circuit when the inductive and capacitive reactances are equal. This condition occurs at only one frequency in a fixed circuit. 2. Resonance exists between an oscillating system and its maintaining periodic agency when a small amplitude of the periodic agency produces relatively large amplitudes of oscillation in the system. 3. Condition existing in a body when the frequency of an applied vibration equals the body's natural frequency.

resonant frequency Electrically, the frequency at which the inductive reactance and capacitive reactance of a series circuit are equal. The resonant frequency is determined by the formula $f = 1/2 \; \pi \sqrt{LC}$, where f is in hertz, L is in henrys and C is in farads. Mechanically, it is the frequency which excites an object at its *natural frequency.*

resonant-reed relay   A relay that operates in response to signals of the proper frequency, power level, and time duration. It consists of a series of reeds, tuned to different natural frequencies. These are positioned over contacts, and their lengths are lying within a surrounding coil. Currents at differing frequencies pass through the coil, and if one is at the natural frequency of a reed, the magnetic field causes the reed to vibrate, thus closing a contact. Usually used for multicircuit control. See also *vibrating reed relays.*

respiration   The process whereby exchange of gas takes place between the organism and its environment. Under normal conditions, an increase in the partial pressure of carbon dioxide in the arterial blood increases the respiration rate and depth of breathing. In a healthy adult, the average frequency of quiet breathing is about 16 times per minute, with an average volume of 500 to 600 cubic centimeters of air entering and leaving the respiratory tract with each breath. Infants have a much higher respiration rate, of about 50 breaths per minute. This decreases in children to a usual rate of 36.

respirator   1. Appliance worn over the mouth and nose to prevent the inhalation of noxious gas. 2. Apparatus used to assist the muscles of respiration.

respiratory acidosis   Acidosis, caused by inability to excrete CO_2 (carbon dioxide), usually the result of chronic lung disease.

respiratory alkalosis   Alkalosis, due to hy-

perventilation with excessive loss of CO_2.

respiratory center   An area of specialization in the medulla oblongata which controls breathing.

respiratory quotient   A ratio of the volume of expired carbon dioxide (CO_2) to the volume of oxygen (O_2) consumed.

respirometer   Instrument for measuring and/or recording respiratory movements. See also *pneumograph* and *spirometer.*

response   1. A quantitative expression of the output of a device in terms of the input under specifically stated conditions. 2. The reaction of an organism to stimulus. 3. The depolarization of a nerve or muscle cell in reaction to stimulus. 4. The fidelity with which the output waveform of a device corresponds to the input waveform. 5. A tabulation of the deviations from perfect fidelity.

response curve   A graphical representation of response. Usually measured in decibels, with reference to a given level on a verical scale. When the response curve of an amplifier, monitor, recorder, pickup, microphone, etc. is accurately plotted, the resulting curve represents the relative levels of amplitude at all frequencies within a specified bandwidth.

response time   Time required for the output of a control system or element to reach a specified fraction of its new value after the application of a step input or disturbance.

resting potential   1. The potential recorded between the inside of a cell (cytoplasm) and the outside (extracellular fluid) in the resting state. Since the potential difference occurs across the cell membrane, the term *membrane potential* is also used. In all cells where this measurement has been made, the inside is found to be negative with respect to the outside, the magnitude being 70 to 100 mV. 2. The ionic potential difference existing between the sodium ion-bearing medium outside the cellular membrane and the potassium ion-bearing cytoplasm within the unstimulated cell. This potential is variable, depending upon the type of cell, from tens of millivolts, up to almost 100 millivolts.

restorer pulses In computers, pairs of complement pulses, applied to restore the coupling-capacitor charge in an alternating current flip-flop.

rest potential Residual potential difference remaining between an electrode and an electrolyte after the electrode has become polarized.

restriking voltage See *reignition voltage*.

resuscitation 1. The act of maintaining respiration and circulation through means external to the patient's body, to prevent death. 2. The process of restoring a normal or near-normal heartbeat after a cardiac emergency has occurred.

resuscitator Apparatus for initiating respiration in cases of asphyxia.

retardation A slowing down of activity; backwardness.

retching Ineffectual efforts to vomit.

retention time Maximum time between writing into a storage tube and obtaining an acceptable output by reading.

retentivity The property of a material measured by the amount of magnetic flux which remains after the material has been saturated with flux, and then the magnetomotive force is removed.

reticular A term applied to tissue meaning resembling a network.

reticular activating system See *brain*.

reticulosis A group of neoplasms arising from lymphoid tissue, e.g., Hodgkin's disease, lymphosarcoma.

retina The sensory membrane of the eye that receives the image formed by the lens, and is connected with the brain by the optic nerve.

retinal Pertaining to the retina.

retinitis Inflammation of the retina.

retinopathy Pathological lesion affecting the retina, e.g., diabetic retinopathy, hypertensive retinopathy.

retrace See *flyback*.

retrace blanking Blanking of a television picture tube during vertical retrace intervals to prevent the retrace lines from being visible on the screen.

retro Prefix meaning backwards or behind.

retrograde amnesia Symptom of concussion. The patient cannot remember what happened immediately before an accident.

return trace On a cathode-ray or television tube, the path of the scanning spot during the interval when it returns to its starting point. During the return interval, its brightness is reduced to a minimum.

return wire Ground wire, common wire, or the negative wire of a direct current circuit.

reverse bias Connecting a voltage to a semiconductor diode, with the P-type material negative and the N-type positive, so that the voltage pulls the carriers away from the junction and thereby stops current flow.

reverse-blocking diode thyristor A two-terminal thyristor which for negative anode-to-cathode voltage does not switch, but exhibits a reverse blocking state.

reverse-blocking triode thyristor A three-terminal thyristor which for negative anode-to-cathode voltage does not switch, but exhibits a reverse blocking state. This device is better known as an *SCR* (*silicon controlled rectifier*).

reverse-breakdown voltage The voltage that produces a sharp increase in reverse current in a semiconductor, without a significant increase in voltage.

reverse-conducting diode thyristor A two-terminal thyristor which for negative anode-to-cathode voltage does not switch, but conducts large currents at voltages comparable in magnitude to the on-state voltage.

reverse-conducting triode thyristor A three-terminal thyristor which for negative anode-to-cathode voltage does not switch, but conducts large currents at voltages comparable in magnitude to the on-state voltage.

reverse current The small value of current which flows when the voltage across a diode is of a polarity reversed from that for normal current flow.

reversed feedback See *negative feedback*.

reverse saturation current The reverse current that flows in a semiconductor due to a specified reverse voltage.

reverse voltage Voltage of that polarity which produces the smaller current.

reversible counter A circuit that accepts two different successive input signals and stores that difference, normally in a coded form. See also *up-down counter*.

RF Abbreviation for radio frequency; extends from approximately 50 kHz to over 30,000 MHz; the frequency range useful for communication.

R/h Abbreviation for roentgens per hour.

rhagades A crack or fissure of skin causing pain; a term especially used of radiat-

ing scars at angle of mouth due to congenital syphilis.

rheobase The minimum potential of electric current necessary to produce stimulation.

rheoencephalography See *electrical impedance cephalography*.

Rhesus factor See *blood groupings*.

rheumatic Pertaining to rheumatism.

rheumatic fever A disorder affecting connective tissue, particularly that of the heart and the joints. The cause is considered to be an allergic reaction to toxins from hemolytic streptococcus.

rheumatic heart disease Chronic rheumatic heart disease which is the result of severe damage and deformation of the valves of the heart due to rheumatic fever.

rheostat An adjustable resistance. One contact is attached to one end of the resistor, the other to a tap which can be moved, thereby increasing or decreasing the amount of resistance in the circuit. A rheostat is physically similar to a potentiometer.

Rh factor See *blood groupings*.

rhinolith A calculus formed in the nose.

rhinorrhea Discharge from the nose.

rhodopsin Visual purple contained in the retina.

rhonchi Abnormal breath sounds, present throughout both phases, but more prominent during expiration. Due to air moving through a narrowed lumen.

rhonchus A roaring bronchial sound heard on auscultation.

rhythm disturbance An aberration of heart rhythm; an arrhythmia. Generally, rhythm disturbances are considered premonitory events, heralding more acute conditions which will occur if the underlying cause goes untreated. For this reason, modern coronary care attaches considerable importance to the treatment of rhythm disturbances.

rickets Generally a disease of infants and young children caused by vitamin D deficiency. There is defective bone calcification that causes skeletal deformities, such as bow-legs, knock-knees, and pigeon chest.

Rigg's disease See *pyorrhea*.

right heart bypass The flow of blood from the entrance of the right atrium directly to the pulmonary arteries, avoiding the right atrium and right ventricle.

rigor mortis The stiffening of the body after death.

rima A fissure.

rima glottidis Slit between vocal cords.

ring counter A special form of counter sometimes called a Johnson or shift counter which has very simple wiring and is fast. It forms a loop or circuits of interconnected flip-flops so arranged that only one is zero and that as input signals are received, the position of the zero state moves in sequence from one flip-flop to another around the loop until they are all zero, then the first one goes to one and this moves in sequence from one flip-flop to another until all are one. It has $2 \times n$ possible counts where n is the number of flip-flops.

ringing Transient decaying oscillation about high or low limit induced by unmatched impedance reflections.

ripple 1. An alternating voltage superimposed on the direct current from a generator or rectifier, caused by a rough commutator on the generator or incomplete filtering in the rectifier. Percent ripple is equal to the ratio of the RMS ripple voltage to the absolute value of the total volt-

age, expressed in percent. 2. The transmission of data serially. It is a serial reaction analogous to a bucket brigade or a row of falling dominoes. 3. The ac component in the output of a dc power supply. 4. The variations above and below the average peak amplitude. 5. The variation of output amplitude in dB across the passband of a filter.

ripple counter A binary counting system in which flip-flops are connected in series. When the first flip-flop changes, it affects the second which affects the third and so on. If there are ten in a row, the signal must go sequentially from the first flip-flop to the tenth.

ripple-through counter See *serial counter*.

rise time (t_r) 1. The time in microseconds required for a pulse or other waveform to rise from 10 percent to 90 percent of its peak value. 2. Of a switching transistor. The time interval between the instants at which the magnitude of the pulse at the output terminals reaches specified lower and upper limits, respectively, when the transistor is being switched from its nonconducting to its conducting state. (The lower and upper limits are usually 10 percent and 90 percent, respectively, of the amplitude of the output pulse.) 3. A measure of the time required for the output voltage of a stage to go from a low voltage level (zero) to a high voltage level (one) once a level change has been started. See also *Step-function response*.

R/m Abbreviation for roentgens per minute.

RMM Abbreviation for read-mainly memory. A nonvolatile memory used much as a ROM or PROM except that the data contained therein may be altered through the use of special techniques (often involving external action) which are much too slow for read/write use.

rms Abbreviation for root-mean-square value.

rms amplitude Amplitude of an alternating current wave which may be used for an accurate computation of power in watts. In communications particularly, with complex waves involved, the root-mean-square amplitude of a current is said to be that current which is computed from a measurement of power dissipated in a known resistance as a result of that current. Often called the *effective value.*

rms value The root-mean-square value of ac voltage, current, or power. Calculated as 0.707 of peak amplitude of a sine wave at given frequency.

rms voltage The effective value of a varying or alternating voltage. That value which would produce the same power as if a continuous voltage were applied to a pure resistance. In sine wave voltages, the rms voltage is equal to 0.707 times the peak voltage.

rochelle salt A crystalline double tartrate of potassium and sodium. It has the strongest piezoelectric effect of any material commercially available, but loses its piezoelectric properties above 68 degrees Fahrenheit and is very sensitive to humidity.

rods Sensitive light receptors in the retina.

roentgen (R) Quantity of X or gamma rays which will produce as a result of ionization one (1) electrostatic unit of electricity (either sign) in one (1) cc of dry air at 0°C and standard atmospheric pressure. One roentgen equals absorption of 83.8 ergs of energy per gram of air.

roentgen densitometer A device for recording changes in concentration of a radiopaque indicator injected into circulation for evaluating circulatory function.

roentgen equivalent man (rem) The quantity of ionizing radiation which produces the same biological damage in man as that caused by the absorption of one roentgen of X-or gamma ray.

roentgen equivalent physical (rep) The quantity of ionizing radiation which, upon absorption in body tissue, deposits 93 ergs of energy per gram of tissue.

roentgen per hour (R/h) The intensity of a field of ionizing radiation which will deliver one roentgen per hour.

roentgen rays Radiant energy in the range of wavelengths between 0.05 and 100 Å. Roentgen rays are also known as X-rays. In diagnostic applications, wavelengths between 0.12 and 0.30 Å are commonly utilized, while in therapeutic applications the range is between 0.05 and 0.12 Å.

roentgens/sec (R/s) The basic exposure dose rate of gamma or X-rays.

role model A person who demonstrates the behavior typical of one in a particular position or situation.

rolloff A gradual attenuation of a range of frequencies. Sometimes called a *slope.*

ROM An acronym for read-only memory, a memory which permits the reading of a predetermined pattern of zeros and ones. This predetermined information is stored in the ROM at the time of its manufacture.

root-mean-square amplitude (rms) The effective value of an alternating wave (current and/or voltage) numerically equal to the direct current value that will produce the same heating effect. Can be computed as the square root of the average of the squares of all instantaneous amplitudes over a complete cycle. For sine waves, the rms value is equal to 0.707 times the peak value.

root-mean-square value Of alternating currents and voltages, the root-mean-square value is the effective current or voltage applied. It is that alternating current or voltage that produces the heating effect the same value of direct current or voltage of an equal value would produce. It is equal to 0.707 times the maximum alternating current value.

rosin joint A poorly soldered connection, in which the wire is held in place by the rosin flux, not by solder. It results from inadequate heating of the joint.

rotary stepping switch (or relay) See *stepping relay*.

rotary switch An electromechanical device which is capable of selecting, making, or breaking an electrical circuit and is actuated by a rotational torque applied to its shaft.

rotators Muscles which cause circular movement.

routine 1. A set of coded instructions arranged in proper sequence to direct the computer to perform a desired operation or sequence of operations. 2. A subdivision of a program consisting of two or more instructions that are functionally related; therefore, a program.

routine library Ordered set of standard and proven computer routines by which problems or parts of problems may be solved.

row 1. The characters, or corresponding bits of binary-coded characters, in a word. 2. Equipment which simultaneously processes the bits of character, the characters of a word, or corresponding bits of binary-coded characters in a word. 3. Corresponding positions in a group of columns.

RPAO Abbreviation for right posteroanterior oblique.

RPO Abbreviation for right posterior oblique.

-rrhage Suffix meaning to erupt or burst forth.

-rrhagia Suffix meaning to erupt or burst forth.

-rrhaphy Suffix meaning to suture.

-rrhea Suffix meaning flow or discharge.

R/s Abbreviation for roentgens/sec.

R-S flip-flop A flip-flop consisting of two cross-coupled NAND gates having two inputs designated R and S. A 1 on the S input and 0 on the R input will reset (clear) the flip-flop to the 0 state, and 1 on the R input and 0 on the S input will set it to the 1 state. It is assumed the zeros will never appear simultaneously at both inputs. If both inputs have ones the flip-flop will stay as it was. One is considered nonactivating. A similar circuit can be formed with NOR gates.

R-S-T flip-flop A flip-flop having three inputs; R, S, and T. It works as the R-S flip-flop, except that the T input is used to cause the flip-flop to change states.

ruby laser An optically pumped solid-state laser, where a ruby crystal produces an extremely narrow and intense beam of coherent red light. Used in surgical procedures for retinal welding and the resection of tumorous tissue.

rugae Wrinkles or creases.

rugose Wrinkled.

run One performane of a given computer routine or program during which no manual interventions are necessary.

rupture A bursting. The breaking of any tissue structure through pressure, as in the bursting of a fallopian tube in an ectopic pregnancy, or the bursting of an aneurysm on a vessel.

R wave The portion of the ECG complex which depicts the gross potentials accompanying depolarization of the ventricular muscle masses of the heart.

sabin A measure of the sound absorption of a surface. It is the equivalent of one square foot of perfectly absorptive surface. Numerically, it is equal to the fraction of the total sound energy which is absorbed by a particular material.

sacculated Bagged or pursed out.

sacral Pertaining to the sacrum.

sacrum A composite bone formed by the union of the five vertebrae between the lumbar and caudal regions, constituting the dorsal part of the pelvis.

sac-type artificial heart A flexible, plastic, inner heart fitted into a larger, rigid housing. Increased air pressure in the space between the rigid wall and the flexible sac compresses the sac and empties its contents.

sadism See *masochism*.

sagittal 1. Relating to the anteroposterior median plane of the body or any plane parallel to it. 2. Arrowlike.

sagittal section Section made by cutting through a specimen from top to bottom so that there are equal right and left halves.

sagittal suture The suture between the parietal bones.

saline A water solution of salt (NaCl). In medical practice this normally refers to physiological or normal saline.

salpingectomy Removal of one or both fallopian tubes, for termination of ectopic pregnancy or sterilization.

salpingography Technique of examination of the fallopian tubes by X-rays.

salpinx A tube, either eustachian or fallopian.

sample 1. An instantaneous value of a variable obtained at regular intervals. 2. To obtain sample values of a complex wave at periodic intervals. See also *sampling*.

sample and hold (S/H) 1. A system in which a sample of an analog input signal is frozen in time (stored in a capacitor) and held while it is converted to a digital representation, or otherwise processed. 2. A circuit that holds or freezes a changing analog input signal voltage. Usually, the volt-

339

age thus frozen is then converted into another form, either by a voltage-controlled oscillator, an analog-to-digital (A/D) converter or some other device.

sampled data That data in which the information content can be or is determined only at discrete intervals of time. Sampled data can be analog or digital in form.

sample rate The rate at which the analog sample is measured and/or displayed per second.

sampling 1. Obtaining the values of a function for discrete, regularly, or irregularly spaced values of the independent variable. 2. In statistics, obtaining a sample from a population. 3. Process of obtaining a sequence of instantaneous values of a wave. 4. In pulse code modulation, the act of selecting samples of an analog wave at recurring intervals such that the original wave can later be reconstructed with reasonable fidelity from the samples.

sampling oscilloscope An oscilloscope which enables a rapid, periodic waveform to be displayed on a cathode-ray tube by stroboscopic sampling of successive waves at different points on the waveform, and representing the sample by a dot on the screen, the accumulation of dots describing the waveform.

sanguine Fullblooded; hopeful; ruddy.

sanguinous Blood-stained. Containing blood.

S.A. node See *sinoatrial node* and *pacemaker*.

saphena varix Saccular enlargement of the termination of the long saphenous vein often without obvious varicose veins.

saphenous nerve Large branch of the femoral nerve.

saphenous opening Below groin near inner side of thigh where superficial saphenous vein passes deep to enter femoral vein.

saphenous veins Superficial leg veins. Long saphenous veins begin on the foot and extend to the groin. Short saphenous veins join the popliteal vein at the knee. Favored sites for introduction of cardiac catheters.

sarcoid Resembling flesh.

sarcology Anatomy of the soft tissues as distinguished from osteology.

sarcoma A tumor that arises from connective tissue, often malignant.

sartorius The long ribbon-shaped muscle of the front of the thigh.

saturable reactor An iron-core reactor having a second control winding carrying a direct current whose value is changed to vary the saturation of the iron core and thereby vary the reactance of the first winding which carries alternating current. With no direct current flow, the reactance is a maximum, and decreases with increasing direct current flow.

saturation 1. The condition which prevails when a further increase in input signal produces no corresponding increase in output—and in particular in the following two cases: (a) When the input is magnetizing current and the output is the resulting flux density within a core material. (b) When further increase in potential produces no corresponding increase in current. 2. The condition in which a further increase in some cause produces no observable change in the resulting effect. 3. A pure color being in a state not contaminated by white. 4. The state reached when sufficient solid or gaseous substance is dissolved in a solution so that no more of that substance can be dissolved.

saturation current 1. Maximum current that can be obtained no matter how much the applied voltage is increased. 2. Of a semiconductor device, that portion of the steady-state reverse current which flows as a result of the transport, across the junction, of minority carriers thermally generated within the regions adjacent to the junction.

saturation value 1. Highest value that can be obtained under given conditions. 2. Value of magnetic flux density beyond which an increase in magnetizing force has no appreciable effect on flux density in a particular sample of magnetic material.

saturation voltage The voltage drop across a switching transistor when it is fully turned on. While a perfect switch has negligible voltage drop, a transistor can only approach that limit. The product of saturation voltage and peak current determines the maximum power which the transistor can handle.

sawtooth Of a waveform, increasing approximately linearly as a function of time for a fixed interval, returning to its original value quite sharply and repeating the process periodically.

sawtooth wave generator Electronic circuit, also called a timebase or sweep generator, which generates a voltage or current with sawtooth characteristics. The voltage generator circuits are fairly simple in design, using a number of arrangements such as the multivibrator and the blocking oscillator.

SB Abbreviation for small bowel.

scaler A circuit with two stable states, which can be triggered to the opposite state by appropriate means. A bistable circuit.

scalp That part of the integument of the head which normally is covered with hair.

scan 1. To sweep an area in the precor-

dium with an ultrasonic transducer, to record an echogram. 2. To sweep an electron beam in a regular to-and-fro pattern, as in a television tube. 3. To periodically sample the voltages that exist at like points in a number of similar circuits. 4. To examine point by point in logical sequence. 5. To sample each of a number of inputs intermittently. A scanning device may provide additional functions such as record or alarm.

scanner An instrument which automatically samples or interrogates the state of various processes, conditions, or physical states and initiates action in accordance with the information obtained.

scanning 1. The successive exposure of small portions of an object to a sensing device of some type. Television, radioactive scanning, facsimile transmission, and photoelectric scanning are all examples of this technique. 2. The physical motion of an ultrasonic probe or array of probes, so as to illuminate and record the echoes from cardiac structures in the time-motion echogram.

scanning yoke See *deflection yoke*.

scaphoid Boat-shaped. The name of a bone of the carpus and of the tarsus.

scapula The shoulder blade.

schematic A graphic representation of a circuit, using symbols to represent components. Also called *schematic circuit diagram* or *schematic drawing*.

schematic circuit diagram See *schematic*.

schematic drawing See *schematic*.

schizophrenia 1. A severe emotional disorder of psychotic depth characteristically marked by a retreat from reality with delusion formation, hallucinations, emotional disharmony, and regressive behavior. Formerly called dementia praecox. 2. The generic term used for a group of disorders characterized by a progressive loss of emotional stability, judgment, and contact with reality.

Schmitt trigger   Bistable pulse generator in which an output pulse of constant amplitude exists only as long as the input voltage exceeds a certain threshold value.

sciatica   A severe pain along the sciatic nerve, which extends from the buttocks along the back of the thigh and leg to the ankle.

sciatic nerve   See *nervous system*.

scientific notation   A system of numeric recording in which a number is expressed in terms of a power of ten. For example, the number 1234 is entered as 1.234×10^3 and the number 0.001234 would appear as 1.234×10^{-3}.

scintillation   1. Random fluctuation, in radio propagation, of the received field about its mean value, the deviations usually being relatively small. 2. A rapid apparent displacement of a blip on a radar display. 3. The flash of light produced by certain crystalline materials when a charged particle is passed through them.

scintillation counter   Type of radiation counter in which incoming particles or photons are counted by means of scintillation induced when the incident radiation strikes a phosphor. Counter comprises phosphor and photomultiplier for detecting and amplifying electrons produced by incident ionizing radiation.

scintillation-counter energy resolution   A measure of the smallest difference in energy between two particles or photons of ionizing radiation that can be discerned by the scintillation counter.

scintillation crystals   Special crystals that emit flashes of light when struck by alpha particles.

scintillation scanner   Scanner with scintillation-type detector for mapping concentration of radioactive isotopes in the body.

scintillator material   A material which emits optical photons in respose to ionizing radiation.

scirrhouse   Hard and fibrous in nature.

sclera   The opaque outer coat of the eyeball, forming the major part of the globe of the eye, the remainder being formed by the cornea.

sclerosis   An induration or hardening.

-sclerosis   Suffix meaning hardening.

sclerotomy   An operation on the sclerotic coat of the eye, for the relief of glaucoma.

-scope   Suffix meaning to observe or to look at.

-scopic   Suffix meaning to observe or to look at.

-scopy   Suffix meaning to view.

scored   Marked with significant notches, lines, or grooves.

scotoma   A blind spot in the field of vision.

SCR   Abbreviation for silicon controlled rectifier.

scratch pad memory   1. A computer information store that interfaces directly with the central processor. It is optimized for speed and has a limited capacity. Its purpose is to supply the central processor with the data for the immediate computation without the delays that would be encountered by interfacing with the main memory. 2. An array of storage elements that is analogous to a pad of paper used for jotting down notes. Scratch pad memories are usually of the very high speed type and are most frequently employed where a small amount of data is temporarily stored and must be rapidly available when needed. Generally, the data stored in interim calculation results or data about to be used. Access is normally random.

screen 1. A metal partition or shield which isolates a device from external magnetic or electric fields. 2. The screen-grid electrode of an electron tube. 3. The chemically coated inside surface of the large end of a cathode-ray tube which becomes luminous when scanned by an electron beam.

screen grid In a tetrode or pentode electron tube, a grid between the control grid and the plate. It shields the control grid from influence by the plate.

SCS Abbreviation for silicon controlled switch. A PNPN structure with all four semiconductor regions accessible, rather than only three as is customary with silicon controlled rectifiers (SCR). Accessibility of the fourth region greatly expands circuit possibilities beyond those of conventional transistors or SCRs. Physically, the SCS is an integrated circuit consisting of a PNP and an NPN transistor in a positive feedback configuration. As such, it offers fewer connections, few parts, lower cost, and better characterization than is available from two separate transistors. Viewed differently, it is an SCR with an extra lead which eliminates rate effect problems. Some prefer to use it as a complementary SCR being triggered by negative-going pulses. Many find the high triggering sensitivity ideal for timing and level sensing applications. By viewing the SCS as a transistor with an additional latching junction, some have developed very useful bistable circuits with high turn-on and turn-off gains.

scurvy Syndrome of extreme vitamin C deficiency, the result of which is hemor-

rhage into the tissues and swelling of mucous membranes.

SDR Abbreviation for skin dose rate.

search time Time required to locate a particular field of data in a computer storage device. Requires a comparison of each field with a predetermined standard until identity is obtained. Contrasted with access time.

sebum The secretion of sebaceous (oil) glands.

secondary 1. A secondary winding: That winding of a transformer in which the current flow is due to electromagnetic induction from current in a primary winding. 2. The lower-voltage conductors of a power distribution system, so called because they are fed from the secondary windings of power transformers. Often 115/230 volts, but sometimes 2,400 volts. 2. The second or backup system to a main system which takes over a function if the main system fails.

secondary cell A voltaic cell which, after being discharged, may be restored to a charged condition by an electric current sent through the cell in a direction opposite that of the discharge current.

secondary disease A disease consequent on another disease already active.

secondary electron An electron driven from a material by bombardment with electrons, photons, or other high-velocity particles. Emission of secondary electrons has an important effect on the operation of electron tubes.

secondary hemorrhage See *hemorrhage.*

secondary hypothyroidism See *hypothyroidism.*

secondary storage In computers, external storage which is linked to and directly controlled by a computer.

secondary voltage 1. The voltage of power distribution circuits, commonly 115/230 volts. 2. The voltage across the secondary winding of a transformer.

secondary winding The transformer winding that receives its energy by electromagnetic induction from the primary winding.

secretion A product of a gland.

section Usually applied to thin slices of tissue cut for microscopic examination.

sedative A drug used to calm patients.

Seebeck effect See *thermoelectric effect.*

selectance A measure of the falling off in response of a resonant device with departure from resonance, expressed as the ratio of the amplitude of response at the resonant frequency to the response at some frequency differing from it by a specified amount.

selectivity 1. Characteristic which determines the extent to which it is possible to differentiate between the desired signal and disturbances of other frequencies. 2. Ability of a circuit (such as a phonocardio amplifier) to reject frequencies other than the one to which it is tuned (usually expressed by a curve in which the input signal voltage required to produce a constant power output is plotted against a frequency). 3. Degree to which a radio receiver can accept the signals of one station while rejecting those of all other stations on adjacent channels.

selenide A compound of selenium with an element or radical.

selenium A nonmetallic element relating to sulphur and tellurium and resembling them chemically.

selenium cell Form of light-sensitive cell. Selenium is an element whose electrical resistance decreases when exposed to light. A thin film of gold is supported between two glass plates and mounted in an evacuated bulb. A very narrow line is scratched in the film and filled with selenium and the current flowing through the cell must cross this. When light falls on the cell, the resistance of the selenium decreases and more current flows. This type of cell is known as a *photoconductive cell.* A second type is the *photoelectric cell.* When selenium is in contact with iron or steel and light falls on the joint between the metals, an electromotive force is developed which produces enough current to operate a small device as required. Most photometers or exposure meters are constructed in this way because the voltage developed is proportional to the amount of illumination.

selenium rectifier Rectifier formed of discs of iron in contact with a layer of metallic selenium. Largely supplanted by the silicon rectifier.

self-bias Bias provided to the control grid of an electron tube by the voltage drop across a resistor in the cathode circuit.

self-generating transducer A transducer which does not require external electrical excitation to provide specified output signals.

self-impedance At any pair of terminals of a network, the ratio of an applied potential difference to the resultant current at these terminals, all other terminals being open.

self-inductance The property of an electrical circuit which determines, for a given rate of change of current in the circuit, the electromotive force which is induced in the same circuit.

self-induction The production of a counterelectromotive force in a conductor when its own magnetic field expands or collapses with a change of current in the conductor.

self-organizing system   Any system that modifies its behavior according to a set of inputs representing the environment within which it operates. Environment includes a signal or energy source that is interpreted as an error indicator. Behavioral modification is generally taken to be accomplished by the system itself, without external intervention in its detailed inner workings.

semantics   The relationships between symbols and their meanings.

semen   The ejaculatory fluid, consisting of sperm cells and secretions of the prostate and bulbourethral glands and seminal vesicles.

semicircular canals   Three canals of the internal ear, the sense organs of equilibrium or balance.

semiconductor   A material whose resistivity is between that of conductors and insulators, and whose resistivity can be altered by light, an electric field, or a magnetic field. Current flow is sometimes by movement of negative electrons, and sometimes by transfer of positive holes. The conductivity characteristics of the device are altered by infusion of dopant impurities. Used in transistors, diodes, photodiodes, photocells, and thermistors. Some examples are: silicon, germanium, selenium, and lead sulfide.

semiconductor acceptor impurity   An impurity that may induce hole induction in a semiconductor. Semiconductor material which contains acceptor impurities is known as P-type, and has a higher conductivity than the intrinsic or pure semiconductor of the same element. It should be noted that the presence of the excess holes does not alter the overall neutral charge of the crystal, since the atoms of the impurity are also neutral, and adding together two neutral charge systems must result in a third neutral charge system.

semiconductor device   Electronic device in which the characteristic distinguishing electronic conduction takes place within a semiconductor.

semiconductor diode   An N-type and P-type material joined together to form a PN junction which allows current to flow in the forward direction from anode to cathode and blocks current in the reverse direction from cathode to anode. High reverse voltages greater than a specified limit, such as transients, can destroy the junction due to excessive reverse leakage currents.

semiconductor donor impurity   An impurity which may introduce electronic conduction in a semiconductor. Semiconductor material which contains donor impurities is known as N-type, and has a greater conductivity than the intrinsic semiconductor of the same element, and slightly greater than that of the corresponding P-type. It should be noted that the presence of the excess electrons does not alter the overall neutral charge of the crystal since the atoms of the impurity are also neutral, and adding together two neutral systems must result in a third neutral charge system. See also *semiconductor acceptor impurity.*

semiconductor integrated circuits   Complex circuits fabricated by suitable and selectively modifying areas on and within a wafer of semiconductor material to yield patterns of interconnected passive as well as active elements. The circuit may be assembled from several chips and uses thin-film elements or even discrete components to achieve a specified performance when the necessary device parameters cannot be achieved by material modification.

semiconductor material A chemical element, like silicon or germanium, which has a crystal lattice whose atomic bonds are such that the crystal can be made to conduct an electric current by means of free electrons or holes. The material is made P-type by infusion of acceptor impurities, or N-type by infusion of donor impurities.

semiconductor memory 1. A memory whose storage medium is a semiconductor circuit. Often used for high-speed buffer memories and for read-only memories. 2. A memory in which semiconductors are used as the storage elements, and characterized by low-to-moderate cost storage and a wide-range-of-memory operating speed, from very fast to relatively slow. Virtually all semiconductor memories are volatile.

semiconductor rectifier diode A semiconductor diode designed for rectification and including its associated mounting and cooling attachments if integral with it.

semilunar valve See *heart valve*.

semipermeable Partially but not freely or wholly permeable.

senescence The normal process of growing old.

senility The feebleness of body and mind incident to old age.

sense wire A wire threaded through a magnetic memory core which detects whether a one or zero is stored in the core when the core is interrogated by a read pulse.

sensitivity 1. The degree of response of an instrument or control unit to a change in the incoming signal. 2. In electrocardiography, a term applied to the response of the ECG writing instrument to extremes of frequency. Synonymous with *frequency response*.

sensor 1. Element or device that detects a change in a selective physical quantity and converts that change into an electrical signal for use as an input to a measuring, recording, or control system. 2. A transducer designed to produce an electrical output proportional to some time-varying quantity, such as temperature, illumination, or pressure. 3. The component of an instrument that converts an input signal into a quantity which is measured by another part of the instrument. 4. The pickup or transducer that senses or responds to the variable to be measured. Sensors are classified into two basic types—active or passive. An active sensor generates its own signal, and thus needs no external excitation; a passive sensor requires external excitation. Many biomedical signals (biopotential) are detected by passive electrodes, as in the ECG, EEG, and EMG. Temperature and pressure measurements are ordinarily made with resistive transducers, requiring external excitation.

sensory nerves Afferent nerves carrying sensory information to the central nervous system.

sensory receptor impulse See *nerve impulse*.

sepsis The condition of being infected by pyogenic bacteria.

septal leaflet The moving flap of the mitral valve which lies most proximal to the interventricular septum.

septic Pertaining to sepsis and to the condition resulting from the infection by pyogenic bacteria.

septicemia The circulation and multiplication of microorganisms in the blood; a very serious condition. Also known as *blood poisoning*.

septum The division between two cavities; such as which the interventricular septum which separates the right ventricle of the heart from the left.

sequential-access storage A form of digital computer storage in which the items of stored information become available only in a one-after-the-other sequence, regardless whether all the information or only some of it is desired.

sequential computer A computer in which events occur in time sequence with little or no simultaneity or overlap of events.

sequential logic Circuitry in which the output depends on previous input states as well as on the present input.

sequential logic element A device having at least one output channel and one or more input channels, all characterized by discrete states such that the state of each output channel is determined by the previous states of the input channel.

serial The technique for handling a binary data word which has more than one bit. The bits are acted upon one at a time, analogous to a parade passing a review point.

serial access 1. Pertaining to the sequential or consecutive transmission of data to or from storage. 2. Pertaining to the process of obtaining data from, or placing data into, storage, where the time required for such access is dependent upon the location of the data most recently obtained or placed in storage. See also *random access*.

serial adder Logical unit which adds two binary words, one binary bit pair at a time. The least significant addition is performed first and progressively more significant additions, including carries, are performed until the sum of the two numbers is formed.

serial bit Digital computer storage in which the individual bits that make up a computer word appear in time sequence.

serial computer A computer having a single arithmetic and logic unit.

serial counter A counter in which the second flip-flop cannot change state unit after the first flip-flop has changed, and the third can only change after the second has changed, etc., so that relatively long delays can occur before all counters have reached their final states following an input pulse to the counter. Also called *ripple-through counter*.

serial memory 1. One in which information is stored in series and is written or read in time sequence, as with a shift register. Compared to a RAM, the advantages of a serial memory are slow to medium speed with lower cost. See also *sequential access memory (SAM)*. 2. A memory whose contained data are accessible only in a fixed order, beginning at some prescribed reference point. Data in any particular location are not available until all data ahead of that location have been read. Such a memory is inherently slow compared with a random access memory.

serial mode Operation in a computer which is performed bit by bit, generally beginning with the least significant bit. Read-in and readout occurs bit after bit by shifting the binary data through the register.

serial operation The handling of information within computer circuits in which the digits of a word are transmitted one at a time along a single line. Although slower than parallel operation, its circuits are much less complex.

serial programming Programming in a digital computer so that only one arithmetical or logical operation can be executed at one time.

series 1. When two devices are connected in such a way that current must pass through both of them in passing through the circuit, they are said to be in series. 2. A mathematical expression of the form x1 + x2 + x3 . . . , where x's are real or complex numbers. 3. A battery of tests or measurements on a specific material, such as blood.

series circuit A circuit whose elements are all connected end to end so that the same current flows through all of the elements in sequence. The voltage across the circuit is the sum of the voltages across the individual elements.

series connection Elements in an electrical circuit are said to be connected in series when the same current flows through all of them in sequence.

series-parallel connection A connection, of cells in a battery or other elements, where several identical series-connected groups are all connected in parallel.

series resonance The condition which exists in a circuit comprising capacitance and inductance in series, when the current through the circuit is in phase with the voltage across the circuit. When this occurs, the impedance of the circuit is equal to its resistance. In a series resonant circuit, the capacitive reactance and the inductive reactance are equal in magnitude. Impedance of the circuit rises when the applied current has a frequency above or below the resonant frequency.

serous Fluid-producing, as in the serous membranes of the thorax and abdomen (pleura and peritoneum, respectively); also pertaining to or resembling serum.

serous membrane A smooth lining of body cavities which are closed, particularly those of the thorax and abdomen. Also called *serosa*.

serum That part of the blood which remains after the cells, platelets, and fibrinogen have been removed, usually by allowing the blood to clot. It consists of a saline solution containing a number of proteins and lipids.

servomechanism A system which responds to a control signal, and in which the difference between the desired state and the actual state is fed back into the control system until continued response eliminates this difference.

servo system An electromechanical system including a feedback loop which can accurately transmit mechanical position information over an electrical circuit.

sessile Having no stem, applied to tumors.

set 1. A permanent change of a given parameter, attributable to any cause. 2. An input on a flip-flop not controlled by the clock and used to effect the Q output. It is this input through which signals can be entered to get the Q output to go to 1. Note it cannot get Q to go to 0. See *asynchronous inputs*.

set input An asynchronous input to a flip-flop used to force the Q output to its high state.

set terminal The flip-flop input terminal that triggers the circuit from its first state to its second state. See *input-output terminal*.

settling time 1. In an operational amplifier, the time elapsed from the application of an ideal instantaneous step input to the time at which the closed-loop amplifier output has entered and remained within a specified error band, usually symmetrical about the final value. Settling time includes a very brief propagation delay, plus the time required for the output to slew to the vicinity of the final value, recover from the overload condition associated with slewing, and finally settle to within the specified error. 2. The time elapsed from the application of a full-scale step input to an amplifier to when the output has entered and remained within a specified error band around its final value. 3. The time it takes for a D/A converter to settle for a full-scale charge. 4. In a feedback control system, the time required for an error to be reduced to a specified fraction, usually 2 percent to 5 percent, of its original magnitude. 5. The time required for the output frequency of a voltage or current-tuned oscillator to change from the initial value to within a specified window around the final value in response to a voltage or current step on the tuning input port.

seven-segment display A display format consisting of seven bars so arranged that each digit from 0 to 9 can be displayed by energizing two or more bars. LED, LCD, and gas discharge displays all use seven-segment display formats.

sexadecimal See *number system.*

S/H See *sample and hold.*

shaker An electromagnetic device capable of imparting known and/or controlled vibratory acceleration to a given object for testing or mixing purposes.

shake-table A laboratory tester in which an instrument component is placed in a vibrator that simulates operating conditions.

shape To alter the waveform of an electric wave by filtering or limiting.

shaping network A network which provides complementary compensation or equalization for the loss-frequency characteristic of a line or its equipment.

shelf life The length of time an item can be stored under specified conditions and still meet specified requirements with a specified level of assurance.

shield 1. A metallic covering over a circuit or equipment component which intercepts electrostatic or electromagnetic fields, and conducts the currents they induce to ground. Magnetic fields are shielded by magnetic material such as iron; electric fields are shielded by materials which are good conductors, such as copper. 2. To protect from stray electric or magnetic fields.

shielded pair Two insulated wires surrounded by an electrostatic shield consisting of braided copper wires or a wrapped metallic foil.

shielding Metal covering used on a cable; also a metal can, case partition, or plates enclosing an electronic circuit or component. Shielding is used to prevent undesirable radiation, pickup of signals, magnetic induction, stray current, ac hum, or radiation of an electrical signal.

shift The process of moving data from one place to another. Generally many bits are moved at once. Shifting is done synchronously and by command of the clock. An eight-bit word can be shifted sequentially (serially)—that is the first bit goes out, second bit takes first bit's place, third bit takes second bit's place, and so on, in the manner of a bucket brigade. Generally referred to as shifting left or right. It takes eight clock pulses to shift an eight-bit word or all bits of a word can be shifted simultaneously. This is called *parallel load* or *parallel shift.*

shift counter See *ring counter.*

shift register 1. A digital storage circuit which uses a chain of flip-flops to shift data from one flip-flop to its adjacent flip-flop on each clock pulse. Data may be shifted several places to the right or to the left, depending on additional gating and the number of clock pulses received. Depending on the number of positions shifted, in a right shift the rightmost characters are lost; in a left shift, the leftmost characters are lost. See also *dynamic shift register* and *static shift register*. 2. An element in the digital family which uses flip-flops to perform a displacement or movement of a set of digits one or more places to the right or left. If the digits are those of a numerical expression, a shift may be the equivalent of multiplying the number by a power of the base. 3. A building block in a digital system consisting of any number of flip-flops connected in series that are controlled by a common clock signal. 4. A particular kind of memory in which information is put in one end, is "bumped" sequentially through the memory, and emerges in a readable form at the other end of the memory. This kind of memory, made from semiconductors, is relatively inexpensive and typically is used as auxiliary memory in video terminals, and nonfade analog displays.

shock 1. Severe circulatory disturbance characterized by fall in blood pressure, weak rapid pulse, thirst, and pallor. The usual cause is rapid diminution in the blood volume. 2. A sudden stimulation of the nerves and convulsive contraction of the muscles caused by a discharge of electricity through the body. The severity depends upon (a) the amount of current, (b) whether the path of the current is through a vital organ, and (c) the duration of the current.

shock excitation The sudden application of a momentary steep-wavefront voltage to

a circuit which causes a damped oscillation. See also *impulse excitation*.

Shockley diode See *four-layer diode*.

shock treatment A form of psychiatric treatment in which electric current, insulin, or carbon dioxide is administered to the patient and results in a convulsive reaction to alter favorably the course of mental illness.

short A short circuit.

short circuit 1. Anastomosis between gut or blood vessels which allows the contents to bypass a section of the normal pathway. 2. A low-resistance connection across a voltage source or between the sides of a circuit or line; usually accidental and usually resulting in excessive current flow which often causes damage. 3. An electrical connection of negligible impedance between two terminals of a circuit. 4. To connect a negligibly small impedance between (a pair or terminals). 5. To cause (a current) to flow into a negligibly small impedance.

shorted Prevented from operating by a short circuit.

shot noise Noise inherent in an electric current, due to the fact that it consists of a stream of finite particles, i.e., electrons.

SHT Abbreviation for secondary of the high-tension circuit.

shunt 1. Precision, low-value resistor placed across the terminals of an ammeter to increase its range. 2. Any part connected, or the act of connecting any part, in parallel with some other part. 3. Branch of an electric circuit having its winding in parallel with the exteral or line circuit. 4. A pathway for blood flow between the atria of the heart, due to atrial-septal defect.

sibilus A hissing sound heard on auscultation of the chest during respiration in bronchitis, etc.

sickle-cell anemia Hereditary anemia found sometimes in blacks. The red blood cells become sickle-shaped or crescentic.

sidebands The bands of spectral energy at frequencies higher and lower than the carrier frequency which result from a modulation process.

side-cutting pliers A pincers having holding jaws which also have a set of cutting blades on the sides of the jaws.

siderosis 1. Inhalation of iron particles causing pneumoconiosis. 2. Excess of iron in the blood.

siemens (S) The international standard unit of conductance which replaces and is identical to the mho. It is the reciprocal of resistance in ohms.

sigmoid Like the Greek letter sigma, applied especially to a bend in the pelvic colon just before it becomes the rectum.

sigmoidoscope An instrument for viewing the interior of the rectum and sigmoid flexure of the colon.

sigmoidoscopy Examination of the sigmoid colon with a lighted instrument.

signal 1. An electrical wave used to convey information. 2. An electric wave accompanying some physiologic action, such as the contraction of heart, movement of the eye, or raising of a member. 3. A transduced voltage representing some physiologic action or function. 4. An alerting signal. 5. An acoustic device such as a bell or a visual device such as a lamp which calls the attention. 6. To transmit an information signal or alerting signal.

signal averaging A technique for extracting a signal waveform (generally a time-varying voltage) from a background of unwanted noise. Simple frequency-domain filtering with passive or active circuit elements is the most widely used method for accomplishing this result. But this type of filtering is effective only when the frequency spectrum of the signal and the frequency spectrum of the noise do not overlap.

A signal averager is a special kind of filter, sometimes referred to as a *comb filter*. It can be used effectively only if the desired signal, with its contaminating noise, can be repreated a number of times. In addition, a synchronization pulse is required.

signal conditioner A device placed between a signal source and a readout device to condition the signal. Examples: damping networks, attenuator networks, preamplifiers, excitation and demodulation circuitry, signal converters (for changing one electrical quantity into another, such as volts to amps), instrument transformers, equalizing or matching networks, and filters.

signal conditioning Any manipulation of transducer or amplifier outputs to make them suitable for input to computer peripheral equipment such as an analog-to-digital converter. Also, operations such as linearizing, and square root extraction performed within the computer.

signal enhancement Ensemble averaging of time-domain signals, where a set of time domain samples are digitized and then averaged. In order to enhance the signal by averaging, the time function must be repetitive, and the start of the ensemble average must have a known relationship to some repetitive event (trigger). Such a repetitive signal is the QRS complex of the electrocardiogram. This has been used to synchronize recordings of electrical activity within the heart's specialized conduction system, where the input source consists *only* of surface ECG electrodes.

signal generator 1. A device that produces a signal of a specific type of audio frequency (AF), radio frequency (RF), etc., for purposes of testing or alignment. 2. A device used to furnish current at a known radio frequency, modulated, and to deliver a measured voltage only at the terminals of the generator without appreciable radiation at any other point. 3. A generator of waveforms which simulate the direct and transduced waves produced by the cardiovascular system, used for educational purposes.

signal-to-noise ratio 1. Ratio of the magnitude of the signal to that of the noise; often expressed in decibels. NOTE: This ratio is expressed in many different ways e.g., in terms of peak values in the case of impulse noise, and in terms of root-mean-square values in the case of random noise, the signal being assumed sinusoidal; in specific cases, other measures of signal and noise may be used if clearly stated. 2. At any point of television transmission, this is the ratio in decibels of the maximum peak-to-peak voltage of the video television signals, including synchronizing pulse, to the root-mean-square voltage of the noise. NOTE: Television transmission signal-to-noise ratio is defined in this way because of the difficulty of defining the root-mean-square value of the video signal or the peak-to-peak value of the random noise. 3. Comparison of the amount of signal to the amount of noise by means of a fraction ratio, measuring both elements in the same units. 4. Ratio of the intensity of the desired signal to that of the undesired noise signal.

signal tracing The process of locating a fault in a circuit by injecting a test signal at the input and checking each stage, usually from the output backwards.

significant digit A digit that contributes to the precision of a number. Significant digits are counted from the first digit on the left that is not zero, and continue to the last accurate digit on the right. (A right-hand zero may be counted if it is an accurate part of the number.) For example: 2,500.0 has five significant digits, 2,500 probably has only two (it is not known that the last two digits are accurate), but 2501 has four, and 0.0025 has two.

sigultus Hiccup.

silica gel A colloidal form of silica which will absorb moisture from the air. Small bags of silica gel are packed with electronic equipment when it is shipped to protect it from moisture.

silicon A dark-gray, hard, crystalline solid, second most abundant among the elements in the earth's crust. Used in making transistors and in alloying with iron in making steel for transformer cores.

silicon controlled rectifier (SCR) Silicon controlled rectifiers (or switches) are usually four-layer diodes. In the forward direction, they are power switches; in the reverse direction, they block current just as any diode. The essential characteristic is a bistable mode of operation. They have two stable conditions: the *on* condition when the device is conducting, and the *off* condition when the device is blocking. They differ among themselves in the type of signal required to make them turn *on* and *off*. Some flip on or off easily. Others need complicated triggering circuitry for turn-off. Switching is controlled by a signal at the gate. Also called *triode thyristor*.

silicon controlled switch (SCS) A four-layer semiconductor device having all layers brought out to terminals. Can be used as a silicon controlled rectifier or as a gate which can be turned on and off like a switch.

silicon diode A semiconductor diode that

uses silicon as the rectifying element.

silicone One of the family of polymeric materials in which the recurring chemical group contains silicon and oxygen atoms as links in the main chain. At present, these compounds are derived from silica (sand) and methyl chloride. The various forms obtainable are resins, oils, greases, plastics, etc., all of which have a high resistance to heat and to water.

silicon power rectifier A rectifier in which a silicon diode is used as the rectifying element.

silicon rectifier Semiconductor diode that converts alternating current to direct current and which can be designed to withstand large currents and high voltages.

silicon solar cell A photovoltaic cell that is composed of a thin sheet of a processed silicon and that is used to convert light energy into power.

silicon-symmetrical switch Thyristor modified by adding a semiconductor layer so that the device becomes a bidirectional switch. Used as an ac phase control, for synchronous switching and motor speed control.

silicosis Occupational disease, usually chronic, causing fibrosis of the lungs. It results from inhalation of the dust of stone, flint, or sand that contains silica (quartz). Called *grinders' disease,* it is observed in workers who have breathed such dust for a period of years.

simulate 1. To represent the function of a device, system, or computer program by another. 2. To represent one computer by another. 3. To represent the behavior of a physical system by the execution of a computer program. 4. To represent a biological system by a mathematical model.

simulator Any device which represents a system or phenomenon and which reflects the effects of changes in the original, so that it may be studied, analyzed, and understood from the behavior of that device.

simultaneous access See *parallel access.*

simultrace recorder A multichannel cathode-ray oscillographic recorder in which direct or transduced physiologic potentials are amplified and simultaneously applied to two cathode-ray tubes: the first is a monitor; the second is optically linked to a photosensitive strip-chart paper. The waves appearing on the monitor are duplicated by the light spots moving on the second tube face. In effect, the monitor gives a presentation simultaneous with, and exactly duplicating that which is being recorded.

sine For any angle in a right triangle, the ratio of the side opposite that angle to the hypotenuse.

sine wave A waveform (often viewed on an oscilloscope) of a pure alternating current or voltage. It is drawn on a graph of amplitude versus time or radial degrees and follows the mathematical rules of sine and cosine values in relation to angular rotation of an alternator. It can be simulated by means of an electronic oscillator.

single-ended amplifier An amplifier with one input terminal and one output terminal tied to a common point and therefore operating at a common potential. This point may or may not be connected to ground.

single-point ground See *uniground.*

single-point grounding A grounding system that attempts to confine all return currents to a network which serves as the circuit reference. The phrase single-point grounding does not imply that the grounding system is limited to one earth connection. To be effective, no appreciable current is allowed to flow in the circuit reference, i.e., the sum of above return currents is zero.

single-pole single-throw (SPST) A switch with only one moving and one stationary contact. Available either normally open (N.O.) or normally closed (N.C.).

single-shot blocking oscillator Blocking oscillator modified to operate as a single-shot trigger circuit.

single-shot multivibrator A monostable multivibrator which, after being triggered to the quasi-stable state, will flop back by itself to the stable state after a certain period of time.

single-shot trigger circuit Trigger circuit in which one triggering pulse initiates one complete cycle of conditions ending with a stable condition. Also called *single-trip trigger circuit*.

single-trip trigger circuit See *single-shot trigger circuit*.

sinistral Pertaining to the left side.

sinoatrial node 1. A microscopic collection of atypical cardiac muscle fibers which is responsible for initiating each cycle of cardiac contraction. 2. A specialized cluster of cells located in the heart at the junction of the right antrium and the superior vena cava. Periodic depolarizations of these cells create the natural pacing impulse which stimulates the heart's musculature into action via the specialized conduction system. See also *bundle of His* and *atrioventricular node*.

sintering The process of bonding metal or other powders by cold-pressing into the desired shape, then heating to form a strong cohesive body.

sinus 1. A passage leading from an abscess, or some inner part, to an external opening. 2. A dilated channel for venous blood. 3. Air sinuses, hollow cavities in the skull bones which communicate with the nose. They are the *frontal, maxillary, ethmoidal,* and *sphenoidal* sinuses. 4. Any irregular cavity, particularly in the circulatory system.

sinus arrythmia Irregular cardiac rhythm due to the controlling effect of the vagus on the sinoatrial node. The heart rate increases in inspiration and slows during expiration.

sinusoid 1. Like a sinus. Channels for small blood vessels as found in the liver, suprarenal glands, etc. 2. The characteristic waveform of a function which rises smoothly and continuously from zero to a positive maximum; then, as smoothly continues back through zero to rise to a negative maximum; then, as smoothly continues back through zero to repeat the cycle. 3. Varying in time or space in accordance with the trigonometric sine function. The movement of a pendulum and the vibration of a tuning fork are examples of sinusoidal motion.

sinus rhythm The normal rhythmic contraction and relaxation of the heart musculature under the control of impulses arising in the sinoatrial node, the heart's natural pacemaker.

skin effect The tendency for a high-frequency current to travel on the outside skin of a conductor, rather than being distributed uniformly throughout the conductor. Skin effect causes an increase in the effective resistance of a conductor at high frequencies.

slave A component in a system that does not act independently, but only under the control of another, controlling component.

sleeping sickness See *encephalitis*.

slewing In random-sampling oscilloscope

technique, the process of incrementally delaying successive samples or a set of samples with respect to the signal being sampled.

slew rate 1. The maximum rate of change of the output voltage of an amplifier while the amplifier is operating in its linear region. Usually given in terms of volts per microsecond. 2. The maximum time rate of change of a closed-loop amplifier output voltage under large signal conditions. The large signal is that ac input voltage which saturates one of the amplifier stages, thus causing current limiting of that stage. Also called *rate limit* or *voltage velocity limit.*

slipped disc An acute or chronic condition, caused by the traumatic or degenerative displacement and protrusions of the softened central core of an intervertebral disc (cartilaginous disc between the spine bones), especially of the lower back. Symptoms are low back pain, which frequently extends to the thigh; muscle spasm; and tenderness.

slow-action relay See *time delay relay.*

slow memory See *slow storage.*

slow-scan television Television system using a slow rate of horizontal scanning suitable for transmitting printed matter, photographs, and illustrations.

slow storage In computers, storage with a relatively long access time. See also *secondary storage.* Also called *slow memory.*

small signal That value of an ac voltage or current which when halved or doubled will not affect the characteristic being measured beyond the normal accuracy of the measurement of that characteristic.

small signal characteristics The characteristics of an amplifier operating in the linear amplification region.

snap-action contacts A contact assembly having two or more equilibrium positions, in one of which the contacts remain with substantially constant contact pressure during the initial motion of the actuating member, until a condition is reached at which stored mechanical energy snaps the contacts to a new position of equilibrium.

snapshot In a computer, a dynamic printout of selected data in storage at breakpoints and checkpoints during the computing operations as contrasted with a static printout, for a dump.

sneak circuit The usually unanticipated part of a complete electrical circuit which carries an unintentional (sneak) current. Sneak currents may prevent proper operation of interconnected equipment.

snr or S/N ratio Abbreviation for signal-to-noise ratio.

socket contact A female contact designed to mate with a male contact. Normally connected to the live side of a circuit.

soft tube An electron tube which does not have a high vacuum, either intentionally or because of occluded gases.

software 1. The internal programs or routines professionally prepared to augment and support programming and computer operations. These routines permit the programmer to use his own language (English) or mathematics (algebra) in communicating with the computer. 2. The totality of programs and routines used to extend the capabilities of computers such as compilers, assemblers, narrators, routines, and subroutines. 3. The package of programming support or utility routines which is provided (or is available with) a given computer. The package generally includes: an assembler, a compiler, an operating system (or monitor), debugging aids, and a library of subroutines. 4. Everything except hardware.

soft X-rays Radiation employing particles of comparatively slow velocity.

solar absorber A surface which converts solar radiation into thermal energy.

solar cell A device which will generate low voltage direct current from the radiant energy in sunlight. Consists of a silicon PN junction in which photons of light energy penetrate a thin layer of N-type silicon and create hole-electron pairs at the PN junction. These travel to the terminals and produce a useful external current.

solar energy conversion The process of changing solar radiation either directly or with a heat engine, to electrical or mechanical power.

solar plexus A plexus of nerves and ganglia in the upper region of the abdomen.

solar radiation Radiation from the sun that is made up of a very wide range of wavelengths, from the long infrared to the short ultraviolet with its greatest intensity in the visible green at about 5,000 angstroms. The solar radiation received on the earth's surface is restricted to the visible and near-infrared, as the atmosphere strongly absorbs the wavelengths located at either end of the spectrum.

solder A low melting-point alloy used to mechanically and/or electrically join two or more metallic portions of a device or circuit. Frequently a tin-lead mixture but may contain many other metallic or semi-metallic elements.

soldering gun A pistol-shaped soldering tool having a trigger switch to turn it on. Operates from 117 volts ac, and has an integral step-down transformer with a single turn secondary which heats the soldering loop.

soldering iron A copper-tipped tool, usually electrically heated, which enables heat-

ing of the work and melting of a solder to make a soldered joint.

solder short A defect which occurs when two or more wires or etched conductors are shorted because of touching solder.

solenoid 1. An electromagnet having an energizing coil which is approximately cylindrical in form, acting on an armature positioned in the center of the coil. 2. An electrical conductor would as a helix with a small pitch, or as two or more coaxial helices. 3. A tubular coil of wire which, when traversed by an electric current, acts as a magnet and tends to pull a movable iron core to a central position.

solenoid valve An hydraulic valve controlled by a relay or solenoid, which is powered by electromagnetic energy. Most hydraulic or gas-driven artificial hearts utilize solenoid valves to feed fluids to the outer heart chamber during the correct time intervals.

soleus A muscle in the calf of the leg.

solid electrolyte fuel cell Self-contained fuel cell in which oxygen is the oxidant and hydrogen is the fuel. The oxidant and fuel are kept separated by a solid electrolyte which has a crystalline structure and a low conductivity.

solidly grounded Grounded through an adequate grounded connection in which no impedance has been inserted intentionally. Also called *directly grounded*.

solid silicon circuit Semiconductor circuit that employs a single piece of silicon material in which the various circuit elements (transistors, diodes, resistors, and capacitors) are formed by diffusion in the planar configuration. By combining oxide masking, diffusion, metal deposition, and alloying, a complex network with active and passive components is made completely within a die that is part of a single semi-conductor wafer. Often, thin-film devices are applied to the surface of the silicon wafer to provide passive circuit elements beyond the range of solid silicon technology. External connections are made through small wires soldered, welded, or thermocompression-bonded to selected points on the surface of the die.

solid state Refers to the electronic properties of crystalline materials, generally semiconductor in type. Vacuum and gas filled tubes function by flow of electrons through space, or by flow of current through ionized gases. Solid-state devices involve the interaction of light, heat, magnetic fields, and electric currents in crystalline materials. The variety of effects obtainable is much larger than in tubes, and less power is generally required.

solid-state component A component whose operation depends on the control of electric or magnetic phenomena in solids, e.g., a transistor, crystal diode, or ferrite.

solid-state computer A computer built primarily from solid-state electronic circuit elements.

solid-state devices Electronic components that convey or control electrons within solid materials—for example, transistors, semiconductor diodes, or magnetic cores.

solid-state laser A laser using a transparent substance (crystalline or glass) as the active medium, doped to provide the energy states necessary for lasing. The pumping mechanism is the radiation from a powerful light source, such as a flashtube. The ruby and Nd-YAG lasers are solid-state lasers.

solid-state physics Dealing with the structure and properties of solids. Includes much research in the structure of metals, alloys, ionic crystals, impurity semiconductivity, lattice defects, and crystal growth.

solid-state relay (SSR) 1. A device in which the control and load currents pass, exclusively, through semiconductors. There are no moving parts or mechanical contacts. 2. An electronically operated switch in which a small control current causes a solid-state device to produce a switching action in the primary current circuit.

solid tantalum capacitor An electrolytic capacitor whose anode is a sintered tantalum powdered slug which has an oxide film dielectric formed on it. The cathode is manganese dioxide, a semiconductor which also acts as the electrolyte. Characteristics: excellent temperature coefficient over a broad range ($-55°C$ to $+125°C$), high capacity per unit volume, low impedance. Generally used in low-voltage applications (up to 125 volts). Has an excellent shelf life.

somatic Pertaining to the nonvisceral portions of the body, such as the skeleton and skeletal muscles.

-some Suffix meaning body.

somnambulism Walking and carrying out other activities while asleep.

sonar A ranging system similar to radar but using sound waves instead of radio waves. Detects presence of objects near a boat in a fog and is also used to find depth of the ocean bottom. (The bat is guided by a system similar to sonar using extremely high pitched sounds that are beyond the range of human hearing.) The sound is sent out at regular intervals and the time required for the sound to be returned by reflection is measured. Knowing the speed of sound, it is easy to determine the distance. The technique has been applied to medical use in echocardiography and in ultrasonic studies of the brain and eye.

sone A unit of loudness. A simple 1000 Hz tone 40 dB above a listener's threshold, produces a loudness of one sone. The loudness of any sound that is judged by the listener to be *n* times that of one-sone tone is *n* sones.

sonic delay line 1. A device using electro-acoustic transducers and the propagation of an elastic wave through a medium so as to achieve the delay of an electrical signal. 2. A form of delay line which uses pulses in the molecules of the medium—in contrast with an electrical delay line, which uses electrical pulses in a wire or in an assembly of coils and capacitors.

sonic surgery The use of focused, ultrasonic waves to produce precisely circumscribed alterations at predetermined sites within tissues. This technique has been used, for example, in treating certain neurologic diseases involving the brain and ear.

sonographic The creation of a visible tracing or pictorial representation from sonic or ultrasonic information. An *echogram* is a pictorial representation of cardiac structures in motion, obtained through echocardiography.

sophisticated vocabulary An advanced and elaborate set of instructions. Some computers can perform only the more common mathematical calculations such as addition, multiplication, subtraction, etc. A computer with a sophisticated vocabulary can go beyond this and perform operations such as linearize, extract square root, select highest number, etc.

soporific Agent causing sleep.

sordes A collection of bacteria, food particles, and epithelial tissue in the mouth.

sort To arrange items of information according to rules dependent upon a key or field contained in the items or records; e.g., to digital sort is to sort first the keys on the least significant digit, and to resort on each higher order digit unit the items are sorted on the most significant digit.

sound Sensation accompanying vibrations of the eardrum at certain frequencies; longitudinal vibrations at those frequencies in any medium. The audible frequencies range from 30 to 40,000 Hz, the middle C of a piano being 256 Hz.

sound energy density Sound energy per unit volume. The commonly used unit is the erg per cubic centimeter.

sound wave A traveling wave in air or other elastic medium produced by audio-frequency vibrations. The velocity of a sound wave in air is about 1,100 feet per second; in water it is about 4,800 feet per second.

source 1. The origination point of electrical energy or signal power supplied to a circuit. 2. In a field-effect transistor, the electrode which serves the same function as a vacuum tube cathode. 3. Supply of energy, or device upstream from a sink (see *sink*). 4. Terminal which usually sources carriers. In MOS devices which are usually symmetrical, it can be interchanged with the drain terminal in a circuit. 5. The working-current terminal (at one end of the channel in an FET) that is the source of holes (P-channel) or free electrons (N-channel) flowing in the channel. Corresponds to emitter in bipolar transistor.

source impedance The impedance which a meter or other instrument "*sees*," i.e., the impedance of the driving circuit when measured from the input terminals of the meter.

source language In a computer, the language from which a statement is translated.

source program A computer program

written in a language designed for ease of expression of a class of problems or procedures, by humans; e.g., symbolic or algebraic. A generator, assembler translator, or compiler routine is used to perform the mechanics of translating the source program into an object program in machine language.

space charge   1. The negative charge produced by the cloud of electrons existing in the space between the cathode and plate in an electron tube, formed by the electrons emitted from the cathode in excess of those immediately attracted to the plate. 2. The region around a PN junction in which holes and electrons recombine leaving no mobile charge carriers and a net charge density different from zero.

spade lug   A solder lug which has an open end so that it can be slipped under the head of a binding screw.

spaghetti   Linen or cotton fabric tubing impregnated with insulating material and baked. Used to insulate short lengths of bare wires.

span   The reach or spread between two established limits such as the difference between high and low values in a given range of physical measurements.

spark suppression   The use of a small capacitance or high resistance across contacts which break currents in inductive circuits, to prevent excessive sparking when the contacts break. A capacitor and resistance in series is also used as a suppressor.

spark suppressor   1. A capacitor and resistor in series, placed across contacts to inhibit sparking and to short-circuit noise voltages from the sparks. 2. A combination of two semiconductor diodes, joined at their cathodes and with their anodes connected across a circuit. Normally nonconducting, these diodes suppress sparks, once

the Zener point of one has been reached so as to permit reverse-mode conduction.

spasm   1. Sudden convulsive involuntary movement. 2. Sudden contraction of a muscle or muscles, especially of the unstriped muscle coats of arteries, intestines, heart, bronchi, etc. The effect of such spasm depends on the part affected; thus, asthma is believed to be due to spasm of the muscular coats of the smaller bronchi; and renal colic is due to spasm of the muscle coat of the ureter.

spasmolytic   Substance which relieves spasm.

spastic   1. In a state of spasm. 2. Popular term for *cerebral palsy*. In patients with this disease the muscles are often spastic, i.e., hypertonic, and there is excessive neuromuscular excitability.

spasticity   State of being hypertonic—so that the muscles are stiff. Spasticity of limb and bladder muscles routinely accompanies transection of the spinal cord.

spastic paralysis (cerebral palsy)   A condition probably stemming from various causes present since birth. Associated with nonprogressive brain damage, cerebral palsy is characterized by spastic, jerky voluntary movements, or constant involuntary and irregular writhing.

spatial   Pertaining to space.

spatula   1. A flat flexible, blunt knife, used for spreading ointments and poultices. 2. A tongue depressor.

specific address   An address that indicates the exact storage location where the referenced operand is to be found or stored in the actual machine-code address numbering system.

specific heat   1. Ratio of the heat capacity of a substance to the heat capacity of an equal mass of water. When so expressed, the specific heat is a dimensionless number. 2. The amount of heat necessary to raise the temperature of a substance by 1°F. Expressed at Btu per pound per degree F or as calories per gram.

specific inductive capacity See *dielectric constant.*

specific insulation resistance See *volume resistivity.*

specimen A fraction or the whole of a sample unit, or a representative example of material evolved from the identical process as the product.

spectral distribution of energy A plot showing the variation of spectral emission with wavelength.

spectrophotometer An instrument capable of measuring transmission or apparent reflectance of visible light as a function of wavelength, thus making possible the accurate evaluation of color, as well as comparisons between two luminous sources with respect to intensity and wavelengths emitted.

spectroradiometer An instrument for measuring the radiant energy from a source at each wavelength through the spectrum. Spectral regions are separated either by calibrated filters or a calibrated monochromator. The detector is usually an energy receiver such as a thermocouple.

spectroscope Any one of a class of instruments used for dispersing radiation visible, or invisible, into its component wavelengths and for observing or measuring the resultant spectrum.

spectroscopy A branch of optics pertaining to radiations which lie in the infrared, visible, ultraviolet and vacuum ultraviolet regions of the spectrum.

spectrum 1. The frequency content of a complex waveform or a graph of a complex waveform represented in the frequency domain. The band of frequencies necessary to pass a certain type of intelligence. 2. The range of wavelengths considered in a system. 3. The distribution of the components of a signal across the frequency range.

spectrum analysis The study of energy distribution across the frequency spectrum for a given electrical signal.

spectrum analyzer A device for sweeping over a spectrum or band of frequencies to determine what frequencies are being produced, and the amplitudes of each component.

speculum A polished instrument for examining the interior cavities of the body, especially the vagina, and rectum, the ear, and the nose.

speech synthesizer System used in research to generate speech from electrical signals in order to study human vocal patterns.

speed of light The speed at which light travels in a vacuum—186,284 miles per second. Its symbol is *c*.

speed of sound The speed at which sound travels—750 miles per hour (1,080 feet per second)—in air at normal sea-level conditions.

sphincter A circular or ringlike muscle surrounding a body orifice; e.g., the anal sphincter muscle.

sphygmocardiograph A device for simultaneous recording of heartbeat and pulse.

sphygmogram A graphic recording of the movements, forms, and forces of an arterial pulse.

sphygmograph An instrument affixed to the wrist, which moves with the beat of the pulse and registers the rate and character of the beats.

sphygmoplethysmograph Device for recording pulse shape and volume.

sphygomanometer An instrument for measuring the arterial tension (blood pressure) of the circulation. It consists of an air-inflatable cuff and pressure indicator. In practice, the cuff is applied to the arm over the brachial artery and pumped to a pressure which occludes the artery, as confirmed by auscultation. This gives the systolic pressure. Air is then bled from the cuff until the blood is just heard passing through the opening artery lumen, creating distinctive Korotkoff sounds. This gives the diastolic pressure.

spike Transient of short duration, comprising part of a pulse, during which the amplitude considerably exceeds the average amplitude of the pulse.

spinal accessory nerve The eleventh cranial nerve, arising in the spinal cord. It innervates the region of the sternum and mastoid process of the temple, and the trapezium.

spinal analgesia or **anaesthesia** Infiltration of local anaesthetic agents into the cerebrospinal fluid by means of a lumbar puncture. This procedure results in a block to the nerves below the level of the injection.

spirograph An instrument for recording respiratory characteristics.

spirometer An instrument for measuring the volume of air entering and leaving the lungs.

spirometry Measurment of lung volumes, capacities, and flow rates at which air enters and leaves the lungs.

split-phase Describing any device, such as an induction motor, which derives a second phase from a single-phase alternating current source by tapping it off through a capacitive or inductive reactor.

split-phase motor A single-phase induction motor having an auxiliary winding connected in parallel with the main winding, but displaced in a magnetic position from the main winding, so as to produce the required rotating magnetic field for starting. The auxiliary circuit is generally opened when the motor has obtained a predetermined speed.

spondylitis Inflammation of a vertebra or vertebrae.

spondylosis Term used to describe non-specific degenerative changes in the intervertebral discs with peripheral ossification. Known as *osteoarthritis of the spine*.

sporadic Occurring occasionally, and at random intervals. Infrequent. A characteristic of impulse noise.

spot See *land*.

sprain The wrenching of a joint resulting in injury to its attachments.

SPST Abbreviation for single-pole single-throw.

SPST contact Single-pole single-throw contact of a switch or relay.

spurious pulse A pulse in a scintillation counter other than one purposely generated, or one due directly to ionizing radiation.

spurious response Response of a frequency-selective system, such as a radio receiver, to an undesired frequency.

sputum Matter ejected from the respiratory tract, often from the lungs.

square wave A waveform which shifts abruptly from one to the other of two fixed values for equal lengths of time. The transition time is negligible when compared with the duration time of each fixed value.

It is the type of wave normally generated for logic operations in a computer.

SSR   Abbreviation for solid-state relay.

stability   The ability to remain stable, without drift, oscillations, or variations. A characteristic shared by organisms in health and by equipment soundly designed, well-maintained, and prudently operated.

stack   A portion of a computer memory and/or registers used for temporarily holding information.

standby   1. A state of equipment nonoperation, but one which will permit complete resumption of operation within a short period of time. 2. A duplicate set of equipment to be used if the primary unit becomes unusable because of malfunction.

standstill   Complete cessation of myocardial contraction.

star connection   Three windings or network elements with one terminal of each connected to a common node.

Starr-Edwards valve   See *heart valve.*

start-stop multivibrator   See *flip-flop multivibrator.*

stasis   A cessation of normal flow; a stagnation of blood or other fluids. Most commonly used to indicate arrest of the circulation of either blood or lymph, but also for intestinal stasis, a holding up of the contents of the bowel.

state   Either of the two conditions of a bistable device, the one state or the zero state. The state of a circuit refers to its output; hence, a flip-flop is said to be in the one state when its output is high, and in the zero state, when its output is low.

state equations   Those equations which express the output of a system and its condition at any time, as a single-valued function of the system's input at the same time

and the state of the system at some known, initial time.

statement   In computer programming, a meaningful expression or generalized instruction in a source language.

state of the art   The level of advancement in an art or science at a given time, such that practicioners in the field have a reasonable probability of successful execution with current knowledge and technique.

static   1. Interference caused by natural electrical disturbances in the atmosphere, or the electromagnetic phenomena capable of causing such interference. 2. A form of information storage in shift registers and memories whereby information will be retained as long as power is applied. 3. Capable of maintaining the same state indefinitely (with power applied) without any change of condition. Not requiring continuous refreshing. 4. An electric charge which does not move, such as the bound charge on a capacitor plate. 5. Crackling or hissing sounds which disturb radio reception. 6. White flecks on a television screen which interfere with television reception. 7. Without movement. 8. Having no moving parts.

static charge   An electric charge held on the surface of an object, particularly on a dielectric.

static check   Of a computer, one or more tests of computing elements, their interconnections, or both, performed under nonfunctional conditions.

static device   As associated with electronics and other control or information handling circuits, the term static refers to devices with switching functions that have no moving parts.

static electricity   Electricity produced when an object such as a glass rod is rubbed with a piece of silk, acquiring the property of attracting small pieces of paper and fluff. The glass rod is said to be electrified, the name being derived from the Greek word for amber, which was found to possess the same property. During the rubbing, electrons are either removed from or added to the rubbed object. The glass is a nonconductor and so the electricity in it does not move—it is thus called static electricity. See also *electricity*.

static field A field that is present between the poles of either a permanent magnet or an electromagent that has a direct current passing through its coils.

static focus The focus attained when the electron beam of a cathode-ray tube is theoretically at rest or is at the position it would occupy if scanning energy were not applied.

staticize 1. In a computer, to convert serial or time-dependent parallel data into static form. 2. To retrieve an instruction and its operands from storage prior to its execution.

staticizer A storage device used in a computer for converting time—sequential information into static parallel information. A bistable switching device.

static memory A type of semiconductor memory where the basic storage element can be set to either of two states in which it will remain so long as the power stays on. See also *dynamic memory*.

static MOS array A circuit made up of MOS devices which does not require a clock signal.

static power conversion equipment Any equipment which converts electrical power from one form to another without the use of moving parts such as rotors or vibrators. Static implies the use of semiconductors.

static (precipitation) Radio static caused by the impact of electrically-charged rain drops or dust particles on a receiving antenna.

static shift register A shift register that uses logic flip-flops for storage. This technique, in integrated form, results in larger storage-cell size and therefore shorter shift register lengths. Its primary advantage is that information will be retained as long as the power supply is connected to the device. A minimum clock rate is not required; in fact, it can be unclocked.

static storage In computers, storage in which the information does not change position, e.g., electrostatic storage, flip-flop storage, binary magnetic core storage. Contrast to dynamic storage.

static switch A semiconductor switching device that has no moving parts.

stationary field Field in which the scalar (or vector) at any point does not change during the time interval under consideration. Also called *constant field*.

steady state The condition of a current or voltage when switching transients have ceased and the current or voltage has reached its regular unvarying level.

stenosis A narrowing or constriction.

step 1. One unit of movement of a stepping switch. A vertical step on a rotary step. 2. A pole step. To move the wipers of a stepping switch from one bank level to another (*vertical step*) on from one contact to another with same bank level (*rotary step*).

step down To decrease the value of an electrical quantity, such as a voltage, usually by means of a transformer or resistance.

step-down transformer A transformer used to reduce an alternating voltage. Its secondary winding has fewer turns; hence, its voltage is less than the primary voltage.

step function A function which is zero for all values of time prior to a certain instant and a constant for all values of time thereafter.

step-function response The characteristic curve or output plotted against time resulting from the input application of a step function. Also called *rise time*.

stepping relay A multiposition relay in which moving wiper contacts mate with successive sets of fixed contacts in a series of steps, moving from one step to the next in successive operations of the relay. Also called *rotary stepping switch* (or relay) or *stepping switch*.

stepping switch See *stepping relay*.

step-recovery diode Varactor in which forward voltage injects carriers across the junction, but before the carriers can combine, voltage reverses and carriers return to their origin in a group. The result is abrupt cessation of reverse current and a harmonic-rich waveform.

step up To increase the value of some electrical quantity, such as a voltage.

step-up transformer Transformer in which the energy transfer is from a low-voltage winding to a high-voltage winding or windings.

steradian Solid angle subtending an area on the surface of a sphere equal to the square of the radius; there are 4Π steradians in a sphere.

stereotaxic Pertaining to or characterized by precise positioning in space.

sterile Free from microorganisms.

sterile field See *field*.

sterilizer A device that destroys micro-

organisms by heat (wet steam under pressure at 120°C for 15 minutes).

sternal puncture Technique employed to obtain sample of bone marrow for investigation. A needle is inserted into the sternum under local anaesthesia, and a small amount of the marrow aspirated.

sternum The breastbone.

stertoric breathing Noisy respiration; snoring; a rattling sound occurring on inspiration, due to mucous obstruction of the larynx.

stethoscope Instrument devised by French physician René T. H. Laennec for the purpose of auscultating heart sounds and respiratory sounds.

sthenic Strong, active. Having or indicating, vigor.

stimulate To excite to functional activity.

stimulation 1. The application of an electric signal, a drug, or other vehicle for activating a tissue, organ, or organ system. 2. Addition of energy to a system, such as a maser or laser.

stimulus 1. Any chemical or physical change in the internal or external environment that produces a response. 2. An input signal to a nerve or muscle cell, or to a group of such cells.

stimulus-response curve (SR curve) A curve of action-potential size versus stimulus intensity.

Stokes-Adams syndrome The occurrence of cerebral symptoms, usually syncope, in patients as a result of a rhythm disturbance. During the attack, the electrocardiogram may reveal either ventricular standstill, ventricular fibrillation, ventricular tachycardia, or the slowing of the idioventricular impulse below a critical rate. Marked sinus bradycardia or atrial fibrillation with high degree of A.V. block may also be implicated. Where heart rate consistently is below 35 contractions per minute, a cardiac pacemaker is usually implanted to elevate rate to 60 beats per minute.

stoma   1. The mouth. 2. A small opening.

stomatitis   Inflammation of the oral mucosa.

-stomy   Suffix meaning to make an artificial opening.

stopcock   A valve for stopping or regulating the flow of fluids or gasses. Used in I.V. (intravenous) systems and in the hydraulic lines leading to pressure transducers.

stopping power   The reciprocal of the thickness of a solid that absorbs the same amount of alpha radiation as one cm of air.

storage   1. Device in which data can be stored and from which it can be obtained at a later time. The means of storing data may be chemical, electrical, or mechanical. 2. Device consisting of electronic, electrostatic, electrical hardware, or other elements into which data may be entered and from which data may be subsequently obtained as needed. Also called *memory*. 3. That part of a control system or computer into which information can be introduced, held, and then extracted at a later time. 4. In an oscilloscope, the ability to retain the image of an electrical event on the cathode-ray tube (CRT) for further analysis after that event has ceased. This image retention may be for only a few seconds with variable persistence storage, or it may be for hours with bistable storage.

storage access time   In a computer, the time that is required to transfer information from a storage location to the local storage register or other location where the information then becomes available for processing.

storage allocation   In a computer, assigning specific sections of memory to blocks of data or instructions.

storage buffer   1. A synchronizing element between two different forms of storage, usually between internal and external. 2. An input device in which information is assembled from external or secondary storage and stored ready for transfer to internal storage. 3. An output device into which information is copied from internal storage and held for transfer to secondary or external storage. Computation continues while transfers between buffer storage and secondary or internal storage or vice versa take place. 4. Any device which stores information temporarily during data transfers.

storage capacity   1. Amount of information that can be retained in a storage (or memory) device, often expressed as the number of words that can be retained (given the number of digits, and the base, of the standard word). When comparisons are made among devices using different bases and word lengths, it is customary to express the capacity in bits. Also called *capacity*. 2. The number of bits that a memory can hold; for example, a 1K semiconductor memory can store 1,000 bits (actually 1,024 bits), a 2K semiconductor memory can store 2,000 bits (actually 2,048 bits). Fixed memories usually contain instructions and, therefore, their capacity is sometimes expressed as the number of words of a certain length that it can hold. For example, 256×4 means that the memory can store 256 four-bit words, which makes it a 1K memory (1,024 bits). The same 1K memory could be a 128×8 memory. In either case however, a 1K memory is specified and not a 256×4 or 128×8 memory.

storage cell An elementary unit of storage, e.g., binary cell, decimal cell.

storage counter A counter in which a series of current pulses charge a capacitor with each pulse raising the voltage to a higher level. A comparator circuit determines when the capacitor voltage reaches a predetermined level. Special techniques are frequently used to linearize the charging curve of the capacitor.

storage cycle 1. Periodic sequence of events occurring when information is transferred to or from the storage device of a computer. 2. Storing, sensing, and regeneration from parts of the storage sequence.

storage dump See *dump*.

storage element Smallest part of a digital computer storage used for storing a single bit.

storage oscilloscope An instrument that has the ability to store a CRT display in order that it may be observed for any required time. This stored display may be instantly erased to make way for storage of a later event.

storage print In computers, a utility program that records the requested core image, core memory, or drum locations in absolute or symbolic form either on the line-printer or on the delayed printer tape.

storage register A special computer register which is used for both arithmetic and control functions. It has a capacity of one word and serves as a buffer between core storage and the central processing unit.

storage time A phenomenon that limits switching speed by increasing the time required to turn off a transistor that has been driven into saturation. It results from the fact that a heavily conducting transistor

has many excess charge carriers moving in the collector region. When the transistor is turned off by a base signal, collector current will continue for some time until all these excess carriers have been removed from the collector region.

store 1. A storage unit. A memory unit in which information can be accumulated until needed for future use. 2. To insert a binary digit into a register or memory unit. See also *memory storage*.

stored program A set of instructions in the computer memory specifying the operations to be performed and the location of the data upon which these operations are to be performed.

stored-program computer 1. A computer in which the instructions specifying the program to be performed are stored in the memory section along with the data to be operated on. 2. A computer which can alter its own instructions in storage as though they were data, and subsequently execute the altered instructions.

strain gauge 1. A device for measuring the expansion or contraction of an object under stress, comprising wires that change resistance with expansion or contraction. See also *load cell*. 2. A pressure transducer, as used in direct blood pressure measurement and mechanics of breathing studies.

stranded wire A conductor composed of a number of wires, electrically parallel, but which may be spirally wound together.

strangury Painful passing of urine.

stray capacitance Any of the small unintentional capacitances which exist between wires or components in a circuit, particularly capacitance to ground. Although not intentional parts of the circuit, they can seriously affect operation at higher frequencies.

stray-current corrosion Corrosion that results when a direct current from a battery or other external source causes a metal in contact with an electrolytic to become anodic with respect to another metal in contact with the same electrolyte. Accelerated corrosion will occur at the electrode where the current direction is from the metal to the electrolyte and will generally be in proportion to the total current.

stray magnetic field Stray magnetic flux from nearby transformers or inductors which can link with conductors or other inductors to produce undesired noise voltages.

stray radiation Radiation not serving any useful purpose. It includes direct radiation and secondary radiation from irradiated objects.

streptomycin An important antibiotic, and one of the primary drugs for treating tuberculosis.

stress A physical, chemical, or emotional factor that causes bodily or mental tension.

stria A narrow streak, stripe, band, or line distinguished by color, texture, or elevation from the tissue in which it is found.

striate Striped. Having fine linear markings.

striation The state or condition of being striated.

striated muscle Striped voluntary muscle.

stricture Abnormal narrowing of a passage, such as a vessel, duct, or the intestine.

stridor A harsh sound during breathing, caused by obstruction to the passage of air.

striking potential 1. Voltage required to start an electric arc or to ionize the gas between electrodes of gas-filled lamp. 2. Smallest grid-cathode potential value at which plate current begins flowing in a gas-filled triode.

-stringent Suffix meaning to draw tight, compress, bind, or cause pain.

strip-chart recorder An instrument which produces a hard-copy graphic record of physiologic waveforms, either bioelectric (such as the ECG, EEG, or EMG) or transduced (such as respiration, cardiovascular blood pressure, or the echocardiogram). Simple recorders, generally limited to uncomplicated tracing, employ a pen or heated stylus, driven by a galvanometer, to inscribe the waveforms on a moving chart paper. More complex multichannel recorders utilize a photo recording process, in which separate amplifier channels drive mirror galvanometers, or control the position and intensity of writing light spots on a cathode-ray tube. In either, the light cast upon the photosensitive paper leaves a latent trace of the waveform which can be made visible by appropriate chemical developing or, in the case of ultraviolet-sensitive paper, exposure to room lighting.

stroboscope 1. A device for producing a flashing light of controlled frequency, used to observe movements, vibrations, rotation, or the like, by synchronizing flash rate to the object's motion rate. 2. A device that can be tuned to the speed of a rotating object such as a motor shaft, causing the rotating member to appear to stand still (stop). The dial, if previously calibrated, reads in revolutions/min, so that speed can be noted instantly.

stroboscopic Utilizing or relating to a stroboscope.

stroke 1. Popular term for cerebral apoplexy; that is, the symptom complex resulting from hemorrhage into or upon the brain, or from embolism or thrombosis of the cerebral vessels. 2. A key-depressing operation in man-machine communication. 3. Straight or cursive portion of a letter or numeral. Also known as a *character stroke*.

stroke volume The quantity of blood pumped during each cardiac contraction. It is arrived at by determining the diastolic volume of the left ventricle, subtracting the volume of blood in the ventricle at the end of systole. In the adult, the average stroke volume is about 60 cc.

stroma The connective and supportive tissues of an organ, including nerve tissue and vascularization.

S-T segment That portion of the surface ECG complex which represents the period wherein all musculature of the ventricles is in a depolarized state.

stupor Unconsciousness, varying from partial to complete.

stylet 1. A thin probe used in surgery. 2. A slender wire inserted into a catheter to lend support during passage, or inserted into a hollow needle to clear the lumen.

styptic Agent which arrests bleeding.

sub A prefix denoting beneath or under.

subacute Less than acute, i.e., gradual in onset.

subassembly Two or more parts which form a portion of an assembly or a unit replaceable as a whole, but having a part or parts which are individually replaceable.

subcarrier A carrier band which is applied as modulation on another carrier wave.

subcarrier oscillator In a telemetry system, the oscillator which is modulated by the measurand or by the equivalent of the measurand in terms of changes in the transfer elements of a transducer.

subchannel In a telemetry system, the route required to convey the magnitude of a single subcommutated measurand.

subclavian Under the clavicle; thus, the subclavian artery and vein are vessels passing under the clavicle.

subclinical Without any obvious signs of the disease.

subcutaneous Beneath the skin.

subdermal electrodes Electrodes placed below the skin.

sublimation A defense mechanism, operating unconsciously, by which instinctual but consciously unacceptable drives are diverted into personally and socially acceptable channels.

subliminal Below the threshold of conscious perception.

sublingual Under the tongue.

subminiaturization The packaging of miniaturized components by special techniques which give increased volumetric efficiency.

subnanosecond Less than a nanosecond.

subprogram In a computer, a part of a larger program which can be compiled independently.

subroutine 1. The set of instructions necessary to direct the computer to carry out a well-defined mathematical or logical operation. 2. A subunit of a routine. A subroutine is often written in relative or symbolic coding even when the routine to which it belongs is not. 3. A portion of a routine that causes a computer to carry out a well-defined mathematical or logical operation. 4. A routine which is arranged so that control may be transferred to it from a master routine and so that, at the conclusion of the subroutine, control reverts to the master routine. (Such a subroutine is usually called a closed subroutine.) 5.

A single routine may simultaneously be both a subroutine with respect to another routine, and a master routine with respect to a third. Usually, control is transferred to a single subroutine from more than one place in the master routine. This avoids having to repeat the same sequence of instructions in different places in the master routine.

substrate 1. That part of an integrated circuit which acts as a support. It may be a ceramic insulator to which an integrated circuit is attached, or a semiconductor chip within which an integrated circuit is fabricated. Typical substrates bearing complete complex integrated circuits are 0.05 inch square. 2. Compound on which an enzyme acts.

subtractor An operational amplifier circuit in which the output is proportional to the difference between its two input voltages or between the net sums of its positive and negative inputs.

suffused Congested. Bloodshot.

sulfonamides Used in the treatment of infections of the respiratory and urinary tracts and in certain types of meningitis.

superciliary Having to do with the eyebrow.

supercilium The eyebrow.

superconduction The property of certain metals which cause them to lose all apparent electrical resistance when cooled to a few degrees above absolute zero (0 degree K or minus 459.7 degrees F). A current started in a superconducting loop will circulate indefinitely without measurable attenuation.

superego See *ego*.

super-emitron See *image iconoscope*.

superior Higher; the upper of two parts; situated above.

superior vena cava A major vein emptying into the right atrium of the heart. It receives and combines the venous bloodflows from several lesser veins serving the tissues and organs of the upper trunk, head, and the arms.

superregeneration Method used to produce greater regeneration, hence, greater amplification, than would otherwise be possible without oscillation. Positive feedback is utilized, to the point where the circuit reaches critical gain and is about to oscillate. At this point, gain is sharply reduced by quenching the stage (either by forced-limiting of stage bias, or by driving down gain with the signal of a low-frequency oscillator).

supersonic Traveling at a speed greater than the speed of sound (1,080 feet per second). Do not confuse this term with ultrasonic which describes sound at frequencies above the audio range.

supervisory program See *executive routine*.

supervisory routine See *executive routine*.

supine Lying face upwards; in the case of the forearm, having the palm uppermost.

supply voltage The voltage at which ac power is supplied for use by the customer. The customer may use it at the supply voltage, or step it down to a lower utilization voltage.

suppuration The formation of pus.

supra Prefix meaning above.

suprarenal Above the kidney. See *adrenal glands*.

supraventricular arrhythmia An arrhythmia originating above the ventricles, that is, either in the atria or in the A.V. nodal region.

supraventricular tachycardia A cardiac arrhythmia having its origin in a focus within the atrioventricular node. Each beat resembles a nodal premature beat, and P waves, if present, may precede, follow, or occur with a QRS complex. Beats are produced by retrograde conduction and are inverted in leads 2, 3, and a VF.

surface leakage The passage of current over the surface of an insulator, rather than through it. Surface leakage in new components is very low, but when a component is installed in equipment and exposed to dust, dirt, moisture, and other degrading environments, leakage current can increase and cause problems.

surface resistivity The resistance between opposite edges of a surface film one cm square. It is measured by determining the resistance between two straight conductors one cm apart, pressed upon the surface of a slab of the material. Water absorbent materials usually show lower resistivity than nonabsorbent ones.

surfactant A substance coating and reducing the surface tension of alveoli, helping to maintain their expansion.

SUS See *unilateral switch*.

sutures 1. Silk, thread, catgut, nylon, etc., used to sew a wound. 2. The union of flat bones by their margins.

Swan-Ganz catheter The multilumen, balloon-tipped catheter which can be advanced from arm vein to pulmonary artery without need for fluoroscopic guidance. It provides a means for measuring pulmonary artery and pulmonary wedge pressure, and for obtaining mixed venous blood. In some versions, the catheter is also equipped with a tiny bead thermistor, so positioned that it can detect blood temperature change, for measurement of cardiac output by the thermal dilution technique.

sweep circuit An oscillator circuit which provides a time base for cathode-ray presentations. It causes the cathode-ray spot to move across the cathode-ray screen as a linear function of time; then returns it abruptly to the starting point at the end of each sweep.

sweep frequency 1. Describes an oscillator whose output frequency is swept over a band of frequencies. 2. An audio or radio frequency which is repetitively varied over a band of frequencies. Provides a test signal for wide-band devices.

sweep-frequency generator A signal source capable of changing frequency automatically and in synchronism with a display device. The frequency sweep can be obtained either by electronic or mechanical means.

sweep generator See *sawtooth wave generator*.

sweep oscillator Oscillator used to develop a sawtooth voltage which can be amplified to deflect the electron beam of a cathode-ray tube. See also *sweep generator*.

switch A device that connects, disconnects, or transfers one or more circuits and is not designated as a controller, relay, or control valve. The term is also applied to the functions performed by switches.

switching diode Diode that provides essentially the same function as a switch. Below a specified applied voltage it has high resistance corresponding to an open switch while above that voltage it suddenly changes to the low-resistance state of a closed switch.

symbolic address In computers, arbitrary identification of a particular word, function, or other information without regard to the location of the information.

symbolic code Code that expresses programs in source language, i.e., by referring to storage locations and machine operations by symbolic names and addresses that are independent of their hardware-determined names and addresses. Also called *pseudocode*.

symbolic coding In digital computer programming, any coding system using symbols other than actual computer addresses so as to make programming easier.

symbolic-language programming In a computer, writing program instructions in a language which facilitates the translation of programs into binary code by making use of mnemonic convention. Also called *assembly language programming*.

symbolic logic A special computer or control system language composed of symbols that the instrumentation can accept and handle. Combinations of these symbols can be fed in to represent many complex operations. See also *mathematical logic*.

sympathectomy Surgical transection of sympathetic nerves usually with excision of part of the sympathetic chain.

symphysis Growing together of bones.

symptom Evidence of a disease or a disturbance of normal body function.

symptomatology A study of the symptoms of disease.

synapse 1. The point at which a nervous impulse is transferred from one neuron to another. See also *nervous system*. 2. Region where nerve cells communicate. There is no continuity between the neurons, and impulses are transmitted from one nerve cell to another by the passage of chemical messengers which stimulate the postsynaptic nerve cell. 3. The junction point of two neurons, across which a nerve impulse passes.

synaptic See *synapse*.

sync Slang expression meaning to synchronize, i.e., to cause to agree in rate or speed.

synchro A wound-rotor magnetic induction transducer capable of electrically transmitting or receiving angular positional information.

synchronism 1. Relationship between two or more periodic quantities of the same frequency when the phase difference between them is constant. 2. Applied to the synchronous motor, the condition under which the motor runs at a speed which is directly related to the frequency of the power applied to the motor and is not dependent upon other variables.

synchronization 1. Precise matching of two waves or functions. 2. In carrier, that degree of matching, in frequency, between the carrier used for modulation and the carrier used for demodulation which is sufficiently accurate to permit efficient functioning of the system. 3. Process of keeping the electron beam on the cathode-ray tube screen in the exact position relative to the scanning beam at the television transmitter.

synchronization pulses Pulses introduced by transmitting equipment into the receiving equipment to keep the two equipments operating in step.

synchronize To make sure that a level or pulse is presented to a system or subsystem at the correct time.

synchronized cardioversion See *cardioversion*.

synchronous 1. In step or in phase as applied to two devices or machines. 2. A term applied to a computer in which the performance of a sequence or operations is controlled by equally spaced clock signals or pulses. 3. Having a constant time interval between successive bits, characters or events.

synchronous computer An automatic digital computer in which all ordinary operations are controlled by equally spaced signals from a master clock.

synchronous generator A circuit designed to synchronize an externally generated signal with a train of clock pulses. The generator produces precisely one output pulse for each cycle of the input signal. The output pulse thus has a width equal to that of the period of the clock pulse train.

synchronous inputs Those terminals on a flip-flop through which data can be entered but only upon command of the clock. These inputs do not have direct control of the output such as those of a gate, but act only when the clock permits and commands. Called *JK inputs* or *ac set* and *reset inputs*.

synchronous logic A digital logic system in which logical operations are performed in synchronism with the clock pulse.

synchronous operation 1. The performance of a switching network, wherein all major circuits in the nework are switched simultaneously by a clock pulse generator. 2. Operation of digital circuits controlled by a clock pulse.

synchrotron A device for accelerating particles, ordinarily electrons, in a circular orbit in an increasing magnetic field applied in synchronism with the orbital motion. See *accelerator*.

synchysis Softening of the vitreous humor of the eye.

syndrome A collective group of signs and symptoms, typical of a particular disease.

synergy The working together of two or more agents, producing an effect which neither can produce alone.

-synia Suffix meaning pain.

synovial Of or pertaining to the viscous fluid secreted by synovial membranes, which serves to lubricate joints.

synovial fluid The liquid which lubricates the joints.

synovial membrane A membrane that lines a joint cavity but not covering articular surfaces.

synthesis The process of putting together parts to form a whole; the building up of complex substances by the union and interaction of simpler materials.

synthetic anticoagulants Used to prevent formation or extension of blood clots within arteries and veins.

syringomyelia Progressive degenerative disease affecting the brainstem and spinal cord in which the tracts of fibers subserving pain and temperature are mainly affected.

system 1. Any combination of components or structures joined together to perform a complete operational function or the functions of two or more subsystems. 2. A group of integrated circuits or other components interconnected to perform a single function or number of related functions. If further interconnected into a larger system, the individual elements are referred to as subsystems. 3. Any of the various organic groupings of the body which are separable by function (i.e., digestive system, cardiovascular system, lymphatic system).

systemic Pertaining to or affecting the body as a whole.

system master tapes Magnetic tapes containing programmed instructions necessary for preparing a computer prior to running programs.

systems flow chart A schematic representation of the flow of information through the components of a processing system.

systems software That collection of programs which controls the computer and helps people use the computer. This includes assemblers, editors, debuggers, operating systems, compilers, loaders, and other utility programs.

systole Period of contraction, of the heart, especially that of the ventricles. It coincides with the interval between the first and second heart sound, during which blood is forced into the aorta and the pulmonary circulation. See also *diastole*.

systolic Referring to that portion of the cardiac cycle when the ventricles are in a state of contraction.

systolic murmur Adventitious sound heard during systole.

systolic pressure The maximum value of blood pressure, occurring during contraction of the ventricles of the heart. See also *diastolic pressure*.

T

T Abbreviation for tera, 10^{12}.

tab See *land*.

table A collection of data in which each item is uniquely identified by a label, by its position relative to the other items, or by some other means.

tachograph An instrument that measures and records rate.

tachy Prefix meaning fast.

tachycardia Abnormally fast heart action (typically in excess of 100 beats per minute).

tachypnea Abnormal increase in the respiratory rate.

tactile Pertaining to the sensation of touch.

talus The ankle.

tamponade Compression. Usually cardiac tamponade, in which the action of the heart is impeded by the presence of fluid in the pericardial sac.

tank circuit A circuit composed of lumped inductance and capacitance capable of storing electrical energy over a band of frequencies centered on the resonant frequency of the circuit.

tap 1. A branch circuit. 2. A connection brought out from an intermediate point of a coil or winding. 3. To make a tapped connection.

tape Thin plastic ribbon of which one side is coated with iron oxide particles to receive magnetic impressions from a recording head; or, if already recorded, to induce signals in a playback head. Tape is regularly wound on reels or packed in magazines or in cartride form.

tape deck 1. An assembly of a magnetic transcription system consisting of a tape-moving mechanism (the tape transport) and a head assembly. Some decks also include recording and playback preamplifiers; these properly are called tape recorders. Some have playback-only preamplifiers; these have no standard name but are often called tape players. 2. A specialized assembly of a digital storage system having components comparable to those in (1), but configured for the storage and accessing of data in bit form, with means for automatic control by a computer.

tape head The transducer on a tape recorder past which the magnetic tape runs during record or replay. It applies a controlled magnetic field to the tape during the recording process and senses the field patterns stored in the tape to provide electrical output during playback.

tape punch Electromechanical device which punches holes in a paper tape corresponding to the electrical signals of a teletypewriter or data code.

taper The type of resistance distribution in a potentiometer, whether linear, left-hand, or right-hand. With linear taper, resistance increases uniformly with rotation of the knob. With left-hand taper, there is little change at the left end, but an increasing rate of change as the knob is turned to the right. A right-hand taper starts with a large change which gets progressively smaller as the knob is turned to the right.

tape recorder A mechanical-electrical device that records signals by selective magnetization of iron oxide particles which are coated on a thin plastic tape.

tape speed The speed, in inches per second, at which magnetic tape is drawn past the recording and playback heads on a tape recorder. Normal speeds are 1-7/8, 3-3/4, and 7-1/2 inches per second, but 15 and 30 inches per second may be used where very high-fidelity reproduction is required.

tape station See *tape unit.*

tape transport See *transport.*

tape unit A device containing a tape drive, together with reading and writing heads and associated controls. Also called *tape station.*

tapped resistor A fixed resistor which has one or more intermediate terminals between those at the two ends. Used as a voltage divider.

target theory The probability of survival for a biological system can be expressed by equations that represent random distribution of ionizations through a sensitive region or target volume of that system. The theory suggests that survival can occur only if ionization has not occurred anywhere within the entire molecule under study. This leads to an exponential dose-response curve. The dose-response curve can be more complex if this requirement is not met.

tarsus 1. The seven small bones of the foot. 2. The cartilaginous framework of the eyelid.

taxis Locomotion of an organism or a cell in response to a directional stimulus.

technology 1. The sum total of all the tools and techniques through which mankind has added leverage to human effort since the beginning of time. 2. Systematic knowledge and its applications to industrial processes.

telautograph A record communication system in which writing movement at the transmitting end causes corresponding movement of a writing instrument at the receiving end.

telecardiophone An amplifying stethoscope which permits heart sounds to be ausculated at a distance.

telecommunication 1. Pertaining to the art and science of telecommunication. 2. Communication over a distance by telephone, radio, telegraph, etc. 3. Any transmission, emission, or reception of signs, signals, writing, images, or sounds, or intelligence of any nature by wire, radio, visual, or other electromagnetic systems.

4. Data transmission between a computing system and remotely located devices via a unit that performs the necessary format conversion and controls the rate of transmission.

telectrocardiograph A device for transmission and remote reception of electrocardiographic signals.

telemeter 1. An electronic instrument that senses and measures a quantity, as that of speed, heart rate, respiration, temperature, pressure, or radiation, then transmits radio signals to a distant station. 2. Measuring, transmitting, and receiving apparatus for indicating or recording the value of a quantity at a distance. 3. To transmit the value of a measured quantity to a remote point by radio signals. See also *telemetering* and *telemetry*.

telemetering A measurement accomplished with the aid of intermediate means which allow perception, recording, or interpretation of data at a distance from a primary sensor. The most widely employed interpretation of telemetering restricts its significance to data transmitted by means of electromagnetic propagation.

telemetry 1. The transmission of measurements, obtained by automatic sensors and the like, over communications channels. 2. The practice of transmitting and receiving the measurement of a variable for readout or other uses. In medicine, term is most commonly applied to systems which makes possible the continuous ECG monitoring of ambulatory postcoronary patients.

telephone Electromechanical device which converts sound waves into electrical signals. These are conveyed over conductors or radio links to the distant point to be reconverted to sound waves which are a replica of the original. The science of telephony deals with the provision of inter-connecting cable networks, junction cables between towns, cities, and continents, and automatic systems of connecting telephone subscribers together.

teleprinter See *teletypewriter*.

teleprocessing A form of information handling in which a data processing system utilizes communication facilities (originally an IBM trademark).

teletype Trademark of Teletype Corporation usually referring to a series of different types of teleprinter equipment, such as tape punches, reperforators, and page printers utilized in communications systems.

teletypewriter A keyboard machine which can transmit and receive alphabetical, numeric, and certain control (nonprinting) characters as trains or electrical pulses on two wires. Attachments can be fitted for punching paper tape and printing on a roll of paper at the same time. Also used for reading punched tape and printing the message that is read. (Also known as a *teleprinter*.)

television standards Television in the United States as defined by the Federal Communications Commission:

Lines per frame	525
Interlace	2 to 1
Frames per second	30
Video bandwidth	4.2 MHz
Channel bandwidth	6.0 MHz
Audio modulation	F3
Polarization	Horizontal
Horizontal scanning (B & W)	15,750 Hz
Horizontal scanning (color)	15,734 Hz
Vertical scanning (B & W)	60 Hz
Vertical scanning (color)	59.94 Hz

Television in Canada, Mexico, Panama, Japan, Korea, Iran, Saudi Arabia, and the Netherlands Antilles uses these same standards.

temperature coefficient A decimal fraction which expresses how much some property of a material, such as expansion, resistance, or capacitance, will change for each degree Centigrade change in temperature.

temperature-compensated zener diode Positive temperature-coefficient reversed-bias zener diode (PN junction) connected in series with one or more negative-temperature coefficient forward-biased diodes within a single package.

temperature compensation 1. The process of making some important characteristic of a circuit independent of temperature changes. 2. The practice of having some component in a circuit change with temperature in such a way as to compensate for temperature changes in other components.

temporal-lobe epilepsy A disease characterized by recurring seizures resulting from lesions within the temporal lobe of the brain.

temporary storage Internal storage locations in a computer reserved for intermediate and partial results.

tenacious Adhesive, hanging-on.

ten basic penicillins Antibacterial drugs that are used primarily to treat infections of the skin, respiratory tract, the genitourinary tract, and other diseases.

tendinitis Inflammation of a tendon.

tendon 1. A cord of fibrous white tissue by which a muscle is attached to a bone or other structure. 2. A steel bar or tensioned wire anchored to formed concrete and allowed to regain its initial length, so as to induce greater strength in the concrete.

tenorrhaphy Operation to suture a tendon.

tenosynovitis Inflammation in the sheath of a tendon.

tenotomy Cutting a tendon.

tension The act of stretching; being stretched.

tensor A muscle which stretches.

tentorium cerebelli The part of the dura mater which separates the cerebral hemispheres from the cerebellum in the brain.

tera (T) Prefix meaning 10^{12} (one million million).

terahertz (THz) One million megahertz, or 10^{12} hertz.

teratogen Agent inducing fetal malformation.

teres Round and smooth.

terminal 1. The point at which electrical conductors are connected to a circuit or to each other. 2. A device for facilitating connections to a circuit or component. 3. A point in a computing system or communication network at which data can either enter or leave. 4. A device that permits gaining access to a central computer—can be a teletype, electric typewriter, a graphic terminal, or other device. 5. Any device capable of sending and/or receiving information over a communications channel; the means by which data are entered into a computer system and by which the decisions of the system are communicated to the environment it affects. A wide variety of terminal devices have been built, including teleprinters, special keyboards, light displays, cathode-ray tubes, thermocouples, pressure gauges and other instrumentation, radar units and telephones.

terminal lug A threaded stud fitted with

nuts under which a wire can be clamped.

terminal point See *land.*

terminated 1. The condition of a wire or cable pair which is connected to (terminated on) binding posts or a terminal block. 2. The condition of a circuit which is connected to a network which has the same impedance the circuit would have if it were infinitely long.

ternary code Code in which each code element may be any one of three distinct kinds or values.

test clip A testing lead device consisting of a pair of spring-closing metal jaws which can be clamped on any terminal to which a temporary test connection is required. Sometimes equipped with needle points for making connection through the insulation.

test lead A flexible insulated lead wire which usually has a test prod on one end. Ordinarily used for making temporary electrical connections between meters or test instruments and a circuit under test. The insulation is normally rubber and standard colors are red and black.

test prod A needle point with an insulating handle, connected to a test lead and used to make temporary test connections.

tetanus Contraction of a muscle without twitching; also, acute infectious disease known as *lockjaw,* manifested by tonic contractions of certain muscles. Caused by the toxin of Clostridium tetani bacillus, and is usually fatal if left untreated.

tetany Tonic spasm of muscles.

tetracycline derivatives Used in the treatment of a wide variety of diseases caused by many different types of microorganisms.

tetracyclines Group of chemically related antibiotics which include Aureomycin (chlortetracycline), Terramycin (oxytetra-

cycline), and Achromycin (tetracycline hydrochloride), et al. Useful in the therapy of infections caused by gram-positive and gram-negative bacteria, as well as ricksettiae and large viruses, such as the PLT–Bedsonia group.

tetrahedron 1. A solid body having four faces. 2. A lead system in vectorcardiography.

TFD Abbreviation for target-film distance.

T flip-flop A type of flip-flop used in ripple counters whose outputs change state each time the input signal voltage falls from 1 to 0 and remains unchanged when the input-signal voltage rises from 0 to 1. There is one change in output state for every two changes in input signal. Thus, the frequency of the output is half the frequency of the input. Also called *binary.*

TGC Abbreviation for time gain compensation. In echographic studies, a means for altering the amplification of echoes as a function of the depth of the interfaces from which the echoes originate. For example, the echo obtained from the anterior pericardium is approximately 20,000 times the amplitude obtained from the posterior pericardium, due to attenuation of the ultrasound beam and echo. TGC allows reduction of gain for echoes near the transducer, while allowing increase of gain for echoes further from the transducer. The effect is a uniform presentation of echoes, undistorted by the effects of attenuation.

thalidomide Nonbarbiturate sedative found to induce fetal malformations if taken by pregnant women.

thalofide cell A photoconducting cell which has thallium oxysulfide as the light sensitive agent.

thaumatrope Instrument demonstrating persistence of visual impressions: images printed on opposite sides of a card appear to fuse when it is rotated rapidly so that each side is presented alternately.

theca A sheath. Examples are the meninges of the spinal cord, and the synovial sheaths of the flexor tendons of the fingers.

thenar Relating to the palm of the hand at the base of the thumb.

therapeutics The branch of medical science dealing with the treatment of disease.

therapy Treatment of a disease. The long history of medicine has established many forms of therapy, including biologic, chemical, electrical, physical, thermal, radiation, shock, endocrine, glandular, psychological, roentgen ray, and many others. Adopting the specific course of therapy for a particular disease is the fundamental concern of the practicing physician.

therm A unit of heat energy equal to 100,000 British thermal units.

thermal agitation 1. Movements of the free electrons in a material, due to heat. In a conductor, they produce minute pulses of current. When these occur in the conductors of a resonant circuit at the input of a high-gain amplifier, the fluctuations are amplified together with signal currents and heard as noise. 2. Minute voltages arising from random electron motion, due to heat, which is a function of absolute temperature, expressed in degrees Kelvin.

thermal breakdown Breakdown caused by decomposition or melting where the temperature rise results from the applied electric stress.

thermal coefficient of resistance The change in the resistivity of a substance due to the effects of temperature only. Usually expressed in ohms per degree change in temperature.

thermal conduction The transport of thermal energy by processes having rates proportional to the temperature gradient and excluding those processes involving a net mass flow.

thermal conductivity The quantity of heat that flows in a unit of time through a unit area of unit thickness having unit difference of temperature between its faces. (The rate can be measured in $Btu/ft^2/s$ in degrees F, or in $gram\text{-}cal/cm^2/s$ in degrees C.)

thermal cutout Heat sensitive switch that automatically opens the circuit of an electrical device when the operating temperature exceeds a safe value.

thermal electrical noise An energy exchange between a thermal environment and an electrical circuit.

thermal EMF The electromotive force generated when the junction of two dissimilar metals is heated. See *thermocouple*.

thermal equilibrium Condition in which a system and its surroundings are at the same temperature.

thermal generation In a semiconductor, the creation of a hole and a free electron by freeing a bound electron through the addition of heat energy.

thermal junction See *thermocouple*.

thermal noise Noise generated by random thermal motion of changed particles.

thermal pneumocardiograph A pneumocardiograph that measures temperature changes betwen air entering and leaving the respiratory passages.

thermal relay A slow-operate relay actuated by the effect of heat from a resistance-wire coil on a bimetallic strip.

thermal resistance The resistance of a substance to conduction of heat.

thermal runaway A regenerative condition in a transistor, where heating at the collector junction causes collector current to increase, which in turn causes more heating, etc. The temperature can rapidly approach levels that are destructive to the transistor.

thermal shock The result occurring when a material is subjected to rapid and wide-range changes in temperature in a testing effort to discover its ability to withstand heat and cold.

thermal time delay relay Relay in which functioning times are determined by the thermal storage capacity of the actuator (usually a bimetal), the critical-operate temperature, the power input, and the thermal insulation.

thermion Ion, either positive or negative, which has been emitted from a heated body. Negative thermions are electrons (*thermoelectrons*).

thermionic Pertaining to the emission of electrons or ions from an incandescent body.

thermionic diode Diode electron tube which has a heated cathode.

thermionic emission Electron emission from a solid body as a result of its elevated temperature.

thermistor 1. An electrical resistor made of a material having resistance which varies sharply in a known manner with temperature. 2. A solid-state semiconducting device used as a temperature and flow sensor, and for temperature compensation. It is made by sintering mixtures of the oxide powders of various metals. It is produced in many shapes, such as beads, discs, flakes, washers, and rods, to which contact wires are at-

tached. As its temperature is increased, the electrical resistance of the thermistor decreases. The associated temperature coefficient of resistance is extremely high, usually nonlinear, and negative.

thermocompression bonding The joining together of two materials without an intermediate material by the application of pressure and heat in the absence of electrical current.

thermocouple 1. A thermoelectric couple (a union of two conductors—as bars of dissimilar metals joined at their extremities—for producing a thermoelectric current) used to measure temperature differences. 2. A temperature-sensing device consisting of two dissimilar metal wires joined together at both ends to deliberately incur the Seebeck effect. One wire of the circuit is opened (both output terminals are the same metal), and a small EMF appears across the terminals upon application of heat at one junction. The magnitude of this EMF is proportional to the difference in temperature between the measuring junction (located at the point of measurement) and the reference junction (usually located in the measuring instrument or in a tumbler of ice water). Special alloys have been developed to serve as the dissimilar metals, such as constantan, alumel, chromel, and platinum-rhodium. These alloys are paired with each other or with pure-element metals such as copper and iron.

thermocouple contact Contact of special material used in connectors employed in thermocouple applications. Materials frequently used are iron, constantan, copper, chromel, alumel, and others.

thermocouple instrument Instrument in which one or more thermojunctions are heated directly or indirectly by an electric current, and supply a direct current which flows through the coil of a suitable direct-current mechanism, such as one of the permanent-magnet, moving-coil type.

thermocouple wire Wire drawn from special metals or alloys and calibrated to established specifications for use as thermocouple pair. For example: iron, constantan, alumel.

thermodynamics The study of the relationship between heat and other forms of energy, as well as the behavior of matter with respect to thermal energy. Generalizations summarized by three laws of thermodynamics include the conservation of energy and the trend found in nature for the availability and distribution of energy. Also encompasses the study of radiation.

thermoelectric 1. Relating to phenomena involving relations between the temperature and the electrical condition in a metal or in contacting metals. 2. Pertaining to the direct conversion of electrical energy to heat or vice versa.

thermoelectric effect A flow of current in a circuit constructed of two metals in a manner that one junction is maintained at a higher temperature than the other. The magnitude and direction of the current depends on the metals and the temperature difference between the junctions. Also called *Seebeck effect*.

thermoelectric generator A semiconductor device using thermocouple action to convert heat directly into electricity. When heat is applied to a PN junction, electrons migrate to the cooler end of the N-type material and holes to the cooler end of the P-type material. To provide a useful voltage, a number of such junctions are connected in series, forming a thermopile.

thermoelectricity The flow of electricity produced when two thermoelectric junc-

tions in the same series circuit are at different temperatures, one heated and one cooled.

thermoelectric junction Any of the junctions between dissimilar wires in a thermocouple or thermopile, where a potential difference is created by the application of heat.

thermoelectromotive force The algebraic sum of the Peltier EMF at thermocouple junction and the Thomson EMFs in the thermocouple metals.

thermoelement Device consisting of a thermocouple and a heating element arranged for measuring small currents.

thermogram The recorded results of a thermographic study, producing a graphic, pictorial plot of heat emitted by areas of the body. High-resolution images are possible from a series of thermal scans using a bolometer as the sensor. See *thermography*.

thermojunction battery Nuclear-type battery which converts heat into electrical energy directly by the thermoelectric or Seebeck effect.

thermolabile A substance which undergoes change with temperature.

thermoluminescence The production of light from heat.

thermomagnetic Pertaining to two effects: (a) the heating of a body by changes in magnetization, and (b) the destruction of magnetization by heating a magnet.

thermometer An instrument for indicating temperatures.

thermopile Battery of thermocouples, consisting of alternate rods of antimony and bismuth suitably joined and connected to a galvanometer. Thermopiles are used for the measurement of heat where thermometers cannot be employed.

thermoplastic The property of resin to soften under application of heat, be rigid at normal temperature, and resoften on each application of heat.

thermoplastic resins Resins that can be remelted and rehardened many times by successive heating and cooling cycles.

thermoregulator A device to maintain or regulate temperature. The words thermostat and thermoregulator are often used interchangeably. It is customary, however, to employ the term thermoregulator in connection with such devices as are extremely sensitive or extremely accurate and, particularly, for devices resembling a mercury-in-glass thermometer with sealed-in electrodes, in which the rising and falling column of mercury makes and breaks an electrical circuit.

thermorelay See *thermostat*.

thermosetting A classification of resin which cures by chemical reaction when heated and when cured, cannot be re-softened by heating.

thermosetting plastic A type of plastic in which an irreversible chemical reaction takes place while it is being molded under heat and pressure. This type of plastic cannot be reheated or softened.

thermosetting resins Resins that cure, set, or harden using heat and/or catalysts to form an end product which cannot be melted by further heating.

thermostat Mechanism to convert expansion of heated metal or fluid into movement and power sufficient to operate small devices, control electric circuits, or small valves, etc. Can be set to operate at definite temperatures. Also called *thermorelay*.

thermostatic switch A temperature-operated switch that receives its operating energy by thermal conduction or convec-

tion from the device being controlled or operated.

thermostat materials Pairs of metals having widely different coefficients of expansion. When they are joined together, a temperature change makes one material expand more than the other, causing a change in shape of the combination that can be used to make or break a circuit as temperature changes. The principal combinations used are nickel and iron, chromium and iron, and pairs of various alloys (bimetals).

thermotherapy Treatment by applying heat.

theta Brain wave signals whose frequency is approximately 3.5 to 7.5 Hz. The associated mental state is fuzzy, unreal, uncertainty, daydreaming, ambiguity.

Thévenin's theorem A fundamental tool for evaluation of network performance; it states: at any given frequency, the current flowing in an impedance connected to two terminals of a linear, bilateral network containing generators of the same frequency is equal to the current flowing in the same impedance when its terminals are connected to a voltage generator of equivalent voltage with the impedance removed, and whose series impedance is the impedance of the network, with all generators replaced by their internal impedances.

thick film A film with a thickness of the order of one one-thousandth of an inch. Generally applied by painting, dipping, silk-screening, or similar processes. The term thick film is used as a specific contrast to the more familiar term, thin film. Metal glaze is a thick film.

thick-film circuit A microelectronic assembly in which the passive circuit elements and their interconnections are defined on a ceramic substrate by using the silk-screen process. The active elements are added as discrete chips.

thick-film hybrid integrated circuit The physical realization of a hybrid integrated circuit fabricated on a thick-film network.

thick-film process A method in which electronic circuit elements, resistors, conductors, capacitors, etc., are produced by applying specially formulated pastes to a ceramic substrate in a defined pattern and sequence. The substrate containing the green pattern is fired at a relatively high temperature to mature the circuit elements and bond them integrally to the substrate. The term thick film distinguishes this technology from thin-film practice in which circuit elements are made by evaporation or sputtering in a high-vacuum environment. Circuits produced by this technology are identified by several descriptive phrases; thick-film passive circuits or thick-film hybrid circuits when active devices are included.

thin film 1. A deposited (normally by sputtering or evaporation) film of conductive or insulating material which may be patterned to form electronic components and conductors on a substrate or used as insulation material between successive layers of components. 2. A coating up to a thickness of about 10^4 angstrom.

thin-film capacitor A capacitor utilizing a metal oxide as the dielectric or insulating material. Both the electrodes and the dielectric are deposited in layers on a substrate. This device is usually associated with microelectronics, integrated, and thin-film circuits.

thin-film circuit Integrated circuit consisting of a passive substrate on which the various passive elements (resistors and capacitors) are deposited in the form of thin patterned films of conductive or non-conductive material. Active components (transistors and diodes) are attached separately as individually packaged devices or in unpackaged (chip) form, or may be formed integrally by thin-film techniques. See also *hybrid thin-film circuit*.

thin-film formation A process that is either additive by pattern formation through masks, or subtractive by selective etching of predeposited films from a substrate.

thin-film resistor 1. Any metal film deposited on a substrate by vacuum evaporating or sputtering techniques. 2. A metal film 500 to 1000 Å thick.

thin-film semiconductor Semiconductor produced by the deposition of an appropriate single-crystal layer on a suitable insulator.

thixotropic A term used to describe materials which are jelly-like at rest but are transformed into a liquid condition when stirred, agitated, or otherwise disturbed.

thixotropy The property by which some compositions become semisolid at rest and liquefy again on agitation.

thoracic Pertaining to the thorax.

thoracic surgery Surgical procedures carried out within the chest.

thoracotomy Operation of opening the thorax.

thorax 1. The part of the body of man and other mammals between the neck and the abdomen. 2. The chest; the cavity which holds the heart and lungs, defined as that volume between the neck and the diaphragm.

three-layer diode See *trigger diode*.

three-wire power system A system of electric supply comprised of three conductors. One of these (known as the neutral wire) is maintained at a potential midway between the potential of the other two conductors which are referred to as the outer conductors. Part of the load may be connected directly between the two outer conductors and the remainder divided as evenly as possible into two parts, each of which is connected between the neutral wire, and one outer conductor. There are thus two distinct supply voltages, one being twice the other.

threshold 1. A term having various interpretations in medical physics, depending upon its application. It implies a minimum value necessary for a certain response (i.e., voltage, frequency, signal). 2. Generally, the minimum value of a signal that can be detected by the system or sensor under consideration, or the minimum stimulus required to effectuate response.

threshold audiogram See *audiogram*.

threshold dose The minimum dose that will produce a detectable effect of any degree.

threshold of sensitivity The smallest stimulus or signal that will result in a detectable output. This phrase is frequently used to describe the voltage point at which a monitor or event marker will trigger.

threshold value The minimum level for which there is a measurable output.

threshold voltage 1. The input voltage level at which a binary logic circuit switches from one logic level (i.e., state) to the other. 2. The voltage at which a **PN** junction begins to pass a current; in a solid-state lamp, the voltage at which light is first emitted.

thrombectomy Removal of a blood clot.

thrombin Essential factor required in the blood clotting mechanism.

thromboembolism An embolism resulting

from a dislodged thrombus or part of a thrombus.

thrombolytic An agency which breaks down clots.

thrombophlebitis Condition caused by the inflammation of a vein complicated by the formation of an intravascular blood clot (thrombus). Circulation is obstructed in the affected area, usually the legs.

thrombosis Formation, development, or presence of a blood clot inside an artery or vein. This condition can be serious if it affects the blood vessels of vital organs, such as the brain, heart, or lungs.

thrombus 1. A clot of blood found in a vessel. It usually occurs on the wall of a vessel where the endothelium is damaged. 2. A clot of blood formed within a blood vessel and remaining attached to its place of origin.

-thymia Suffix meaning thymus gland or mind.

thyratron A hot-cathode gas tube in which one or more control electrodes can initiate a current flow, but thereafter cannot control it.

thyristor A name given to many semiconductor devices including: reverse blocking triodes (SCRs), bidirectional triodes (Triacs), four-layer diodes, gated bilateral switches, gate turn-off switches, and light activated SCRs. A bistable semiconductor device of three or more junctions of which at least one can switch from a blocking state to a conducting state within a single quadrant of its current voltage characteristics.

thyristor rectifier A power rectifier which uses a thyristor as the rectifier and voltage controlling element.

thyrite A material used in voltage-dependent resistors. The resistance of thyrite decreases as the voltage applied to it increases, making it useful for protecting devices from high-voltage transients.

thyroid cartilage The large cartilage of the larynx forming the Adam's apple.

thyroidectomy Operative removal of the thyroid gland.

thyrotoxic myopathy A chronic disease resulting in muscular atrophy. It is associated with hyperthyroidism.

thyroxine The active secretion of the thyroid gland. A substance rich in iodine. Chemically it is $C_{15}H_{11}I_4NO_4$. Used in the form of sodium salt for replacement therapy in cases of hyperthyroidism or absent thyroid function.

THz Abbreviation for terahertz (10^{12} hertz).

tibia The inner and usually larger of the two bones of the vertebrate hind limb between the knee and ankle.

tibial See *tibia*.

tidal air That air which is inspired and expired during normal breathing.

tidal volume The volume of gas inspired or expired during each quiet respiration cycle.

time base 1. A reference time signal recorded at given intervals with the information signal. 2. The axis of chart motion. 3. The abscissa of many plotted curves.

timebase See *sawtooth wave generator*.

time constant 1. The time required for an exponential quality to change by an amount equal to 0.632 times the total change required to reach steady state. 2. The number of seconds required for a capacitor in an RC circuit to reach 63.2 percent of full charge after the application of a step-function voltage.

time constant of fall Time required for a pulse to fall from 70.7 percent to 26.0 percent of its maximum amplitude, excluding spike.

time constant of rise Time required for a pulse to rise from 26.0 percent to 70.7 percent of its maximum amplitude, excluding spike.

time delay 1. Total elapsed time or lag required for a given command to be effected after giving the command. 2. The time required for a signal to travel between two points, on a circuit or through space.

time delay relay Relay in which there is an appreciable interval of time between energizing of the coil and movement of the armature, or between deenergizing of the armature. Also called *slow-action relay*. (Examples are slow-operating relays and slow release relays.)

time division multiplex A system which enables the transmitting of a number of signals over a single common path by transmitting them sequentially at different instants of time.

time division multiplexer A device which samples all data input from different low-speed devices, and retransmits all the samples in an equal amount of time.

time division multiplexing 1. A signaling method characterized by the sequential and noninterfering transmission of more than one signal in a communication channel. Signals from all terminal locations are distinguished from one another by each signal occupying a different position in time with reference to synchronizing signals. 2. A system of multiplexing in which channels are established by automatically connecting terminals, one at a time, at regular intervals.

time frame 1. In telemetry, the time

period containing all elements between corresponding points of two successive reference markers. 2. In data recording, the events lying between two recorded time reference markers.

time gate   A transducer which has an output only during selected time intervals.

time lapse VTR   A videotape recorder which provides a continuous record of events over a period of from 12 to 48 hours on one reel of videotape.

time-sharing   1. A method of operation in which a computer facility is shared by several users for different purposes at (apparently) the same time. Although the computer actually services each user in sequence, the high speed of the computer makes it appear that the users are all handled simultaneously. 2. A means of making more efficient use of a facility by allowing more than one using activity access to the facility on a sequential basis.

timing signal   Any signal recorded simultaneously with data to provide a time index.

tinnitus   A ringing, humming, or hissing sound in the ears, subjective or functional in origin.

-tion   Suffix meaning act or state of.

tissue   An aggregation of similarly specialized cells united in the performance of a particular function.

tissue turgor   The condition of normal tissue fullness and resilience.

titration   Quantitative analysis by volume by means of standard solutions.

-tocia   Suffix meaning birth.

tocography   Method of recording alterations in the intrauterine pressure.

toggle   1. Another term for a flip-flop. It implies that the flip-flop will change state

at the receipt of a clock pulse. Mainly used to describe flip-flops connected as a counter, where the process of changing state is described as toggling. 2. Using switches to enter data into the computer memory. 3. To switch between two states, as in a flip-flop. See *flip-flop.*

toggle rate   Twice the frequency at which a flip-flop completes a full cycle encompassing both states. Usually used to denote the maximum input frequency that a flip-flop can follow.

toggle switch   A switch with a projecting lever whose movement through a small arc opens or closes one or more electric circuits.

tolerance   1. An amount of permissible variation from a standard, often expressed in percent. 2. The ability to endure without ill effect.

tolerance dose   Maximum level that can be tolerated by an individual or material within a specific period of time with the expectation of no harmful effects to the person or material.

-tome   Suffix meaning a cutting instrument.

tomography   Technique of radiography whereby a uniplanar focus is obtained by means of differential movement of the tissue in relation to the X-ray beam. The effect is like that of rotating the body about a point so that only one region is in focus. In practice, the patient remains still while the apparatus moves.

-tomy   Suffix meaning cutting into, or incision.

tone   1. A sound wave capable of exciting an auditory sensation having pitch. 2. An electrical wave of audio frequency. 3. A sound that is distinct and identifiable by its regularity of frequency, or constant pitch.

It may be a single frequency or several frequencies which have a fixed harmonic relation. 4. The condition of a muscle tissue.

tonometer Instrument for measuring tension such as that used to measure intraocular tension.

tonsillectomy Operative removal of the tonsils.

tonus 1. The degree of muscular contraction when not undergoing shortening. 2. The state of slight normal muscle contraction due to motor nerve reflexes of low frequency.

top hat A form of semiconductor package which resembles a formal or top hat.

topical Pertaining to local external application.

toroid An inductor or transformer wound on a doughnut-shaped magnetic core. All loading coils are wound as balanced toroids.

torso The human trunk.

total emissivity The ratio of radiation emitted by a surface to the radiation emitted by the surface of a black body under identical conditions. Important conditions which effect emissivity of a material are surface finish, color, temperature, and wavelength or radiation. Emissivity may be expressed for radiation of a single wavelength (*monochromatic emissivity*), for total radiation of a specified range of wavelengths (*total spectral emissivity*), or for total radiation of all wavelengths (total emissivity).

total-harmonic distortion Ratio of the power at the fundamental frequency, measured at the output of the transmission system considered, to the power of all harmonics observed at the output of the system because of its nonlinearity when a single frequency signal of specified power is applied to the input of the system. It is expressed in decibels.

total lung capacity The volume of gas contained within the lungs at the end of a maximum inspiration.

touch voltage The potential difference between a grounded metallic structure and a point on the earth's surface equal to the normal maximum horizontal reach (approximately three feet).

tourniquet A device to compress blood vessels in order to inhibit circulation.

toxic Pertaining to a poison; harmful to the body. Combining form meaning poison.

toxin 1. A poison, usually of bacterial origin. 2. Any poisonous substance.

toxoid A nonpoisonous modification of a toxin. Sometimes used to immunize against disease.

trabecula A septum extending into an organ from its capsule or wall.

trace 1. Pattern that appears on the screen of a cathode-ray tube. 2. Visible line or lines appearing on the screen of a cathode-ray tube as a result of the deflection of the electron stream. 3. The written record appearing on a strip chart.

trace elements Mineral substances whose presence in minute amounts in the diet is necessary for the maintenance of health, e.g., cobalt, copper, manganese.

tracer Radioactive isotope or substance containing a radioactive isotope which enables the substance to be traced in metabolic systems.

trachea 1. The main trunk of the system of tubes by which air passes to and from the lungs. 2. The windpipe; the air passage from the larynx to the bronchi. See *bronchi*.

tracheostomy The formation of an opening into the trachea to provide an accessory airway.

tracheotomy A surgical incision into the trachea.

tracks See *channels*.

track width The width of the track on a recording tape, corresponding to a given record gap. The most common track widths encountered in longitudinal recording are 48 to 50 mils, several such tracks being accommodated on a half-inch-wide tape.

traction The act of drawing, as in applying a force along the axis of a bone.

trans Prefix meaning across.

transceiver A combination of transmitting and receiving equipment in the same housing, in which some or all of the components are used jointly for both transmitting and receiving.

transconductor An active or passive network whose short-circuit output current is a specific, accurately-known, linear or nonlinear function of the input voltage, thereby establishing a predetermined relationship between input voltage and output current.

transducer A device for converting one form of energy into another form of energy, e.g., a pneumatic or hydraulic signal into an electric signal.

transfer characteristic 1. Relation, usually shown by a graph, between the voltage of one electrode and the current to another electrode, with all other electrode voltages being maintained constant. 2. Function which, multiplied by an input magnitude, will give a resulting output magnitude. 3. Relation between the illumination on a camera tube and the corresponding output-signal current, under specified conditions of illumination.

transfer constant Of an electric transducer, the arithmetic mean of the natural logarithm of the ratio of input to output voltages, and the natural logarithm of the ratio of input to output currents when the transducer is terminated in its image impedances. Also called *image transfer constant, transfer factor*.

transfer factor See *transfer constant*.

transfer function 1. A relationship between one system variable and another that enables the second variable to be determined from the first. 2. A mathematical expression that describes the relationship between the values of a set of conditions at two different times, typically, at the beginning and end of a process. 3. The mathematical expression that relates the output and input of any characteristics of a filter as a function of frequency. The function usually is complex and therefore usually contains components corresponding to both attenuation and phase.

transfer impedance Between any two pairs of terminals of a network, the ratio of a potential difference applied at one pair of terminals to the resultant current at the other pair of terminals, all terminals being terminated in any specified manner.

transfer rate The rate at which data may be transferred between the computer registers and storage, input, or output devices. Usually stated in number of characters per second.

transformer A device with two or more windings linked by a common magnetic circuit. Variations of current in a primary winding are converted by mutual induction into variations of current and voltage in a secondary winding.

transformer coupling Coupling of circuits by means of a transformer.

transformer loss Ratio of signal power that an ideal transformer of the same impedance ratio would deliver to the load impedance to signal power that the actual transformer delivers to the load impedance. This ratio is usually expressed in decibels.

transformer voltage ratio Ratio of the root-mean-square primary terminal voltage to the root-mean-square secondary terminal voltage under specified conditions of load.

transient 1. Instantaneous surge of voltage or current which occurs as the result of a change from one steady-state condition to another. 2. Phenomenon which takes place in a system owing to a sudden change in conditions and which persists for a relatively short time after the change has occurred. 3. Distinct line or series of lines perpendicular to the direction of scanning produced in the recorded copy immediately following a sudden change in density. 4. Damped oscillatory quantity occurring in the output of a system as a result of a sudden change in input.

transient distortion See *distortion*.

transient response The response of an amplifier or circuit to high-frequency components of a signal. Evaluated by the response to a square-wave test signal.

transistance The characteristic of an electric element which controls voltages or currents so as to accomplish gain or switching action in a circuit. Examples of the physical realization of transistance occur in transistors, diodes, saturable reactors, etc.

transistor A semiconductor device which provides solid-state amplification, switching, and rectification of electrical current as a result of effects occurring at junctions between N-type and P-type material. An N-type semiconductor material is one with an excess of free electrons in its structure. A P-type has a deficiency of electrons, termed an excess of holes. Silicon is the main material used—with impurities introduced to determine the conductivity type.

transistor base Region which lies between an emitter and a collector of a transistor, into which minority carriers are injected.

transistor chip Unencapsulated transistor element of very small size used in microcircuits.

transistor dissipation The power dissipated in the form of heat by the collector. The difference between the power supplied to the collector and the power delivered by the transistor to the load.

transistor resistor logic (TRL) An early form of logic circuit design that was employed prior to the advent of monolithic circuits. The input elements are resistors; however, unlike RTL there is only one active output transistor. In discrete circuit designs, this logic form was the least expensive since it required a minimum number (1) of active devices.

transistor transistor logic (TTL or T²L) A logic circuit design similar to DTL, with the diode inputs replaced by a multiple emitter transistor. In a 4-input DTL gate, there are 4 diodes at the input. A 4-input TTL gate will have 4 emitters of a single transistor as the input element. TTL gates using NPN transistors are positive level NAND gates or negative level NOR gates. Also known as *multi-emitter transistor logic*.

transition region Region around a PN junction in which the majority carriers of each side diffuse across the junction to recombine with their respective counterparts.

transit time In an electron tube, the time required for an electron to pass from the cathode to the plate.

translate To change information from one computer language to another.

translation The changing of programmed information from one form to another.

transmigration The passage of cells through a membrane.

transmission line Any circuit (open wire, paired cable, or coaxial cable) used to transfer energy from one point to another.

transmittance The ratio of the radiant power emitted by a body to the total radiant power received by the body.

transmitter A device which converts a signal into some transmittable form of energy and transmits it from one location to another. In the field of instrumentation, the signal converted is usually the measured value of some physical quantity such as temperature, pressure, flow, or liquid level.

transmutation The process in which one species of atom is transformed into another by a nuclear reaction.

transplantation Operation to remove a portion of tissue from one part of the body to another, or from a donor's body to that of a recipient.

transponder 1. An electrical device which, when it receives a signal, automatically transmits a corresponding signal. 2. A combined radio receiver and transmitter which automatically broadcasts identifiable signals when it receives the proper signal or interrogation.

transport Sometimes referred to as *tape transport*, the mechanical drive system of the tape deck.

transudation Oozing of fluid through a membrane or from a tissue.

transvenous pacemaker The introduction of a catheter electrode into the right ventricle from a vein, usually the right external jugular, for purposes of cardiac pacing.

trap 1. An unprogrammed conditional jump to a known location, automatically activated by hardware, with the location from which the jump occurred, recorded. 2. A tuned circuit that eliminates a given signal or keeps an unwanted signal out of a given circuit without blocking the desired signal.

trapezium First bone in second row of carpal bones.

trapping In a computer, instructions that cause a central processing unit to initiate an internal interrupt which transfers control to a subroutine which stimulates the desired operation of the instruction. Also, the subroutine can be changed, thereby changing the operation on the instruction.

trauma Injury, inflicted more or less suddenly, by some physical agent.

tri Prefix meaning three.

triac Formal name is bidirectional triode thyristor. A thyristor that can be triggered into conduction in either direction. Terminals are main terminal 2, main terminal 1, and gate.

triad 1. Group of three bits or three pulses, usually in sequence on one wire or simultaneously on three wires. 2. One of thousands of phosphor-dot groupings on the screen of a color CRT.

tribo A prefix meaning, or resulting from, or pertaining to, friction.

triboelectricity Electrostatic charges generated as a result of friction between two types of material. See also *electricity*.

triboelectric series A listing of materials which can be electrified by rubbing one with the other. Also called *electrostatic series*.

triceps Certain muscles with three heads, especially the one at the back of the arm which extends the elbow.

trickle charge   A continuous charge of a storage battery at a very low rate.

tricuspid valve   The valve which admits blood from the right atrium into the right ventricle of the heart, just as contraction commences. See *heart valve*.

trigeminal nerve   The fifth cranial nerve, arising in the medulla oblongata, spinal cord, and pons. It has three divisions: mandibular, maxillary, and optholmic, and innervates the face, providing sensory and motor functions. See *cranial nerves*.

trigeminal neuralgia   Pain in the face, of unknown cause. The distribution is confined to branches of the trigeminal nerve. The pain is paroxysmal and precipitated by mild stimuli such as washing the face or eating.

trigger   1. To start action, by means of one circuit, in another circuit which then functions for a period of time under its own control. 2. Short pulse, either positive or negative, which can be used to set into motion a chain of events. 3. A timing pulse which initiates logic circuit operations.

trigger action   Initiation of main current flow instantaneously by a weak controlling impulse in a device.

trigger diode   1. Symmetrical three-layer avalanche diode used in activating silicon controlled rectifiers. It has a symmetrical switching mode, and hence fires when the breakover voltage is exceeded in either polarity. Also called *diac*. 2. A two-terminal voltage-controlled device exhibiting a bilateral negative-resistance characteristic. The device has symmetrical switching voltages ranging from 20 to 40 volts and is specifically designed for use as a trigger in ac power control circuits such as those using triacs.

triggered blocking oscillator   A blocking oscillator that can be reset to its starting condition by the application of a trigger voltage.

triggering level   The instantaneous level of a triggering signal at which a trigger is to be generated. Also, the name of the control which selects the level.

triggering signal   The signal from which a trigger is derived.

trigistor   A PNPN device with a gating control acting as a fast-acting switch; similar in nature to a thyratron.

trimester   A three-month period.

trimmer   A small adjustable circuit element connected in series or parallel with a circuit element of the same kind that its adjustment sets the combination of the two to a desired value.

trimming potentiometer   An electrical mechanical device with three terminals. Two terminals are connected to the ends of a resistive element and one terminal is connected to a movable conductive contact which slides over the element, thus allowing the input voltage to be divided as a function of the mechanical input. It can function as either a voltage divider or rheostat.

triode   Three-electrode thermionic tube invented in 1907 by Lee De Forest (1873–1961) and used as a signal amplifier. It consists of a cathode, anode, and control grid (a screen of open mesh or a spiral of wire arranged between these) in an evacuated envelope. The grid is quite near the cathode and the electron current flowing from it to the anode can be controlled by the application of a small electrical (negative) potential. The amplifying property of the triode is due to the largeness of the changes in anode current caused by small changes in grid potential. The amplification factor (μ) of the tube is measured by the ratio of the anode and grid-voltage changes required to produce the same changes in anode current. When signals are applied to the grid, the changes in anode current flowing through the tube's load resistance cause voltage changes across it which are amplified replicas of the input signals.

triode laser Gas laser, the light output of which may be modulated by signal voltages applied to an integral grid.

triode thyristor See *silicon controlled rectifier*.

trocar A surgical instrument for puncturing cavities to remove fluid entrapped within. Consists of a hollow tube (or *cannula*) with a concentric bayonet-pointed *stylet*. After insertion of the trocar, the stylet can be removed so that fluid can be withdrawn through the tube.

trochlear nerve The fourth cranial nerve, arising in the midbrain. It is a paired nerve, which innervates the musculature of the eyes.

-trophic Suffix meaning nourishment.

-trophy Suffix meaning nourishment.

-tropia Suffix meaning turn or react.

troubleshooting Locating and diagnosing malfunctions or breakdowns in equipment by means of systematic checking or analysis.

truth table 1. A tabulation relating all output logic levels of a digital circuit to all possible combinations of input logic levels in such a way as to completely characterize the circuit functions. 2. A table used either to describe or design a simple logic element. A list of the input conditions possible is made; the output conditions which will result, or which are required, are arranged alongside. The logic expressions may then be read off.

TSD Abbreviation for target-skin distance.

T²L Abbreviation for transistor transistor logic.

TTL Abbreviation for transistor transistor logic.

TTL (or T²L) compatibility A term used to describe the capabilities of a MOS device to drive, or be driven by, biopolar circuitry.

tubal Relating to a tube, and especially to an oviduct.

tube An elongated, hollow, cylindrical structure.

tubercle 1. A small eminence. 2. The small grayish nodule which is the specific lesion of the tubercle bacillus.

tumor A swelling or growth of new tissue; it develops independently of surrounding structures and serves no specific function of its own. Also called *neoplasm*.

tuned circuit A circuit which will resonate at a selected frequency. The circuit contains both capacitance and inductance, one of which is adjustable.

tuned filter An arrangement of electronic components which either attenuates signals at a particular frequency and passes signals at other frequencies, or vice versa.

tuned reed frequency meter See *vibrating-reed meter*.

tuner A component that receives radio broadcasts and converts them into audio signals. It may be built on a separate chassis, or as part of a receiver.

tunica A term applied to several membranes, e.g., tunica vaginalis, the serous coat of the testicle.

tunnel diode A semiconductor PN diode in which, under proper conditions, electrons tunnel across the junctions at the speed of light, contrasted to the slow motion of the electrical charge carriers in conventional transistors. The tunnel diode exhibits negative resistance characteristics at one point in its operation; that is, as the voltage rises the current falls.

tunnel effect   Piercing of a rectangular potential barrier in a semiconductor by a particle that does not have sufficient energy to go over the barrier.

tunneling   See *tunnel diode.*

tunnel rectifier   Tunnel diode having a relatively low peak-current rating as compared with other tunnel diodes used in memory-circuit applications.

turbinectomy   Operation to excise a turbinate bone.

turgid   Swollen; distended.

turn-on time   The time required for an output to turn on (sink current, to ground output, to go to 0 volts). It is the propagation time of an appropriate input signal to cause the output to go to 0 volts.

turns ratio   In a transformer, the number of turns in the secondary winding divided by the number of turns in the primary winding. Under normal circumstances, it is also the voltage ratio.

T wave   That portion of the surface ECG waves which follows the QRS complex, signaling that the ventricles have repolarized and are again ready to contract.

twisted pair   A pair of insulated conductors spirally twisted together. Has many uses, such as for jumpers or cross connections.

two-address   In a computer, having the property that each complete instruction includes operation and specifies the location of two registers usually one containing an operand and the other the result of the operation.

two-conductor jack   Receptacle having two through circuits, tip and sleeve.

two-phase   Describing an alternating current circuit in which there are two sinusoidal voltages differing in phase by 90 electrical degrees, or one-quarter of a cycle.

two's complement arithmetic   A method in which subtraction is performed by means of adding the two's complement of one number to the number it is to be subtracted from. Two's complement is formed by adding one to the one's complement of the given binary number.

two-track recording   One quarter-inch wide tape, the arrangement by which only two channels of sound may be recorded, either as a stereo pair in one direction or as separate monophonic tracks (usually in opposite directions).

two-track tape   See *half-track tape.*

tympanic cavity   See *tympanum.*

tympanic membrane   The membrane separating the middle from the external ear, commonly called the *eardrum.*

tympanites   A distended state of the abdomen caused by gas in the intestines.

tympanoplasty   Operation to reconstruct sound-conducting mechanism in the middle ear.

tympanum   Also called tympanic cavity. A part of the middle ear, and comprises a cavity in the temporal bone deep to the tympanic membrane.

u Abbreviation for the twelfth letter, μ (mu) of the Greek alphabet. 2. Abbreviation for micro (one-millionth).

UGI Abbreviation for upper gastrointestinal.

UJT Abbreviation for unijunction transistor.

UL Abbreviation for Underwriters' Laboratories, Inc.

ulcer An open lesion on the surface of the skin or mucous membrane.

ulcerative Pertaining to ulceration.

ulcerative colitis A disease with inflammation and ulceration of the colon. There is diarrhea, and mucus and blood are passed in the stools. The patient is anemic. The disease may be mild or severe and pathogenic organisms appear not to cause it, though emotional stress seems to precipitate it.

ulna The inner bone of the forearm.

ulnar An artery, a vein, and a nerve running beside the ulna.

ultra Prefix meaning beyond.

ultrasonic Describing frequencies higher than can be heard by the human ear, therefore above 20,000 hertz. Ultrasonic waves are used in ultrasonoscopes and echocardiographs to obtain sonar-like images of such structures as the eye, brain, and heart. The importance of these applications in medicine is growing, because they are noninvasive, and, unlike fluoroscopy and X-ray, do not subject the patient to hazardous radiation.

ultrasonic cleaning The cleaning of durable articles by immersing them in a solvent which is violently agitated by a sound wave having a frequency of 20 to 30 kilohertz.

ultrasonic diagnosis A method of obtaining information from within the body in a visual presentation without employing ionizing radiation. It differs from X-ray in that the form of energy used in high-frequency sound, which is inaudible. The sound is transmitted in very brief pulses followed by relatively lengthy silent intervals. Also, in contrast to X-ray techniques, in which the film is placed behind the tissue being examined, ultrasonic information is picked up at the original point of transmission in the form of echoes from internal structures. These returning echoes are converted to electrical energy and displayed or recorded in either a static or dynamic pattern, as desired, either on a cathode-ray tube screen or on a graphic recorder chart.

Diagnostic ultrasound equipment detects differences in acoustic impedance at tissue interfaces along its route of travel. Echoes are received not only from soft-tissue-to-bone interfaces but also from many soft-tissue-to-soft-tissue and soft-tissue-to-fluid interfaces, which is one of the unique capabilities of ultrasonic diagnosis. Some applications of ultrasonics in diagnosis have been visualization of tumors, detection of diseased heart valves, echoencephalography, detection of pericardial effusion, fetal cephalometry, detection and localization of renal calculi, localization of foreign bodies in soft tissue, detection of common-bile-duct stones, and detection (with extraction) of foreign bodies in the eye.

ultrasonic disintegrator An apparatus for using pressure waves produced by an ultrasonic generator to tear cells apart.

ultrasonic generator Device that produces signals of ultrasonic frequency.

ultrasonic reflectoscope A trade name for a medical ultrasound instrument used in ultrasonic imaging and echocardiography.

ultrasonic scanner Ultrasonic diagnostic instrument for visualizing tissue/organ interfaces and presenting cross-sectional images.

ultrasonograph Instrument which displays an echosonogram or transonogram on a permanent record form, such as a chart.

ultrasonography A medical diagnostic technique in which ultrasonic energy is directed into the body and returning

echoes are detected in sequence. See *echoencephalography*.

ultrasonoscope Instrument which displays an echosonogram on an oscilloscope display, usually with auxiliary output to a chart recording instrument, for M-mode recording.

ultraviolet (UV) Zone of invisible radiations beyond the violet end of the spectrum of visible radiation. Since these wavelengths are shorter than the visible, their photons have more energy, enough to initiate some chemical and biological reactions, and to degrade most plastics.

ultraviolet sources Source of radiation of frequencies above that of visible violet (from 4,000 angstroms to about 400 angstroms, the beginning of X-ray region).

unbalanced line A line or circuit which is asymmetric with respect to ground and/or other conductors, usually having ground serve as one of the circuit conductors; e.g., a coaxial line.

unblanking pulse Voltage applied to a cathode-ray tube to overcome bias and cause trace to be visible.

unbonded strain gauge transducer A transducer made up of resistance strain gauges arranged in a Wheatstone bridge with two or four active arms which respond to an applied force. The unbonded strain gauge filaments are suspended in the air and are activated by a mechanism attached to the diaphragm column or other force-responding element.

unciform The hook-shaped bone of the wrist.

unconditional jump In a digital computer, instruction which interrupts the normal process of obtaining instructions in an ordered sequence, and specifies the address from which the next instruction must be taken.

unconscious That part of the mind the content of which is only rarely subject to awareness. It is the repository for knowledge that has never been conscious or that may have been conscious briefly and was then repressed.

Underwriters' Laboratories, Inc. (UL) A nonprofit laboratory sponsored by the National Board of Fire Underwriters which examines and tests devices, materials, and systems whose action may affect casualty, fire, and life hazards for the purpose of establishing safety standards on types of equipment or components.

undulant Wavelike. Increasing and decreasing in a cyclic fashion.

uni Prefix meaning single or one.

unidirectional transducer Transducer that measures stimuli in only one direction from a reference zero or rest position. Also called *unilateral transducer*.

uniground A single point in an electrical system connected to ground to eliminate noise currents. Also called *single-point ground*.

unijunction transistor (UJT) Formerly called a double base diode. A three-terminal semiconductor device which exhibits a stable negative resistance characteristic between two of its terminals. It is this negative resistance feature that makes the UJT suitable for the applications with which it is associated. Thyristor trigger circuits, oscillator circuits, timing circuits, bistable circuits, etc.

unilateral Denoting that a certain organ or point is located on only one side of the body.

unilateral switch A semiconductor device much like a miniature SCR; switches at a fixed voltage determined by its internal construction. Also called *SUS*.

unilateral transducer See *unidirectional transducer*.

uninterruptable power systems A solid-state power conversion system to provide regulated ac power to critical loads. System provides uninterrupted power even during brownouts and blackouts.

unipolar 1. Having but one pole, polarity, or direction. When applied to amplifiers or power supplies, it means that the output can vary in only one polarity from zero and, therefore, must always contain a dc component. 2. In electrocardiography, the technique devised by Wilson and his associates, in which the potential at an exploring electrode is differentially compared to that of an indifferent point, obtained by resistively summing the potentials at the right arm, left arm, and left leg. 3. Refers to transistors in which the working current flows through only one type of semiconductor material, either N-type or P-type. In unipolar transistors, the working current consists of either positive or negative electrical charges, but never both. All MOS IC transistors are unipolar. Unipolar (MOS) IC transistors operate slower than bipolar IC transistors, but take up much less space on a chip and are much more economical to manufacture.

unipolar pulse A pulse which has appreciable amplitude in one direction only.

unipolar transistor See *FET*.

unit The lowest standard quantity in any system of measurement. The unit of electrical energy, for example, is the kilowatt-hour.

unit charge Electrical charge which will exert a repelling force of one dyne on an equal and like charge one centimeter away in a vacuum, assuming that each charge is concentrated at a point.

unit magnetic pole Two equal magnetic poles of the same sign have unit value when they repel each other with a force of one dyne if placed one centimeter apart in a vacuum.

unity coupling The condition existing in a perfect transformer, where all of the magnetic flux generated by the primary winding passes through the secondary winding.

unloading circuit In an analog computer, a computing element or combination of computing elements capable of reproducing or amplifying a given voltage signal while drawing negligible current from the voltage source, thus decreasing the load errors.

unwind In computers, to rearrange and code a sequence of instructions to eliminate excessive operations. See also *automatic coding*.

update 1. To modify a master file in a computer with current information according to a specified procedure. 2. To refresh a display from a memory, so as to create the impression of a continuous presentation. 3. To periodically replace displayed or stored information with new information.

up-down counter A counter which can count in an ascending or descending order depending upon the logic at the up-down inputs. Also called *reversible counter*.

uremia An elevation of the urea concentration in the blood above its normal value of about 30 mg/100 ml. Also used to describe a syndrome resulting from impaired renal function which may be due to renal or extrarenal causes, e.g., dehydration. There are disturbances of salt and water balance and acid-base equilibrium in addition to elevation of the blood urea.

ureter The vessel between the kidney and the bladder, down which the urine passes.

urethra The vessel betwen the bladder and the exterior through which the urine is discharged.

urethrography X-ray examination of the urethra by means of retrograde injection of a radiopaque dye.

urethroscope An instrument for viewing the interior of the urethra.

urethrotomy Incision of the urethra to remedy stricture; the instrument used being a urethrotome.

-uria Suffix meaning urine, or condition of urine.

urinalysis (urine analysis) Examination of urine constituents, both normal (urea, uric acid, total nitrogen, ammonia, chlorides, phosphate, and others) and abnormal (albumin, glucose, acetone, bile, blood, cells, and bacteria).

urticaria Raised, itching, white patches on the skin; commonly called *nettlerash* or *hives*.

USASCII Abbreviation for USA Standard Code for Information Interchange. The standard code, using a coded character set consisting of 7-bit coded characters (bits including parity check), used for information interchange among data processing communication systems and associated equipment. The USASCII set consists of control characters and graphic characters. Synonymous with *ASCII*.

utero 1. A combining form from uterus, as uterovaginal. 2. The Latin dative of uterus, as in utero, in the uterus.

uterus The hollow muscular organ in females which is the place of nourishment of the embryo and fetus.

uuf Abbreviation for the obsolete term micromicrofarad, superseded by picofarad (pf).

UV Abbreviation for ultraviolet.

V Abbreviation for volts.

vaccines Immunizing preparations for the prevention or modification of infectious diseases, such as poliomyelitis, measles, and mumps when injected into a subject.

vacuole Specialized region within a cell surrounded by plasma membrane.

vacuum 1. A space devoid of molecules, thus representing an area of negative pressure. 2. The condition of a space, such as within an electron tube, from which air has been pumped until the small amount of gas remaining will not adversely affect operation of the device which is in the vacuum.

vacuum-tight See *hermetic*.

vagina The passage leading from the cervix to the vulva. The lower end of this canal is formed by the hymen.

vagotomy Surgical division of the vagus nerve sometimes performed on patients with peptic or duodenal ulcers.

vagus nerve The tenth cranial nerve, arising in the medulla oblongata. It is also called the *pneumogastric nerve*. The vagus nerve provides motor and sensory functions for the many structures which it innervates in the neck, abdomen, and thorax. It is associated with breathing, swallowing, digestion, cardiac, and hepatic functions.

valence A measure of the ability of an atom to combine directly with other atoms. Dependent upon the number and arrangement of the electrons in the outermost shell of the atom.

valence band In a semiconductor, the band of energies just below the conduction band, separated from it by the forbidden gap.

valence electron 1. An electron in the conduction band of a semiconductor, where it is free to move under the influence of an electric field. 2. Any of the electrons which comprise the outer shell of an atom, which can enter into chemical reactions and can move through influence of an electric field. It is the movement of valence electrons which constitutes the current flow in an N-type semiconductor.

validity check 1. A check that a code group is actually a character of the particular code in use. 2. Computer check of input data, based on known limits for variables in given fields.

valvotomy Incision into a valve, especially heart valve. The purpose of the operation is to widen the orifice of a stenosed valve.

Van de Graaff generator Generator of very high voltage (to millions of volts), commonly used for particle acceleration in nuclear studies. See *accelerator*.

varactor A PN junction semiconductor diode which functions as a voltage-variable capacitor. The capacitance of a varacter is dependent upon applied voltage amplitude and polarity. This property is useful in the construction of voltage-controlled oscillators and tuned amplifiers.

variable A condition which is subject to change and which is assigned a set of values during an experiment.

variable capacitor A capacitor whose capacitance can be varied by varying the separation between a pair of plates, or by varying the depth of insertion of interleaved plates. Most widely used for tuning radio frequency circuits.

variable-frequency oscillator (VFO) A stable oscillator whose frequency may be changed or adjusted.

variable-reluctance Describing any of a number of transducers, such as pressure transducers, microphones, phonograph pickups, etc., in which the input is made to vary the reluctance of a magnetic path.

variable-resistance transducer A transducer that produces an output related to an internal change in resistance. The internal resistance change is secondary to some variation in the properties of the received energy. Flood-flow, temperature, and respiration-monitoring units often contain variable-resistance transducers as sensors.

varices Dilated, twisted veins.

varicose veins Abnormally distended and lengthened superficial veins caused by slowing and obstruction of the normal blood backflow. Varicose veins are most commonly observed in the legs, anus and rectum (hemorrhoids), and scrotum (varicocele).

varicotomy Excision of varicose vein.

varistor A two-terminal device containing a pair of diodes, either copper-oxide or silicon, connected in parallel, but with opposing polarities. Used as a voltage limiting device since its resistance drops as the applied voltage is increased.

varix An enlarged and tortuous vein.

varmeter Volt-ampere reactive meter; instrument designed to measure the reactive component of power (volts times amperes, reactive).

vas A vessel, or duct of the body; as vas deferens, the duct of the testis.

vascular 1. Possessing a blood supply. 2. Concerning the blood vessels and the supply of blood.

vasectomy Removal of a part of the vas deferens.

vasoconstriction A reduction in the caliber of the lumens of the blood vessels, owing to a reaction of the vasomotor nerves.

vasodilation An increase in the caliber of the lumens of the blood vessels by vasomotor action.

vasomotor Any agent that affects the caliber of a blood vessel.

vasomotor nerves The nerves which control the caliber of the lumens in the blood vessels, through control of musculature which acts upon the vessels.

vasospasm Spasm of the blood vessels.

vector 1. A quantity that has magnitude, direction, and sense and that is commonly represented by a directed line segment whose length represents the magnitude and whose orientation in space represents the direction. 2. Any living organism which transmits a pathogen from one individual to another.

vectorcardiograph An instrument that measures both magnitude and direction of heart signals by displaying ECG signals on any desired set of axes, usually X, Y, and Z. In practice, electrodes may be placed so as to provide three-dimensional ECG planes on the body: frontal, transverse, and saggital. Vectorcardiographic plots of these ECG planes provide a means for localizing the site of an infarct, after the method of Grant.

vectorcardiography A systematic method for registering the magnitude and direction of cardiac electrical forces using a cathode-ray oscilloscope, or oscillographic recorder, which converts two scalar ECG inputs from differently situated electrodes on the subject into a loop (x-y) plot, representing the instantaneous electrical forces accompanying a cardiac contraction. The loop plot is highly sensitive to phase differences between the two ECG leads, thus providing a sensitive indicator of very small changes in the ECG wave which would be difficult to detect in the traditional scalar record.

vectorscope 1. A device utilized in viewing vectorcardiograms. 2. An oscilloscope having a circular timebase which has extreme stability. Time delay between two signals can be checked, because the phase difference at any particular frequency can be related to time difference.

vein A vessel carrying the blood to the heart.

velocity of light The velocity of light in a vacuum is 2,997,925 meters per second or 186,280 miles per second. For rough calculations, the figure of 3,000,000 meters per second is generally used.

velocity of sound The velocity of sound varies with temperature and the transmission medium. Here are several examples:

Medium	Degrees Centigrade	Velocity ft/second
Air	0	1088
Air	20	1129
Water	15	4714
Lead	20	4026
Steel	20	16360
Glass	20	18000
Tissue	37	5052

velocity transducer A transducer which generates an output proportional to imparted velocities.

venepuncture Inserting a needle into a vein.

venereal Relating to sexual intercourse.

venereal diseases Infectious diseases transmitted during sexual intercourse, e.g., *gonorrhea, syphilis.*

venereology The study of venereal disease.

venesection Blood-letting. A vein is opened and blood drained off from it. Once used as a panacea for almost any ailment. There are very few present-day indications for venesection.

venography X-ray examination of veins.

venous Relating to the veins.

venous hemorrhage See *hemorrhage.*

venous thrombosis The formation of a thrombus within a vein.

ventilation 1. The supply of fresh air. 2. The process of breathing.

ventral A position more toward the front

of some point of reference (opposite of dorsal).

ventricle Small cavity or pouch. The cavities in the interior of the brain of higher animals are termed ventricles, and the heart contains two ventricles.

ventricular See *ventricle*.

ventricular aneurism An abnormal, localized dilation of a wall, occurring through weakening of the surrounding tissue.

ventricular fibrillation Rapid and uncoordinated contraction of the heart's ventricular muscle tissues (typically in the range of 300 to 500 per minute). A principal causative agent is aberrant conduction through infarcted tissue, which causes depolarization during the heart's vulnerable period. Also, an irritable focus may cause spontaneous depolarization leading to fibrillation. Because the ventricular wall is quivering rather than contracting forcefully, cardiac output (bloodflow) decreases practically to zero. This arrhythmia is fatal unless terminated within a matter of minutes.

ventricular premature beat (VPB) A contraction of the heart's ventricular muscles, occurring prior to the normal time. The usual cause is an irritable focus in either ventricle, which produces excitatory waves triggering spontaneous contraction of surrounding musculature. Because this upsets the usual, coordinated sequence of contraction, there is some risk that the muscles will continue to depolarize at random, in the lethal arrhythmia termed ventricular fibrillation. Modern coronary care techniques are aimed at reducing the incidence of VPBs, to lower the risk of fibrillation in the patient recuperating from acute myocardial infarction.

ventricular systole The portion of the

cardiac cycle during which the ventricles reach peak contraction and discharge the blood under maximum pressure. Electrically, it is reported by the occurrence of the QRS complex in the electrocardiogram. However, peak pressure occurs some 20 to 200 milliseconds later.

ventricular tachycardia A rapid series of impulses originating in an ectopic focus in either ventricle of the heart. The paroxysms typically have a rate of 150 to 200 beats per minute and start and stop abruptly. The ECG taken after a paroxysm will typically show evidence of myocardial ischemia.

ventriculography Fluoroscopic examination of the ventricles of the heart. A radiopaque dye is introduced into the ventricles enabling their size and position to be observed. The use of video equipment and on-line computing equipment makes possible the determination of ventricular dimensions and volume changes throughout the cardiac cycle, for dynamic evaluation of the heart's pumping efficiency.

ventriculostomy Operation to open a ventricle of the brain, usually in order to construct a bypass when the flow of cerebrospinal fluid is obstructed.

venule A small vein; especially one of the tiny veins connecting the capillary bed with the larger systemic veins.

verifier Of computers, a device on which a record can be compared or tested for identity character by character with a retranscription or copy as it is being prepared.

vernier Any device, control, or scale used to obtain fine adjustment or more accurate measurement than the main measuring scale.

vertebrae The thirty-three small bones

which form the backbone, or spinal column.

vertex The top of the head.

vertigo Dizziness or giddiness.

vesical Relating to the bladder.

vesicle A small bladderlike cavity.

vestibule 1. A small cavity of the ear into which the cochlea opens. 2. The space between the labia minora.

vestigial Rudimentary. Bearing a trace of something now vanished or degenerate.

viability See *viable.*

viable Capable of living; especially, born with such form and development of organs as to be normally capable of living.

vibrating reed meter Frequency meter consisting of a row of steel reeds, each having a different natural frequency. All are excited by an electromagnet that is fed with the alternating current whose frequency is to be measured. The reed, whose frequency corresponds most nearly with that of the current, vibrates. The frequency value is read on a scale beside the row of reeds. Also called *reed frequency meter, tuned reed frequency meter.*

vibrating reed relays A type of relay that is actuated by sound frequency. Can be triggered by an electrical resonant circuit, or simply by a mechanically induced sound vibration. See also *resonant reed relay.*

vibrator 1. Electromagnetic device which is used to change a continuous steady current into a pulsating current. 2. Vibrating reed with contacts arranged to supply direct current to two windings of a transformer so that alternating current is supplied from another winding to the load.

vibrocardiography The measurement of the acceleration produced at the chest wall by the pumping motion of the myocardium.

This technique is also known as kinetocardiography, vibrocardiography, and phonocardiography depending primarily upon the frequency range and choice of motion derivative to be displayed. The vibrocardiogram can be obtained by differentiating the apexcardiogram twice (and vice versa) if comparable frequency response transducers are used. The information is essentially the same, but the acceleration display delineates certain periods of the cycle more clearly. The duration of the isometric contraction and the ejection phase can be accurately measured, and it is believed that these two parameters are valuable in estimating the strength of the heart.

vibrophonocardiograph An instrument for recording heart vibrations and sounds.

video fluoroscopy The use of a television camera and video monitors to input, process, and display the images produced by a fluoroscope. The method permits easy distribution of fluoroscopic images to strategically located monitors, allows minimum radiation to be used, and permits electronic enhancement of the fluoroscopic image after conversion into a video signal.

video signal In television, the signal that conveys all of the intelligence present in the image together with the necessary synchronizing and equalizing pulses.

video tape recorder A tape recorder capable of recording a video (television) signal having a bandwidth of 4 megahertz. Uses a revolving recording head which laterally scans across a moving one-inch magnetic tape.

virtual memory 1. A technique that permits the user to treat secondary (disc) storage as an extension of core memory, thus giving the virtual appearance of a larger core memory to the programmer. 2. The use of techniques that permit the computer programmer to use the memory as though both the main memory and mass memory were available simultaneously.

virulence Ability of an organism to overcome the resistance of the host.

virus A submicroscopic pathogen.

viscid Sticky, thick, adhesive. Also called *viscous*.

viscosity The quality of stickiness.

viscous See *viscid*.

viscus Any large interior organ such as the stomach, heart, lung, etc.

visible spectrum That region of the electromagnetic spectrum to which the retina is sensitive and by which the eye sees. It extends from about 400 to about 750 millimicrons in wavelengths of radiation.

vital capacity That volume of air which can be exhaled after the deepest possible inhalation.

vivisection Scientific examination of a living animal.

vocabulary A list of operating codes or instructions available to the programmer for writing the program for a given problem for a specific computer.

voice analyzer An electronic instrument for printing out waveforms corresponding to vocal characteristics. An aid in identifying speech problems as well as speakers.

voice frequency 1. The frequency range of ordinary speech, usually considered to be a band from 100 Hz to 3 kHz. 2. Any of the frequencies in the band 300 to 3400 Hz which must be transmitted to reproduce voice with reasonable fidelity.

voiceprint A speech spectrograph sufficiently sensitive and detailed to identify individual human voices.

volatile memory In computers, any mem-ory which can retain information only as long as energizing power is applied. Contrast to nonvolatile memory.

volatile storage In computers, storage media which have volatile memory. Contrast to nonvolatile storage.

volt (V) 1. A measure of electrical pressure or potential. It can be likened to the pressure used to push water through a pipe. The water would be analogous to current, and the pipe walls and size of opening would offer resistance. Hence, one volt is the pressure required to force one ampere of current through a resistance of one ohm. 2. The unit measurement of electromotive force.

Volta effect Difference of potential that exists when dissimilar metals are placed in contact.

voltage 1. Term used to signify electrical pressure. Voltage is a force that causes current to flow through an electrical conductor. 2. Voltage of a circuit: the greatest effective difference of potential between any two conductors of the circuit concerned. 3. The term most often used in place of electromotive force, potential, potential difference, or voltage drop, to designate electric pressure that exists between two points and is capable of producing a flow of current when a closed circuit is connected between the two points.

voltage amplification The ratio of the signal voltage across a specified load impedance to the signal voltage across the input of a transducer.

voltage amplifier An amplifier designed to greatly increase the voltage of a signal, but having a high output impedance and supplying little power.

voltage attenuation Ratio of the input signal voltage to the voltage delivered to a specified load impedance in a transducer.

voltage comparator An amplifying device with a differential input that will provide an output polarity reversal when one input signal exceeds the other. When operating with open loop and without phase compensation, operational amplifiers make fast and accurate voltage comparators.

voltage divider Resistor or reactor connected across a voltage and tapped to make available a fixed or variable fraction of the applied voltage.

voltage doubler Voltage multiplier which separately rectifies each half cycle of the applied alternating voltage and adds the two rectified voltages to produce a direct voltage, the amplitude of which is approximately twice the peak amplitude of the applied alternating voltage.

voltage drop 1. The decrease in voltage as a current traverses a resistance. 2. The voltage measured across a resistance through which a current is flowing.

voltage gain Of a circuit, difference between the output signal voltage level in decibels and the input signal voltage level in decibels. This value is equal to 20 times the common logarithm of the ratio of the output voltage to the input voltage. The voltage gain is equal to the amplification factor of the tube or transistor only for a matched load.

voltage level At any point in a circuit, the ratio of the voltage existing at that point to an arbitrary value of voltage used as a reference.

voltage multiplier 1. Rectifying circuit that produces a direct voltage having an amplitude approximately equal to an integral multiple of the peak amplitude of the applied alternating voltage. 2. Series arrangement of capacitors charged by rapidly rotating brushes in sequence, giving a high direct voltage equal to the source voltage multiplied by the number of capacitors in series. 3. Precision resistor used in series with a voltmeter to extend its measuring range.

voltage reference circuit A circuit which provides an extremely stable voltage to which an unknown voltage may be compared.

voltage regulation Ratio of the difference between no-load and full-load output voltage of a device to the full-load output voltage. Expressed in percent.

voltage regulator 1. A circuit which keeps an output voltage at a predetermined value, or which varies the voltage according to a predetermined plan, regardless of normal voltage variations to its input, or impedance changes in the output load. 2. A zener diode or a gas-filled electronic tube which has the property of maintaining a nearly constant voltage across its terminals over a considerable range of current through the tube. Used in electronic voltage regulators.

voltages comparator A circuit that compares two analog voltages and generates a logic output when the two voltages are equal or one is greater or lesser than the reference.

voltage to ground The highest voltage existing between any conductor of a circuit and the earth.

voltaic cell An electric cell having two electrodes of unlike metals immersed in a solution that chemically affects one or both of them, thus producing an electromotive force. The name is derived from Count Alessandro Volta, an early Italian physicist who discovered this effect.

voltammeter A measuring instrument having terminals and scales permitting it to be used either as a voltmeter or as an ammeter.

volt-ampere A unit of apparent power equal to the product of volts times amperes without taking into account the power factor. Volt-amperes times power-factor equals power in watts.

voltmeter Instrument for measuring potential difference; may be calibrated in volts, microvolts, millivolts, or kilovolts.

volt-ohm-milliammeter A multipurpose test meter having a single meter element with a number of scales, and a rotary switch with which one can select any of a number of voltage, current, and resistance ranges.

volume resistivity The electrical resistance between opposite faces of a 1-cm cube of insulating material, commonly expressed in ohm-centimeters. Also called *specific insulation resistance*.

volvulus Intestinal obstruction due to twisting of the bowel.

vomer A bone of the septum of the nose.

vomitus Emesis; matter ejected from the stomach via the mouth.

vulva Female external genitalia.

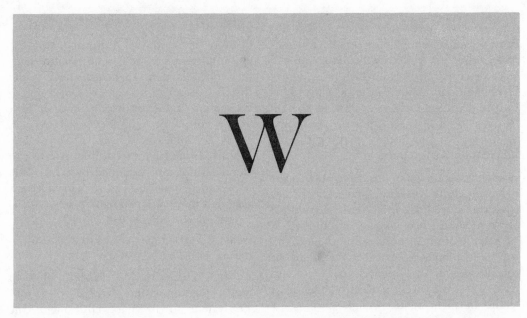

W Abbreviation for watt.

wafer 1. A thin piece of semiconductor crystal upon which several devices are formed. The wafer is then cut into smaller pieces (dice) each containing one device. The dice are then packaged for use. 2. A single section of a wafer switch.

wafer switch A switch with fixed terminals on the periphery of a flat piece of insulating material (wafer), and a movable terminal (or terminals) attached to a shaft in the center which can be rotated to cause the movable terminal to contact one or more of the fixed terminals in succession. Two or more wafers can be stacked to increase the switch capacity.

watt (W) 1. Practical unit of electrical power. It is the power required to do work at the rate of one joule per second. In a direct current circuit, the power in watts is equal to volts multiplied by amperes. In an alternating current circuit, the true power in watts is effective volts multiplied by the circuit power factor. There are 746 watts in one horsepower. 2. A measure of acoustical power. The electrical wattage rating of an amplifier describes the power it can develop to drive a loudspeaker. Acoustical wattage describes the actual sound produced by a loudspeaker in the given environment. (The two figures, in any given amplifier-speaker system are necessarily very widely divergent, inasmuch as the low efficiency of a speaker necessitates their receiving relatively large amounts of amplifier power in order to produce satisfactory sound levels.)

watt-hour A unit for measuring electrical energy. It is the work done by one watt acting for one hour. The kilowatt-hour, one thousand watt-hours, is the unit commonly used.

watt-hour meter A totalizing meter which registers the total electrical energy used, usually in kilowatt-hours.

wattless power See *reactive volt-amperes.*

watt second 1. The amount of energy corresponding to one watt acting for one second. Equal to one *joule.* 2. An energy-rating scale for defibrillators of the type described by Bernard Lown.

watt-seconds into 50-ohm load The energy rating scale for defibrillators of the type described by Lown, which presupposes a 40-ohm load as the discharge path. This rating accounts for energy lost in the resistance of the series inductor prescribed by Lown, for purposes of shaping the defibrillation energy pulse.

wave 1. A periodic variation of an electric voltage or current, bioelectric potential, transduced potential, sonic, ultrasonic, or radio-frequency voltage or current. 2. A wave motion in any medium: mechanical as in blood, acoustical as sound in air or ultrasound in tissue, electrical as current waves on wires, or electromagnetic as radio and light waves through space.

waveform 1. The graphical representation of a wave, formed by plotting amplitude versus time. Most scalar biophysical parameters are recorded as plots of amplitude versus time, as in the ECG, EEG, EMG, phonocardiogram, blood pressure waveform, and cardiac output curve. 2. A graphical representation of the relationship between voltage, current, or power against time. It also provides a picture of the behavior of signals at given frequencies.

waveform analyzer A frequency-selective voltmeter which can be used to determine the frequency and amplitude of each sine-wave component of a complex wave.

waveform generator Electronic instrument for production of graphic waveshapes that show variations in voltage magnitude versus time.

wavelength 1. The distance between the crests of two waves. This definition applies just as well to sound, light, electrical, and mechanical waves as it does to waves in the ocean. Very short wavelengths are commonly measured in *angstroms* (A), with one angstrom being equal to 10^{-7} mm. 2. The

distance between the beginning and the end of a complete cycle of any spatial periodic phenomenon. In acoustics, it is the distance occupied by one cycle of a repetitive sound traveling through the air at a velocity of about 1,100 feet per second. (A 1,100 Hz tone has a wavelength of one foot.) 3. In magnetic recording, it refers to the length of tape occupied by a full cycle of recorded signal (at 7½ ips tape speed, a recorded frequency of 1,000 Hz has a wavelength, on the tape, of 0.0075 inch).

weal 1. Raised patch on skin due to intradermal effusion. 2. A welt.

weber (Wb) Unit of magnetic flux in meter-kilogram-second system of units. Equivalent to 10^8 maxwells and is the amount of flux which, when linked to a single turn of wire, generates a force of one volt in the turn as it decreases uniformly to zero in one second.

wedge pressure monitoring The pulmonary artery pressure, as measured using a flow-directed balloon-tipped catheter. The measurement accurately reflects the mean left atrial pressure. Therefore, in the absence of mitral valve disease, it is a useful indication of left ventricular dynamics. The catheter is introduced into the venous system through the antecubital area. The balloon tip is partially inflated and is carried by the bloodstream from the insertion site to the great intrathoracic veins, into the pulmonary artery, where the measurement is made.

wet cell Cell whose electrolyte is in liquid form and free to flow and move.

wet circuit 1. A circuit which carries direct current. 2. Circuit having current flow to melt (microscopically) contact material at point of contact, thereby dissolving and evaporating away contaminants.

Wheatstone bridge Arrangement of resistances, electric source, and galvanometer used for the measurement of unknown resistances. The circuit has four arms consisting of resistors. Three are known. One of these three is variable, and connects to the terminal where a fourth, unknown resistance is applied. The two known fixed resistor arms establish a reference potential, and the unknown and variable resistor must be brought to equal that potential. The circuit depends for its operation on the fact that no current flows between points at the same potential, and that when the same current flows through two resistances, the voltage drop across each is proportional to their resistance. After the bridge has been balanced by adjustment of the variable resistance shown by the zero reading on the galvanometer, the unknown resistance X is given by the equation:

$$X = \frac{R_1}{R_2} \times R_3$$

white noise Random acoustic or electric noise having equal energy per cycle over a wide total frequency band.

white room An area in which the atmosphere is controlled to eliminate dust, moisture, and bacteria. Used in the production and assembly of components and systems whose reliability or functions might be adversely affected by the presence of foreign matter.

wicking 1. The tendency for flux and solder to run in under the insulation of a wire when its end is being soldered to a terminal. 2. The longitudinal flow of water inside a cable due to capillary action.

Wien bridge A four-arm alternating current bridge which can be used to measure capacitance or inductance in terms of resistance and frequency. Two adjacent arms have noninductive resistors. The other two arms both have resistance and capacitance,

if used to measure capacitance, or resistance and inductance, if used to measure inductance.

Wien bridge oscillator A type of phase-shift oscillator which uses resistance and capacitance in a bridge circuit to control frequency.

winding The many turns of insulated wire that form a coil used in a relay, transformer, or other electromagnetic device. See also *primary winding* and *secondary winding*.

Wilson center terminal See *indifferent electrode*.

wiper 1. A moving contact which makes contact with a terminal in a stepping switch or relay. 2. The movable contact in a rheostat, potentiometer, variable resistor, or variable capacitor.

wire 1. A conductor of round, square, or rectangular section, either bare or insulated. 2. A slender rod or filament of drawn metal. The term is a generally used one, which may refer to any single conductor. If larger than 9 AWG or multiple conductors, it is usually referred to as cable.

wired AND Externally connecting separate circuits or functions so that the combination of their outputs results in an AND function. The point at which the separate circuits are wired together will be a 1 if all circuits feeding into this point are 1. Also called *dot AND,* or *implied AND.*

wired OR Externally connecting separate circuits or functions so that the combination of their outputs results in an OR function. The point at which the separate circuits are wired together will be 1 if any of the circuits feeding into this point are 1. Also called *dot OR,* or *implied OR.*

wirewound resistor A resistor in which the resistance element is a length of high-resistance wire or ribbon wound onto an insulating form.

wirewound trimming potentiometer A trimming potentiometer characterized by a resistance element made up of turns of wire on which the wiper contacts only a small portion of each turn.

wire-wrap A solderless terminating process wherein a bared end of insulated solid wire is wrapped tightly around a terminal post having sharp corners. Low interface resistance is achieved by the virtual fusing of the metals at points of contact. Connection is held together thereafter by the elastic strains left in the two members. Wire-wrap is a registered trademark of the Gardner-Denver Co.

wiring diagram A circuit diagram that shows electrical components and all of the wires which interconnect them. The diagram shows the designation of the terminals, color coding of the wires, and whether the wires are singles, pairs, triples, or quads.

word 1. A set of characters or symbols which is treated as a unit. 2. A group of binary digits containing sufficient information to direct a logical operation. 3. Either two or four sequential bytes. In most computers, the shortest instruction usually requires one word. Most computers operate most rapidly when doing arithmetic operations on numbers which are one word long. A computer's size is really the number of bytes or words in its memory. 4. An ordered set of characters which comprises the normal unit in which information may be stored, transmitted, or acted upon. This unit is treated and transported by the computer circuits as an entity, by the control unit as an instruction, and by the arithmetic unit as a quantity.

word generator An instrument that gen-erates a data stream of ones and zeros under complete control of the operator with regard to bit position, bit frequency, etc. It may be considered to be read-only memory, a paper-tape-reader substitute, a computer simulator, a data-transmission-line tester, a programming device, or even a programmable pulse generator.

word length The number of bits in a sequence that is treated as a unit and that can be stored in one computer location. Longer words imply higher precision and more intricate instructions.

work function 1. General term applied to the energy required to transfer electrons or other particles from the interior of one medium across a boundary into an adjacent medium. 2. Photoelectric work function applies to the transfer of electrons from a metal to a vacuum under the action of light, while the thermionic work function covers the same transfer under the influence of heat.

working voltage 1. The voltage rating of a fixed capacitor. It is the recommended maximum voltage at which the unit should be operated. 2. The rated voltage which can be applied to a device or conductor continuously without danger of break-down.

write 1. To impart or introduce information, usually into some form of storage device. 2. To transfer computer information to an output device; to copy from internal storage to external storage; and to record information in a register, location, or other storage device. See also *read*.

write head Device that stores digital information as coded electrical pulses on a magnetic drum or tape.

write pulse In a computer a pulse that is used to enter information into one or more magnetic cells for storage purposes.

writing rate In a graphic recorder, the maximum linear speed at which the writing source (cathode-ray tube, mirror galvanometer, heated stylus, or pen) can produce a satisfactorily visible trace.

writing speed In a cathode-ray tube, the maximum linear speed at which the electron beam can produce a satisfactorily visible trace.

writing time/div. The minimum time per unit distance required to record a trace. The method of recording must be specified.

WVDC Abbreviation for working voltage, direct current, which is the maximum safe dc operating voltage that can be applied across the terminals of a capacitor at its maximum operating temperature.

X-Y-Z

xenon flashtube A high intensity source of incoherent white light in which a capacitor is discharged through a tube of xenon gas. This is often used as a pumping source in solid-state lasers.

xeroderma Excessive dryness of the skin.

xerographic printer Device for printing an optical image on paper, in which light and dark areas are represented by electrostatically charged and uncharged areas. Powdered ink, dusted on the paper, adheres to the charged areas and is subsequently melted into the paper by the application of heat, to form a fused, permanent image.

xerography A printing process of electrostatic photography that uses a photoconductive, insulating medium, in conjunction with infrared, visible, or ultraviolet radiation, to produce latent electrostatic charge patterns for achieving an observable record.

xeroradiography equipment Equipment employing principles of electrostatics and photoconductivity to record X-ray images on a sensitized plate in a short time after exposure.

xerosis Abnormal dryness, e.g., of the conjunctiva or the skin.

xistor Abbreviation for transistor.

X-ray fluorescence spectroscope Analytical apparatus using fluorescence induced in sample irradiated by X-rays for sample analysis.

X-rays Electromagnetic radiation of very short wavelength ranging from 0.01 angstrom unit to 10 angstrom units. It was discovered by W. K. Röntgen in 1895. X-rays penetrate matter that is opaque to ordinary visible light, the degree of penetration depending on the density. They affect a photographic plate. The shadow picture of a complex object penetrated by X-rays shows the disposition and shape of materials of different densities in the object, and so, X-rays are widely used in medical practice for locating bone fractures and dislocations and for locating tumors and lesions. X-rays are generated when electrons strike a metal target, usually tungsten. See *roentgen ray*.

X-ray spectrometer, X-ray spectrograph Instrument for analyzing X-ray spectrum of crystals, incorporates several techniques —diffraction, fluorescence, absorption, etc.

X-ray synchronizer System that times X-ray exposure to coincide with other physiologic activity such as R-wave in cardiac cycle.

X-ray therapy apparatus High-energy X-ray equipment used for radiation—destruction of malignant tumors or for radiation —treatment of skin conditions, etc. Also called *radiotherapy apparatus.*

X-ray viewer Light box or panel for viewing one or several X-rays.

xy plotter A computer-driven device which draws coordinate points in graph form.

x-y recorder 1. A graphic recording device which resolves inputs to its vertical and horizontal axes into records which represent the magnitude, direction, and phase differences between the two. 2. A plotting device in which x and y inputs are continuously resolved to give a pen-position signal.

yoke 1. A piece of ferromagnetic material without windings, which permanently connects two or more magnet cores. 2. A core assembly which is used to produce electromagnetic deflection of the electron beam in a cathode-ray tube.

Z Symbol for impedance.

zapping Slang for burning out.

z-axis 1. Property of some cathode-ray tube (CRT) circuits and some recorders which use CRTs as the writing elements, by which a voltage applied to the z-axis terminals can be used to brighten the CRT trace for a selected portion of the displayed waveform. 2. In a solid model, the axis

mutually perpendicular to the x and y axes.

Zeeman effect The shifting of the energy levels of an atom molecule, or nucleus, together with the removal of some degeneracy that takes place when the system is placed in a magnetic field. As a consequence of these effects, the spectrum of the system is altered, and lines which were single in the absence of the field, split up in the presence of it.

zener breakdown Conduction in a semiconductor when electrons are pulled out of their convalent bonds by a high electric field.

zener breakdown theory A critical current density is attained by internal field emission, or tunneling, due to the high applied field.

zener diode A two layer device that, above a certain reverse voltage (the zener value), has a sudden rise in current. If forward biased, the diode is an ordinary rectifier, but when reverse biased, the diode exhibits a typical knee, or sharp break in its current-voltage graph. The voltage across the device remains essentially constant for any further increase of reverse current up to the allowable dissipation rating. The zener diode is a good voltage regulator, overvoltage protector, and voltage reference.

zener effect A reverse-current breakdown at the junction of a semiconductor or insulator caused by the presence of a high electric field. See *avalanche.*

zener voltage The voltage at which zener breakdown occurs.

zero-access storage Computer storage for which waiting time is negligible.

zero-address instruction In computers, an instruction consisting only of an operation part. The locations of the operands are defined by the computer code.

zero adjustment Bringing the pointer or pen of an x-y recorder to zero when the input signal is zero.

zero beat The condition achieved, when adjusting two oscillatory circuits to the same frequency by listening to the two tones, when the beat note disappears.

zero bias Of an electron tube, a condition where there is no potential difference between the cathode and the control grid.

zero-input terminal See *reset terminal*.

zero suppression In computers, the elimination of nonsignificant zeros to the left of significant digits, usually before printing.

zona A zone; a segment; any encircling or beltlike structure, either external or internal, longitudinal or transverse.

zoning Purifying a metal by passing it through an induction coil. The impurities are swept ahead of the heating effect. Specifically used in purifying semiconductor crystals.

zoology That part of biology which deals with the study of animal life.